AF618808

Advanced Concepts of Crystal Structures

Advanced Concepts of Crystal Structures

Edited by **Rupert Preston**

New York

Published by NY Research Press,
23 West, 55th Street, Suite 816,
New York, NY 10019, USA
www.nyresearchpress.com

Advanced Concepts of Crystal Structures
Edited by Rupert Preston

International Standard Book Number: 978-1-63238-012-8 (Hardback)

Printed in the United States of America.

Contents

Preface

Almost every element has a specific arrangement, which is expressed through its crystal structure. Any crystal structure refers to the unique and highly ordered arrangement of atoms or molecules in any liquid or solid. The structure is usually a repeating array of 3D 'boxes', also known as unit cells. To understand the structure of any element, the simplest possible representation of molecules, also known as the asymmetric unit is considered. On the basis of the physical and chemical properties which they exhibit, crystals are divided into four categories, namely: Covalent Crystals, Metallic Crystals, Ionic Crystals and Molecular Crystals.

A crystal structure describes a highly ordered structure, occurring due to the intrinsic nature of molecules to form symmetric patterns. The final crystal lattice is constructed through repetition of the unit cell in a three-dimension form. The symmetry properties of the crystal are located in its space group, whereas the angles between the lengths of the edges of a unit cell are called the lattice parameters.

The intrinsic symmetry of any element plays a crucial role in determining its physical properties. The patterns are generally found upon the points of a lattice. Optical transparency, cleavage, electronic band structure are some of the properties which are ultimately defined by the crystal structure.

I would like to thank all the researchers who have contributed their researches and data in this book. I would also like to thank the publishing house for give me this unmatched opportunity.

Editor

1

Crystal Structure, Thermal Behaviour and Vibrational Spectra of Tetraethylammonium Dihydrogenmonophosphate Bis Phosphoric Acid

Ikram Dhouib[1*], Salih Al-Juaid[2], Tahar Mhiri[1], Zakaria Elaoud[1]
[1]Laboratoire de l'Etat Solide, Département de Chimie, Faculté des Sciences de Sfax, Sfax, Tunisia
[2]Chemistry Department, Faculty of Science, King Abdulaziz University, Jeddah, KSA

ABSTRACT

Single crystals of the tetraethylammonium dihydrogenmonophosphate bis trihydrogenmonophosphate $[CH_3CH_2]_4 N^+ (H_2PO_4^-)(H_3PO_4)_2$ (TEP), were grown by slow evaporation solution technique at room temperature. The compound was characterised by IR, Raman, differential thermal analysis (TG-DTA) and single crystal X-ray diffraction. It crystallizes in the monoclinic system (space group $P2_1/c$) with the following unit cell dimensions: a = 7.765 (2) Å, b = 16.531 (4) Å, c = 14.843 (2) Å, β = 100.99 (2)°, $Z = 4$, $D_x = 1.67 \mathrm{Mg \cdot m^{-3}}$, $Dm = 1.532\ \mathrm{Mg \cdot m^{-3}}$, λ (*MoKa*) = 0.71073 Å, $\mu = 0.384\ \mathrm{mm^{-1}}$, $F(000) = 991$, T = 20 (2)°. The structure was solved by the direct method and refined to final R value of 0.0342 and $R_w = 0.107$ for 3239 independent reflections. The structure consists of infinite parallel two-dimensional planes built of mutually $H_2PO_4^-$, H_3PO_4 tetrahedra and $[(CH_3CH_2)_4 N]^+$ cations connected by strong O-H···O and C-H···O hydrogen bonding. There are no contacts other than van der Waals interactions between the layers.

Keywords: Organic Phosphate; Crystalline Structure; Vibrational Spectra; Thermal Behaviour; Differential Thermal Analysis.

1. Introduction

In organic-cation monophosphates, the phosphate anions generally observed are the acidic ones $[HPO_4]^{2-}$ or $[H_2PO_4]^-$. Such anions are interconnected by strong hydrogen bonds so as to build infinite networks with various geometries: ribbons [1], chains [2,3], two-dimensional network [4-6], and three-dimensional network [7]. These entities can be associated to organic molecules to produce compounds having a particular interest as non-linear optical materials.

During a systematic investigation of interactions between monophosphoric acid and organic molecules containing one or more nitrogen atom different structures of monophosphate salts have been described:
$NH_3(CH_2)_2NH_3(H_2PO_4)_2$ [8],
$NH_3(CH_2)_2NH_3(H_2PO_4)_2H_3PO_4$ [9],
$NH_3(CH_2)_2NH_3HPO_4$ [2],
$NH_3(CH_2)_3NH_3HPO_4 \cdot H_2O$ [10],
$NH_3(CH_2)_3NH_3(H_2PO_4)_2$ [11],
$NH_3(CH_2)_4NH_3HPO_4 \cdot 2H_2O$ [12],
$NH_3(CH_2)_4NH_3(H_2PO_4)_2$ [13],
$[NH_3(CH_2)_2NH_2(CH_2)_2NH_2(CH_2)_2NH_3]$ $(HPO_4)_2 \cdot 2H_2O$ [14],
$[C_6H_5CH_2CHCH_2CH_2NH_2CH_2CH_2] \cdot H_2PO_4$ [15],
$[C_6H_5CH_2NH_2CH_3] \cdot H_2PO_4 \cdot H_2O$ [16],
$[NH_3(CH_2)_2]_2NH \cdot HPO_4 \cdot 2H_2O$ [17,18].

In the present work we describe the chemical preparation, the crystal structure, calorimetric measurements and Fourier transform infrared (FTIR) spectral characterization of a new compound: the tetraethylammonium dihydrogenmonophosphate bis trihydrogenmonophosphate $[CH_3CH_2]_4 N+(H_2PO_4^-)(H_3PO_4)^2$, here after abbreviated to (TEP), which is a close analogue of previously ob-

*Corresponding author.

tained $[CH_3CH_2]_4N^+(H_2AsO_4^-)(H_3AsO_4)_2$ [19].

2. Experimental

2.1. Synthesis

The crystal of TEP is easily prepared by slow evaporation at room temperature of an aqueous solution of H_3PO_4 and $[(CH_3CH_2)_4NOH]$. Schematically the reaction is:

$$[(CH_3CH_2)_4NOH]+3H_3PO_4 \rightarrow$$
$$[CH_3CH_2]_4N^+(H_2PO_4^-)(H_3PO_4)_2+H_2O$$

After some days of evaporation, colorless needle-shaped monocrystals appear in the solution. The chemical analysis of phosphorus and acidic proton has been carried out [20].

Density was measured at room temperature by flotation in toluene. The average value of density,

$$Dm = 1.532\ \text{Mg}\cdot\text{m}^{-3},$$

is agreement with that calculated,

$$Dx = 1.598\ \text{Mg}\cdot\text{m}^{-3}.$$

The cell contains two formula units of

$$[CH_3CH_2]_4N(H_2PO_4)(H_3PO_4)_2.$$

2.2. Characterizations

The infrared spectrum was recorded in the range 400-4000 cm^{-1} with a "Perkin Elmer FTIR-1000" spectrophotometer using a sample dispersed in a KBr pellet. Back scattering Raman spectra were obtained under microscope with a Horiba Jobin Yvon Raman spectrometer (Lab RAM HR 800 λ = 633 nm) in the 50 - 4050 cm^{-1} range.

Setaram thermoanalyser, TG-DTA92, was used to perform thermal treatment on samples of

$$[CH_3CH_2]_4N(H_2PO_4)(H_3PO_4)_2.$$

TG-DTA thermograms were obtained with 34.35 mg sample in an open platinum crucible, heated in air with 5˚C·min^{-1} heating rate, from room temperature to 275˚C, an empty crucible was used as reference.

2.3. X-Ray Single Crystal Structure Determination

Single-crystal X-ray data

$$[CH_3CH_2]_4N(H_2PO_4)(H_3PO_4)_2$$

were collected at room temperature on a Nonius Kappa-CCD diffractometer using Mo-K$_\alpha$ radiation λ = 0.71073 Å through the program COLLECT [21]. Correction for Lorentz-polarisation effect, peak integration and background determination were carried out with the program DENZO [22]. Frame scaling and unit cell parameters refinement were performed with the program SCALE-PACK [22].

Table 1. Crystal data and summary of intensity data collection and structure refinement.

Compound	$[CH_3CH_2]_4NH_2PO_4\ (H_3PO_4)_2$
Color/Shape	Colourless/Parallelepiped
Molecular weight, g·mol^{-1}	450.23
Space group	$P2_{1/c}$
Temp. Deg ˚C	20 (2)
Unit cell dimensions (Å)	
a, Å	7.765 (2)
b, Å	16.531 (4)
c, Å	14.843 (2)
β, ˚	100.99 (2)
Cell volume, Å^3	1870.33 (8)
Formula units/unit cell	4
D_{calc}, g·cm^{-3}	1.598
μ_{calc}, cm^{-1}	0.384
Diffractometer/scan	KAPPA CAD4. Enraf-Nonius
Radiation, graphite monochromater	Mo-K$_\alpha$ (λ = 0.71073)
Max. crystal dimensions, mm	0.7 × 0.3 × 0.3
Reflections measured	3239
Range of h, k, l	−9/3, −18/19, −17/17
Reflections with $F_0 > 4\sigma(F_0)$[b]	2506
Weights	$1/\sigma^2(F_o^2) + (0.0655P)^2 + 0.00P]$ avec $P = (F_o^2 + 2F_c^2)/3$
GOF	1.157
$R=\sum\lVert F_o\rvert-\lvert F_c\rVert/\sum\lVert F_o\rVert$	0.0342
Rw	0.1072
T_{min}	0.7749
T_{max}	0.8935
Computer programs[c]	SHELXS-97(Sheldrick, 1990) SHELXL-97(Sheldrick, 1997)

Pertinent details of the crystal structure of $[CH_3CH_2]_4N(H_2PO_4)(H_3PO_4)_2$ are listed in **Table 1**. The crystal structure has been solved and refined in the monoclinic symmetry, space group $P2_1/c$, using the WINGX environment [23] and based on SHELXS97 [24] and SHELXL97 [25] softwares. All the hydrogen positions of the diprotonated cation were placed geometrically and held in the riding mode, the C-H bonds were fixed and affined at 0.78 and 1.01Å, respectively). Interatomic distances, bond angles and the hydrogen bonds scheme are listed in **Tables 2** and **3** respectively.

3. Results and Discussion

3.1. Structural Analysis

The asymmetric unit is composed of one $[H_2PO_4]^-$ anion, two neutral phosphoric acid $[H_3PO_4]$ and one $[(CH_3CH_2)_4N]^+$ cation. The structural arrangement of $[CH_3CH_2]_4N(H_2PO_4)(H_3PO_4)_2$ can be described as an alternation of organic $[N(C_2H_5)_4]^+$ groups and inorganic $(H_2PO_4)(H_3PO_4)_2$ is illustrated in **Figure 1**, viewed in projection along the direction of its two-fold screw axes.

One can distinguish the chains of the acid molecules and the chains of the dihydrogenphosphates anions running parallel to the Z axis in each layer (see **Figure 2**).

The molecules/anions in the chains are related to each

Table 2. Principal intratomic distances (Å) and bond angles (°) in $[CH_3CH_2]_4N^+(H_2PO_4^-)(H_3PO_4)_2$.

Tetrahedron around P		$[CH_3CH_2]_4N^+$ cations	
P1-O1	1.552 (2)	N-C1	1.512 (3)
P1-O2	1.496 (2)	N-C2	1.522 (3)
P1-O3	1.554 (2)	N-C3	1.526 (3)
P1-O4	1.544 (2)	N-C4	1.522 (3)
O1-H1	0.78 (5)		
O4-H4	0.84 (4)		
O3-H3	0.78 (4)		
O2-P1-O4	110.0 (1)	C1-C6	1.517 (4)
O2-P1-O1	111.54 (1)	C2-C8	1.509 (4)
O4-P1-O1	108.1 (1)	C3-C5	1.516 (3)
O4-P1-O3	108.1 (1)	C4-C7	1.517 (3)
O1-P1-O3	104.5 (1)	C1-N-C4	110.4 (2)
O2-P1-O3	113.5 (1)	C1-N-C2	107.0 (2)
P2-O5	1.511 (2)	C4-N-C2	110.9 (2)
P2-O6	1.505 (2)	C1-N-C3	111.1 (2)
P2-O7	1.56 (12)	C4-N-C3	106.5 (2)
P2-O8	1.571 (2)	C2-N-C3	111.0 (2)
O6-H6	0.82 (1)		
O7-H7	0.82 (4)		
O6-P2-O5	114.63 (9)	C8-C2-N	114.9 (2)
O6-P2-O7	106.19 (9)	C7-C4-N	114.8 (2)
O5-P2-O7	110.64 (1)	C5-C3-N	114.4 (2)
O6-P2-O8	107.63 (9)	C1-C6-N	115.1 (2)
O5-P2-O8	109.45(1)		
O7-P2-O8	108.05 (1)		
P3-O9	1.547 (12)		
P3-O10	1.540 (2)		
P3-O11	1.505 (2)		
P3-O12	1.554 (2)		
O9-H9	0.86 (4)		
O10-H10	0.78 (4)		
O12-H12	0.83 (4)		
O11-P3-O10	113.71 (9)		
O11-P3-O9	108.93 (9)		
O10-P3-O9	105.66 (9)		
O11-P3-O12	111.7 (1)		
O10-P3-O12	106.7 (1)		
O9-P3-O12	109.97 (9)		

Table 3. Principal interatomic distances (Å) and bond angles (°) and detailed of the hydrogen bonding scheme.

D-H···A	d (D-H) (Å)	d (H···A) (Å)	d (D···A) (Å)	D-H···A (angle °)
O1-H1···O2^{v}	0.78 (5)	1.81 (5)	2.578 (2)	170 (5)
O3-H3···O11	0.78 (4)	1.83 (4)	2.598 (2)	174 (5)
O4-H4···O6	0.84 (4)	1.72 (4)	2.554 (2)	171 (4)
O9-H9···O5^{i}	0.86 (4)	1.68 (4)	2.521 (2)	165 (4)
O7-H7···O2^{ii}	0.82 (4)	1.83 (4)	2.617 (2)	162 (4)
O10-H10···O6	0.78 (4)	1.76 (4)	2.539 (2)	177 (4)
O12-H12···O5^{iii}	0.83 (4)	1.75 (4)	2.584 (2)	178 (4)
C1-H1B···O8^{iv}	1.01 (3)	2.37 (3)	3.315 (3)	154 (2)

Symmetry codes: ^{i}–x + 1, –y + 1, –z + 1; ^{ii}–x + 1, –y + 2, –z + 1; ^{iii}–x + 2, –y + 1, –z + 1; ^{iv}x, y – 1, z; ^{v}–x + 2, –y + 2, –z + 1; ^{vi}x + 1, y, z.

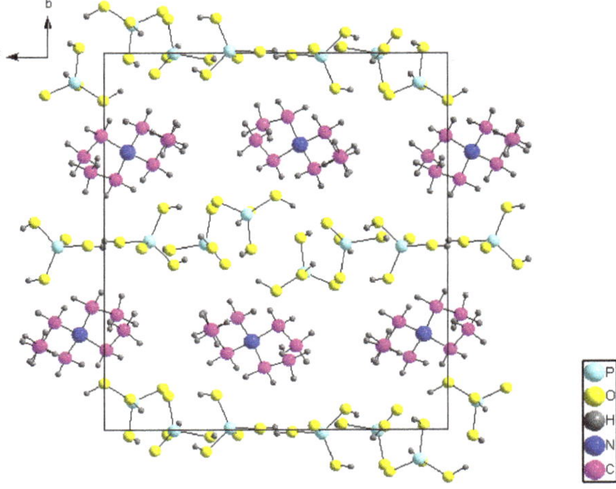

Figure 1. Projection along the a axis of the atomic arrangement of $[CH_3CH_2]_4N^+(H_2PO_4^-)(H_3PO_4)_2$.

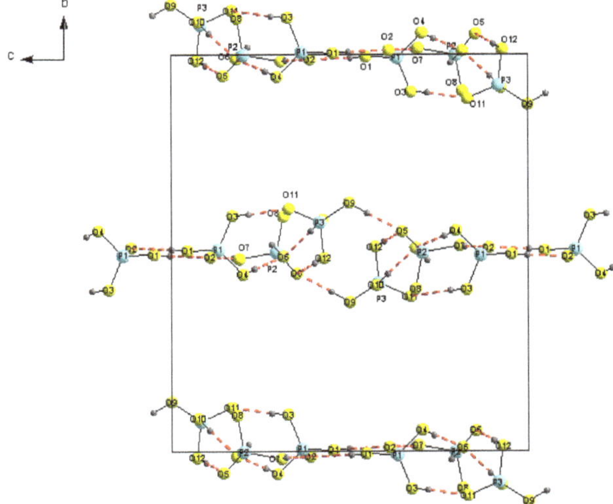

Figure 2. Projection along the axis of the inorganic arrangement.

other by the two-fold screw axis. The distance (P (1) – P (3) = 4.313 (1) Å) between two H_3PO_4 groups linked by hydrogen bond is shortened when compared to the remaining P-P bond lengths P (1) − P (2): 4.499 (1) Å and P (2) − P(3): 4.369 (1) Å between H_3PO_4 and $H_2PO_4^-$ groups. The $H_2P(2)O_4^-$ anions, $H_3P(1)O_4$ and $H_3P(3)O_4$ molecules are joined into trimmers by the hydrogen bonds (see **Table 3**). The dihydrogenphosphate anions $H_2P(2)O_4^-$ and the acid $H_3P(3)O_4$ molecules do not form any straight contact between themselves. All their of $H_2P(2)O_4^-$ one O-H groups are involved in the hydrogen bonds with the O(7)atoms of the acid $H_3P(1)O_4$ molecules and all their of $H_3P(3)O_4$ three O-H group are involved in the hydrogen bonds with the O(9),O(10) and O(12) of the dihydrogenphosphate anions $H_2P(2)O_4^-$.

The acid $H_3P(1)O_4$ molecules form strong $(O1\text{-}H1\cdots O2^{v})$ hydrogen bonds between themselves.

Note that the O(5) atoms of the dihydrogenphosphates anions participate in two hydrogen bonds with the OH groups (O(9)-H(9) and (O(12)-H(12)) of the neighbour orthophosphoric acid molecules of the same chain.

The structure is based on sheets of $H_2PO_4^-$ and H_3PO_4 tetrahedra fused together by strong intralayer O-H···O hydrogen bonds, giving to trimmers, d(O) − O < 2.73Å) [26,27]. The midplanes of these clusters, are located at $z = 0.25$ and $z = 0.75$.

The organic group of tetraethylammonium monocation structure is similar to that in the TESe and TEAs analogue [19,28]. Each tetrapropylammonium cation makes three short contacts with trimmers of the dihydrogenarsenate through the C-H···O hydrogen bonds: $C1\text{-}H1B\cdots O8^{iv}$ (see **Table 3**). The H···O distance is equal to 2.37 (3) Å, respectively. The C···O distance is equal to 3.315 (3) Å, showing that those hydrogen bonds are very weak [29]. The C-H···O angle is equal to 154(2). The structure of the tetrapropylammonium cation is similar to that in the crystals of tetraethylammonium chloride, is in the tetraethylammonium chloride monohydrate [30], tetraethylammonium chloride tetrahydrate [31] and anhydrous tetraethyl- ammonium chloride [32]. The lengths of the N-C bonds are in the range between 1.512 (3) (N-Cl) and 1.516 (3) Å (N-C3). The lengths of the other two N-C bonds are equal to 1.522 (3) Å. The C-N-C angles are in between 106.5(2) to 111.1(2)°. The C-C bonds lengths are in the region between 1.509(4) and 1.517(4). Thus, they are very similar. The final coordinates and U_{eq} or U_{iso} of TEP are given in **Table 4**.

3.2. Vibrational Investigations

The FT-IR and FT-Raman spectra of the title crystal were measured for the powder sample at room temperature. The bands observed in the measured region arise from the vibrations of hydrogen bonds, vibrations of the tetrapropylammonium cation, phosphate groups, and lattice vibrations. Vibrational spectra are shown in **Figures 3** and **4**.

3.2.1. The Hydrogen Bond Vibrations

The valence vibrations of O-H groups interconnected by a system of hydrogen bonds in the crystal appear in IR spectrum as broad bands in the 3500 - 1800 cm^{-1} region [15]. As can be seen from **Table 3**, these consist of hydrogen bonds of O-H···O type with length ranging from 2.521(2) to 2.617 (2) Å. X-ray data show that H atoms of OH groups generate hydrogen bonds with the oxygen atoms of the PO_4 group.

The stretching type of vibrations of hydrogen bonds νOH displays a well defined and medium intense band in IR with the AB structure in the region between 3500 and 2000 cm^{-1}. In IR spectrum the frequency of A band is 2750 cm^{-1} and B is 2460 cm^{-1}. For Raman experiment the broad maxima are found at 2822 and 2550 cm^{-1}. The intense IR bands correspond to a strong hydrogen bond with the O···O distance in the region 2.6 Å - 2.7Å [33]. The in-plane OH bending mode δOH gives rise to a medium band in infrared spectrum without counterparts in Raman spectrum at 1252 cm^{-1}. The out-of-plane bending

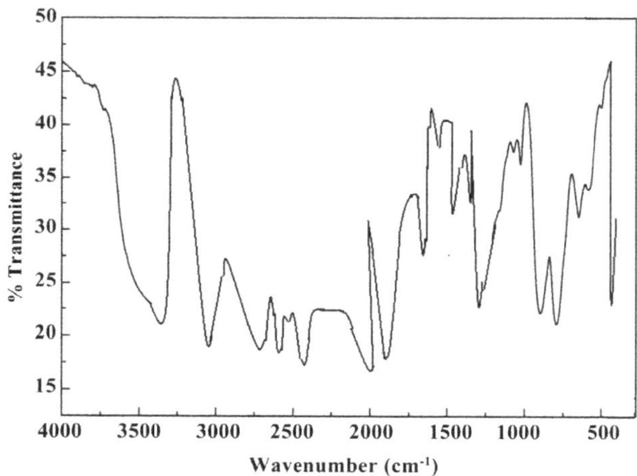

Figure 3. IR spectrum of $[CH_3CH_2]_4N\ (H_2PO_4)(H_3PO_4)_2$.

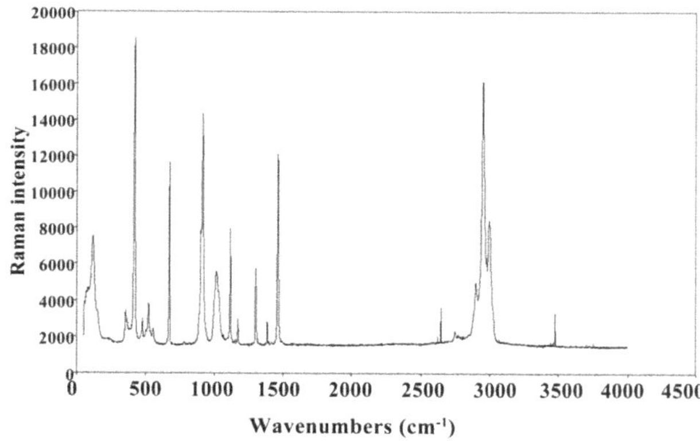

Figure 4. Raman spectrum of $[CH_3CH_2]_4N\ (H_2PO_4)(H_3PO_4)_2$.

Table 4. Principal intratomic distances (Å) and bond angles (°) in $[CH_3CH_2]_4N^+(H_2PO_4^-)(H_3PO_4)_2$ $U_{eq}=\frac{1}{3}\sum_i\sum_j U_{ij}a_i^*a_j^*a_ia_j$ **.**

Atoms	x/a	y/b	z/c	Ueq or Uiso*
P1	0.15869 (7)	0.49303 (4)	0.13317 (4)	0.0127 (2)
P2	0.70980 (7)	0.49978 (4)	0.2963 (4)	0.0113 (2)
P3	0.20672 (7)	0.42355 (4)	0.4170 (4)	0.0114 (2)
O (1)	0.2430 (2)	0.4957 (1)	0.0464 (1)	0.0196 (4)
O (2)	−0.0332 (2)	0.5117 (1)	0.1105 (1)	0.0198 (4)
O (3)	0.1999 (2)	0.4066 (1)	0.1727 (1)	0.0181 (4)
O (4)	0.2539 (2)	0.5555 (1)	0.2025 (1)	0.0174 (4)
O (5)	0.8479 (2)	0.5531 (1)	0.3526 (1)	0.0139 (4)
O (6)	0.5279 (2)	0.5109 (1)	0.3156 (1)	0.0149 (4)
O (7)	0.6960 (2)	0.5166 (1)	0.1917 (1)	0.0238 (5)
O (8)	0.7616 (2)	0.4086 (1)	0.3146 (1)	0.0194 (4)
O (9)	0.1654 (2)	0.3748 (1)	0.4992 (1)	0.0150 (4)
O (10)	0.4068 (2)	0.4171 (1)	0.4239 (1)	0.0185 (4)
O (11)	0.1023 (2)	0.38909 (1)	0.3296 (1)	0.0153 (4)
O (12)	0.1651(2)	0.51456 (1)	0.4281 (1)	0.0165 (4)
H1	0.171(6)	0.492 (3)	0.002 (3)	0.065 (15)*
H3	0.165(6)	0.404 (3)	0.218 (3)	0.073 (14)*
H4	0.350(5)	0.541 (2)	0.235 (3)	0.051 (11)*
H6	0.4600(2)	0.4802 (3)	0.2829 (2)	0.022 (11)*
H7	0.793(5)	0.514 (2)	0.178 (3)	0.048 (11)*
H9	0.180(5)	0.401 (3)	0.550 (3)	0.064 (13)*
H10	0.446(5)	0.446 (2)	0.392 (2)	0.045 (11)*
H12	0.063(5)	0.527 (2)	0.403 (3)	0.042 (10)*
n	0.6747(2)	0.2404 (1)	0.5702 (1)	0.0159 (4)
C (1)	0.8236 (3)	0.2084 (2)	0.6421 (2)	0.0200 (5)
C (2)	0.5598 (3)	0.1684 (2)	0.5342 (2)	0.0217 (5)
C (3)	0.7439 (3)	0.2823 (2)	0.4926 (2)	0.0180 (5)
C (4)	0.5705 (3)	0.3036 (2)	0.6116 (2)	0.0172 (5)
C (5)	0.8522 (4)	0.2285 (2)	0.4423 (2)	0.0251 (6)
C (6)	0.9576 (3)	0.2711 (2)	0.6843 (2)	0.0259 (6)
C (7)	0.4887 (4)	0.2736 (2)	0.6904 (2)	0.0250 (6)
C (8)	0.4006 (4)	0.1887 (2)	0.4623 (2)	0.0314 (7)
H (1A)	0.879 (4)	0.169 (2)	0.6136 (19)	0.024 (7)*
H (1B)	0.768 (3)	0.182 (2)	0.6914 (18)	0.022 (7)*
H (2A)	0.627 (4)	0.127 (2)	0.5076 (19)	0.023 (7)*
H (2B)	0.527 (3)	0.143 (2)	0.5870 (18)	0.018 (7)*
H (3A)	0.811 (4)	0.324 (2)	0.517 (2)	0.028 (8)*
H (3B)	0.641 (3)	0.301 (2)	0.4522 (18)	0.015 (6)*
H (4A)	0.653 (3)	0.348 2)	0.632 (2)	0.017 (7)*
H (4B)	0.483 (3)	0.321 (2)	0.564 (2)	0.012 (6)*
H (5A)	0.948 (4)	0.207 (2)	0.482 (2)	0.023 (7)*
H (5B)	0.787 (4)	0.185 (2)	0.411 (2)	0.029 (8)*
H (5c)	0.897 (4)	0.263 (2)	0.396 (2)	0.025 (7)*
H (6A)	1.020 (4)	0.292 (2)	0.641 (2)	0.029 (8)*
H (6B)	0.903 (4)	0.312 (2)	0.715 (2)	0.041 (9)*
H (6C)	1.040 (4)	0.244 (2)	0.729 (2)	0.035 (8)*

Continued

H (7A)	0.576 (4)	0.258 (2)	0.737 (2)	0.031 (8)*
H (7B)	0.406 (4)	0.233 (2)	0.671 (2)	0.027 (8)*
H (7C)	0.420 (4)	0.318 (2)	0.708 (2)	0.030 (8)*
H (8A)	0.429 (4)	0.212 (2)	0.409 (2)	0.036 (9)*
H (8B)	0.348 (5)	0.137 (3)	0.448 (3)	0.062 (1)*
H (8C)	0.311 (5)	0.225 (3)	0.483 (3)	0.064 (1)*

* is relatif to Uiso of hydogen.

γOH mode appears in the region 900 and 700 cm^{-1}, but is not observed in spectra measured at ambient temperature.

3.2.2. The Tetraethylammonium Cation Vibrations

The frequencies observed in the infrared spectra at 2977 and 2957 cm^{-1} are assigned to the δ symmetric CH_2 and δasymmetric CH_3 absorptions; the different modes of asymmetric (δas) and symmetric (δs) deformation of the methyl groups are found at 1495 and 1397 cm^{-1}, respectively. A weak band which appeared at 1323 cm^{-1} is related to the N-C vibration [34,35]. Besides, sharp bands at 1196, 1172 and 1102, 1116 cm^{-1} are associated to CH_3 and CH_2 rocking vibration modes in IR and spectively. The two bands observed at 952 cm^{-1} in IR and 915 cm^{-1} in Raman were assigned to $\nu_1(NC_4)$ stretchRaman, reing modes [36-38]. The deformation mode $\nu_2(NC_4)$ appears at 744 and 672 cm^{-1} in IR and Raman spectrum, respectively. The splitting ν (C-C-C-N) bending mode at 772 cm^{-1} may correspond to different conformers of the organic chains. The band observed at 463 cm^{-1} in infrared and 478 cm^{-1} in Raman spectrum arises from the deformation vibration $\nu_4(NC_4)$ of the TEP entity.

3.2.3. Internal Vibrations of the Phosphate Groups

The unperturbed PO_4^{3-} ion is a tetrahedron with a point group symmetry T_d. The υ_1 (A)and υ_3 ($F2$) symmetric and asymmetric stretching modes are observed in 1000 - 700 cm^{-1} region, whereas the υ_2 (E) and υ_4 ($F2$) symmetric and asymmetric bending modes are distinguished in the 500 - 400 cm^{-1} domain [39,40].

The interpreting of the IR and Raman spectrum is made in terms of internal modes of two atomic groups, PO_2 and $P(OH)_2$, included in $H_2PO_4^-$ anion. The two stretching vibrations, asymmetric and symmetric of PO_2 group, are observed in the region 1065 - 895 cm^{-1}; while those related to $P(OH)_2$ group occur as two intense bands in the domain 1000 - 861 cm^{-1}. The splitting of F2 stretching mode of PO_4, into three intense components at 1059, 1032, 889 cm^{-1} is a strong evidence of the symmetry lowering of H_2PO_4 in the solid state. On the other hand, bending modes of H_2PO_4 group are observed at lower frequencies. The bands located in the region 487 - 400 cm^{-1}, are respectively attributed to the torsion $\rho(PO_2)$, to the wagging $\omega(PO_2)$, and to the bending $\delta[P(OH)_2]$ vibrations.

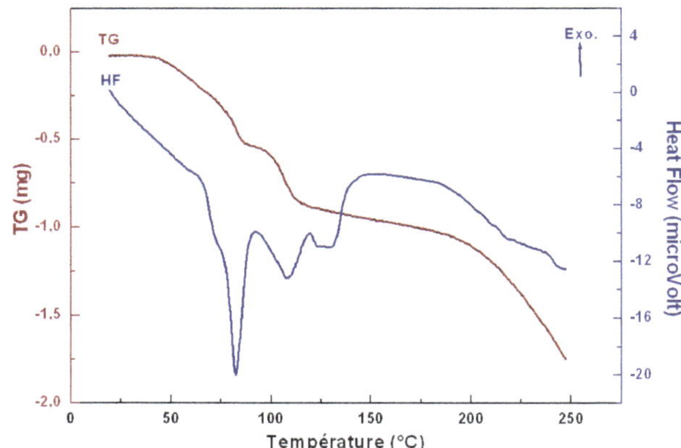

Figure 5. The TG-DSC curve of $[CH_3CH_2]_4N$ $(H_2PO_4)(H_3PO_4)_2$.

3.3. Thermal Analysis (DSC and ATG)

DSC-ATG of TEP was done in air at the rate of 5˚C/min. One more characteristic feature of the room temperature TEP is its high hygroscopicity (**Figure 5**), which illustrates the calorimetric (DSC) and thermogravimetric (TGA) results. This compound is stable until 320 K, above this temperature, or loss of weight appears at 355 K and 381 K. It is due to the departure of adsorbed water. The endothermic peak observed at T = 401 K is attributed to the melting of TEP.

4. Conclusions

Crystals of a hybrid material,

$$(CH_3CH_2)_4N(H_2PO_4)(H_3PO_4)_2,$$

have been prepared by slow evaporation of aqueous solution $(CH_3CH_2)_4NOH$ and H_3PO_4 at room temperature.

The structure consists of strong two dimensional character based on sheets of $H_2PO_4^-$ and H_3PO_4 tetrahedra fused together by strong intralayer O-H···O hydrogen bonds, giving to trimmers. The planes of inorganic groups alternate with planes of organic cations. In addition, the two layers spreading in this network are themselves interconnected by strong O-H···O and C-H···O hydrogen bonding.

Vibrational study recorded is of great interest as it

verifies the dependence of groups constituting our material. The hydrogen bonds confirmed by IR and X-Ray diffraction explain the stability of our compound.

REFERENCES

[1] L. Baouab and A. Jouini, "Crystal Structures and Thermal Behavior of Two New Organic Monophosphates," *Journal of Solid State Chemistry*, Vol. 141, No. 2, 1998, pp. 343-351.

[2] M. T. Averbuch-Pouchot and A. Durif, "Structures of Ethylenediammonium Monohydrogentetraoxophosphate(V) and Ethylenediammonium Monohydrogentetraoxoarsenate(V)," *Acta Crystallographica*, Vol. C43,1987, pp. 1894-1896.

[3] M. T. Averbuch-Pouchot, A. Durif and J. C. Guitel, "Structures of -Alanine, DL-Alanine and Sarcosine Monophosphates," *Acta Crystallographica*, Vol. C44, 1988, pp. 1968-1972.

[4] M. T. Averbuch-Pouchot, A. Durif and J. C. Guitel, "Structures of Glycine Monophosphate and Glycine Cyclo-Triphosphate," *Acta Crystallographica*, Vol. C44, 1988, pp. 99-102.

[5] M. Bagieu-Beucher, "Structure of Cytosinium Dihydrogenmonophosphate," *Acta Crystallographica*, Vol. C46, 1990, p. 238.

[6] R. H. Blessing, "Hydrogen Bonding and Thermal Vibrations in Crystalline Phosphate Salts of Histidine and Imidazole," *Acta Crystallographica*, Vol. B42, 1986, pp. 613-621.

[7] M. T. Averbuch-Pouchot, A. Durif and J. C. Guitel, "Structure of Ethylenediammonium Dihydrogentetraoxophosphate(V) Pentahydrogenbis[Tetraoxophosphate(V)]," *Acta Crystallographica*, Vol. C45, 1989, pp. 421-423.

[8] S. Kamoun, M. Kamoun, A. Jouini and A. Daoud, "Structure of Ethylenediammonium Bis(Dihydrogenmonophosphate)," *Acta Crystallographica*, Vol. C45, 1989, pp. 481-482.

[9] M. Bagieu-Beucher, A. Durif and J. C. Guitel, "Structure of Ethylenediammonium Dihydrogentetraoxophosphate(V) Pentahydrogenbis[tetraoxophosphate(V)]," *Acta Crystallographica*, Vol. C43, 1989, pp. 421-423.

[10] S. Kamoun, A. Jouini and A. Daoud, "Structure du Propanediammonium-1,3 Monohydrogénomonophosphate Monohydrate," *Acta Crystallographica*, Vol. C47, 1991, pp. 117-119.

[11] S. Kamoun, A. Jouini, A. Daoud, A. Durif and J. C. Guitel, "Structure du Diammonium-1,3 Propane Bis(Dihydrogénomonophosphate)," *Acta Crystallographica*, Vol. C48, 1992, pp. 133-135.

[12] S. Kamoun and A. Jouini, "Etude Calorimétrique et Structure Cristalline du Putrescinium Monohydrogénomonophosphate Dihydrate $NH_3(CH_2)_4NH_3HPO_42H_2O$," *Journal of Solid State Chemistry*, Vol. 89, No. 1, 1990, pp. 67-74.

[13] F. Takusagaxa and T. F. Koetzle, "A Study of the Charge Density in Putrescine Diphosphate at 85 K," *Acta Crystallographica*, Vol. B35, 1979, pp. 867-877.

[14] Z. Elaoud, S. Kamoun, T. Mhiri and J. J. Jaud, "Crystal Structure of Triethylentetraammonium Bis Monohydrogenmonophosphate Dihydrate, $[NH_3(CH_2)_2NH_2(CH_2)_2NH_2(CH_2)_2NH_3](HPO_4)_2{\cdot}2H_2O$," *Journal of Chemical Crystallography*, Vol. 29, 1999, pp. 541-545.

[15] Z. Elaoud, S. Kamoun, T. Mhiri, F. Romain and H. J. Burzlaff, "Crystal Structure and Phase Transitions in N-benzyl Piperidinium Dihydrogenmonophosphate, $C_6H_5CH_2CHCH_2CH_2NH_2CH_2CH_2{\cdot}H_2PO_4$," *Journal of Solid State Chemistry*, Vol. 155, 2000, pp. 298-304.

[16] Z. Elaoud, S. Kamoun, J. J. Jaud and T. Mhiri, "Crystal Structure of N-benzylmethylammonium Dihydrogen-Monophosphate Monohydrate, $[C_6H_5CH_2NH_2CH_3]H_2PO_4{\cdot}H_2O$," *Journal of Chemical Crystallography*, Vol. 28, 1998, pp. 313-315.

[17] S. Kamoun, A. Jouini and A. Daoud, "Structure du Aza-3 Pentanediyle-1,5 Diammonium Monohydrogénomonophosphate Dihydrate," *Acta Crystallographica*, Vol. C46, 1990, pp. 1481-1483.

[18] S. Kamoun, A. Daoud, A. Elfakir, M. Quarton and I. Ledoux, "Linear and Nonlinear—Optical Properties on N-Diethylendiammonium Monohydrogenophosphate Dihydrate," *Journal of Solid State Chemistry*, Vol. 94, 1995, pp. 893-896.

[19] C. Lee and W. T. A. Harrison, "Tetraethylammonium Dihydrogenarsenate Bis(Arsenic Acid) and 1,4-Diazoniabicyclo[2.2.2]octane Bis(Dihydrogenarsenate) Arsenic Acid: Hydrogen-Bonded Networks Containing Dihydrogenarsenate Anions and Neutral Arsenic Acid Molecules," *Acta Crystallographica*, Vol. C63, 2007, pp. m308-m311.

[20] G. Charlot, "Chimie Analytique Quantitative," Vol. 2, Masson and Cie, Paris, 1974.

[21] E. A. Muller, R. J. Cannon, A. N. Sarjeant, K. M. Ok, P. S. Halasyamani and A. J. Norquist, "Directed Synthesis of Noncentrosymmetric Molybdates," *Crystal Growth & Design*, Vol. 5, No. 5, 2005, pp. 1913-1917.

[22] Nonius, "Kappa CCD Program Software," Nonius BV, Delft, 1998.

[23] L. J. Farrugia, "WinGX Suite for Small-Molecule Single-Crystal Crystallography," *Journal of Applied Crystallography*, Vol. 32, 1999, p. 837.

[24] G. M. Sheldrick, "SHELXS-97 Programs for Crystal Solution," University of Göttingen, Göttingen, 1997.

[25] G. M. Sheldrick, "SHELXL-97 Programs for Crystal Structure Refinement," University of Göttingen, Göttingen, 1997.

[26] I. D. Brown, "On the Geometry of O-H...O Hydrogen Bonds," *Acta Crystallographica*, Vol. A32, 1976, pp. 24-31.

[27] R. H. Blessing. "Hydrogen Bonding and Thermal Vibrations in Crystalline Phosphate Salts of Histidine and Imidazole," *Acta Crystallographica*, Vol. B42, 1986, pp. 613-621.

[28] J. Baran, M. Śledź, M. Drozd, A. Pietraszko, A. Haznar and H. Ratajczak, "Structural, Vibrational and DSC Investigations of the Tetraethylammonium Hydrogenselenate Crystal," *Journal of Molecular Structure*, Vol. 526, 2000, pp. 361-371.

[29] E. Steinwender, E. T. G. Lutz, J. H .van der Maas and J. A. Kanters, "2-Ethynyladamantan-2-ol: A Model Compound with Distinct OH···π and CH···O Hydrogen Bonds," *Vibrational Spectroscopy*, Vol. 4, No. 2, 1993, pp. 217-229.

[30] J. H. Loehlin and A. Kvick, "Tetraethylammonium Chloride Monohydrate," *Acta Crystallographica*, Vol. B34, 1978, pp. 3488-3490.

[31] T. C. W. Mak, H. J. B. Slot and P. T. Beurskens, "Tetraethylammonium Chloride Tetrahydrate, a 'Double Channel' Host Lattice Constructed from $(H_2O)_4Cl^-$ Tetrahedra Linked between Vertices," *Journal of Inclusion Phenomena*, Vol. 4, No. 3, 1986, p. 295.

[32] R. J. Staples and Z. Kristallogr, "Crystal Structure of 4,4-Dimethyloxazolidine-2-Thione, C_5H_9NOS," *NCS*, 1999, pp. 214- 231.

[33] A. Novak, "Hydrogen Bonding in Solids Correlation of Spectroscopic and Crystallographic Data," *Structure and Bonding*, Vol. 18, 1974, pp. 177-178.

[34] M. Gosniowska, Z. Ciunik, G. Bator, R. Jakubas and J. Baran, "Structure and Phase Transitions in Tetramethylammonium Tetrabromoindate(III) and Tetraethylammonium Tetrabromoindate(III) Crystals," *Journal of Molecular Structure*, Vol. 555, 2000, p. 243.

[35] M. Karbowiak, J. Hanuza, J. Janczak and J. Drozdzynski, "Synthesis, Structural and Spectroscopic Properties of Tetra(Tetraethylammonium) Heptaisothiocyanato Uranate (III) and Neodymate(III)," *Journal of Alloys and Compounds*, Vol. 225, 1995, p. 338.

[36] H. G. Heddrich and C. E. Blom, "Flame Diagnostics and Molecular Constants of CaO by Tunable Diode Laser Spectroscopy," *The Journal of Chemical Physics*, Vol. 90, 1989, p. 4660.

[37] H. G. Heddrich and C. E. Blom, "The Infrared Spectrum of Barium Oxide in the Gas Phase," *Journal of Molecular Spectroscopy*, Vol. 140, 1990, p. 103.

[38] A. D. Kirkwood, K. D. Bier, J. K. Thompson, T. L. Haslett, A. S. Hubber and M. Moskovits, "Ultraviolet-Visible and Raman Spectroscopy of Diatomic Manganese Isolated in Rare-Gas Matrixes," *The Journal of Chemical Physics*, Vol. 95, 1991, p. 2644.

[39] G. Ma, T. Zhang, K. Yu and J. Braz, "Synthesis, X-Ray Crystal Structure and Thermal Decomposition Mechanism of $[Zn(MCZ)_3](NO_3)_2{\cdot}H_2O$ (MCZ=Methyl Carbazate)," *Journal of the Brazilian Chemical Society*, Vol. 16, 2005, p. 796.

[40] M. Drozd, "The Equilibrium Structures, Vibrational Spectra, NLO and Directional Properties of Transition Dipole Moments of Diguanidinium Arsenate Monohydrate and Diguanidinium Phosphate Monohydrate: The Theoretical DFT Calculations," *Spectrochimica Acta Part A*, Vol. 65, No. 5, 2006, pp. 1069-1086.

Supplementary Material

CCDC831725 contains the supplementary crystallographic data for 1. These data can be obtained free of charge via www.ccdc.cam.ac.uk/data_request/cif or by emailing data_request@ccdc.cam.ac.uk, or by contacting the Cambridge Crystallography Data Centre 12, Union Road, Cambridge CB2 1EZ, UK [Fax: +441223336030].

Growth Optimization, Strain Compensation and Structure Design of InAs/GaSb Type-II Superlattices for Mid-Infrared Imaging

Yuxin Song[1*], Shumin Wang[1,2*], Carl Asplund[3], Rickard Marcks von Würtemberg[3], Hedda Malm[3], Amir Karim[4], Xiang Lu[5], Jun Shao[5]

[1]Department of Microtechnology and Nanoscience, Photonics Laboratory, Chalmers University of Technology, Gothenburg, Sweden

[2]State Key Laboratory of Functional Materials for Informatics, Shanghai Institute of Microsystem and Information Technology, Chinese Academy of Sciences, Shanghai, China

[3]IRnova AB, Kista, Sweden

[4]Acreo AB, Kista, Sweden

[5]National Laboratory for Infrared Physics, Shanghai Institute of Technical Physics, CAS, Shanghai, China

ABSTRACT

InAs/GaSb type-II superlattce (T2SL) photodetector structures at the MWIR regime were grown by molecular beam epitaxy. The growth temperature and group-V soaking times were optimized with respect to interface and transport quality. Novel strain compensation schemes with insertion of InSb layers were proposed and tested to be efficient to tune the overall strain between tensile and compressive without degradation of interface and optical quality. The effect of the proposed methods is modeled by analytic functions. Band structure calculations were also carried out for the proposed T2SL structures to assist optimizing sample designs. Single pixel photodiodes with a low dark current were demonstrated.

Keywords: InAs/GaSb; Type-II Superlattce; Molecular Beam Epitaxy; Strain Compensation

1. Introduction

In the two atmospheric transmission windows at a wavelength range of 3 - 5 and 8 - 14 μm [1], there is a large demand for imaging within industrial, biomedical, civilian-surveillance and military applications. The most developed materials are $Cd_xHg_{1-x}Te$ (MCT), InSb and quantum well infrared photodetectors (QWIPs) [2-4]. Type-II superlattices (T2SL) comprising combinations of (Al, Ga, In)Sb and InAs layers have received increased interest the last decade as a promising candidate in photodetector applications. InAs/GaSb T2SL structure was first proposed by Sai-Halasz *et al.* in 1977 [5] and suggested to be a candidate for infrared detection with potentials for superior performance and production yield compared with MCT by Smilth and Mailhiot in 1987 [6]. This is especially true for applications in the long-wave infrared (LWIR) atmospheric transmission window of 8 - 12 μm, where the difficulty of maintaining sufficient composition control of MCT leads to low production yields, and where the low quantum efficiency of quan- tum-well infrared (QWIP) detectors preclude their use for certain critical applications. For imaging in the mid- wave infrared (MWIR) transmission window of 3 - 5 μm, on the other hand, there are already two well established, competing technologies with rather high manufacturability: large-area, high quality focal plane detectors of InSb are available, but they require cooling typically to 80 K. MCT detectors operate with similar signal-to-noise ratios at temperatures beyond 100 K, which means longer cooler lifetime and reduced power consumption. However, the use of the new material T2SL can be motivated also for this wavelength range, since it already surpasses performance of InSb by a wide margin, and is compatible with standard III-V processes and processing tools, unlike MCT. Furthermore, T2SL holds-at least theoretically-promise of even higher operating temperatures than

*Corresponding authors.

MCT from strongly reduced Auger recombination. The provision is that the density of the as-yet unidentified Shockley-Read-Hall centers can be decreased significantly.

InAs and GaSb are well known III-V materials with established processing technologies. The closely matched lattice constants make T2SL materials possible to be grown by epitaxy without introducing large extra strain. By changing the layer thicknesses, the effective band gap can be tuned from 0.5 eV to 0 eV (semi-metallic) [7]. Detectors with a cutoff wavelength as long as 32 μm have been demonstrated [8]. The close lattice constants also allow for integrating multiple structures with different wavelengths in the same device, resulting in multi-color detectors and cameras [9]. Tunneling currents in T2SLs can be reduced due to a large electron effective mass. Moreover, Auger recombination can be suppressed to a large extent owing to the large valence band splitting [10]. T2SL structures can also provide high quantum efficiency. The responsivity of T2SL detectors has already proved to be similar to those made of MCT [11]. Thanks to modern molecular beam epitaxy (MBE) technology, T2SL materials can be grown with a very low defect density and high uniformity, making it especially suitable for fabrication of FPAs [12]. During the period 1996 to 2005, significant developments in growth and fabrication of T2SL structures were achieved, leading to the first demonstration of a T2SL based FPA in 2005 [13]. Since then, interest has grown significantly in development of T2SL based detectors and FPAs [9,14-17].

One major challenge for successful growth is the control of overall strain, primarily caused by the small lattice mismatch between InAs and GaSb [18,19] and the chemical nature at InAs/GaSb interfaces. Strain relaxation and formation of dislocations can occur if the overall strain is not effectively compensated in a several micrometers thick detector structure. The InAs/GaSb interface can be of GaAs type (tensile strain) or InSb type (compressive strain). Some strain compensation methods were tested. The main idea is to increase the percentage of InSb at interfaces to introduce compressive strain, thus compensating the accumulated tensile strain caused by InAs and by GaAs-type interfaces. Plis *et al.* studied the strain compensation by using Sb soaking after the growth of InAs layer to covert part of the InAs into InSb [20]. The method of intentionally inserting a thin InSb layer after the growth of InAs layer was utilized by Rodriguez [21], Haugan [22], Plis [23] *et al.* Interface quality also plays a very important role for T2SL-based device performance, and therefore large efforts were made toward understanding and improving the quality of interfaces [18] [24-26]. The effects of soaking using group-V elements at interfaces on interface quality and overall strain were studied [20,27]. Moreover, control of background doping level [28-30] and influence of substrate quality on device performance were also investigated [31,32].

In this paper, we present optimization of growth parameters and strain compensation of InAs/GaSb T2SLs grown by MBE aiming for fabricating mid-wavelength infrared (MWIR) FPA devices. The MBE system was designed for growth of both As- and Sb-compounds and a heavy background As pressure is expected. This will make it more difficult for effective strain compensation compared with a dedicated MBE system with a minimum effect of background As. Novel strain compensation schemes with insertion of one monolayer InSb layer inside the GaSb layer, at the InAs/GaSb and GaSb/InAs interfaces and their combinations are proposed, modeled, analyzed and tested, and their influence on structural and optical qualities is compared. We find that proper combinations of the proposed schemes can control interface quality of InAs/GaSb T2SLs and tune the overall strain between tensile and compressive. Band structure calculations for the proposed structures were carried out to assist understanding and future improvements. Finally, single pixel photodetectors with a low dark current were demonstrated using the optimized growth conditions.

2. Experiments

2.1. Sample Growth

All the samples were grown by a Riber Compact21 MBE system with a cluster tool, equipped with dual filament effusion cells for In and Ga and valved crackers for As and Sb. To lower the substrate cost at the initial optimizations of growth parameters, semi-insulation GaAs(100) substrates were employed. The GaAs substrates were first de-oxidized at 660°C (all growth temperatures used in this paper were read by a thermocouple) followed by a 100 nm GaAs buffer layer grown at 620°C to smoothen the growth front. The substrate temperature was then ramped down to 510°C with the As valve closed. An atomic monolayer of Ga was deposited to achieve group-III-rich surface reconstruction and then a 1 μm GaSb buffer layer was grown using the interfacial misfit array (IMF) method [33] to accommodate the 7.8% lattice mismatch between GaAs and GaSb and to obtain smooth surface with a low defect density. Latter samples were grown on n-type GaSb (100) substrates. The GaSb substrates were de-oxidized at 580°C followed by a 100 nm GaSb buffer layer grown at 510°C. The T2SL structures were grown on the buffer layer at varied, but much lower growth temperatures (T_g). Reflection high-energy electron diffraction (RHEED) was used for in-situ monitoring of the growth process.

2.2. Structural Property Characteriztion

X-ray diffraction (XRD) based on PANalytical X'Pert

PRO X-ray diffraction system was employed as the main tool to characterize the structural properties. As most physical processes occur at, or close to, the interfaces between InAs and GaSb layers in the superlattice, the interface quality is one of the major parameters for structural quality and can be quantified by measuring the full width at half maximum (FWHM) of the diffraction fringes from the superlattice in XRD rocking curves. The narrower the diffraction peaks, the smoother the interfaces. Moreover, the separation between the GaSb substrate (or GaSb buffer layer) peak and the 0th order diffraction peak of the superlattice in XRD rocking curves can be used to measure the overall lattice mismatch of a T2SL to GaSb. Although the lattice constants of InAs and GaSb are very close (6.0960 vs 6.0583 Å), there is still 0.6% lattice mismatch (tensile strain) when growing InAs on GaSb. Different types of interfaces discussed in detail below may also bring in additional strain. Strain relaxation may occur for micrometer thick devices. An example of this can be seen in the XRD rocking curves of a sample grown on GaAs in **Figure 1(a)**, where each superlattice peak is split into two. A cross-sectional transmission electron microscopy (XTEM) image shown in **Figure 1(b)** confirms that the upper part of the T2SL structure is partially relaxed by nucleation of misfit dislocations (MDs). These MDs are usually accompanied by threading dislocations (TDs) terminated on the sample surface. The two sets of diffraction peaks correspond to X-ray interference from the upper relaxed and the lower strained superlattices, respectively. Simulation and fitting of the XRD rocking curves for T2SLs is possible, but trivial. As the total thickness of one period of a T2SL structure for MWIR detection is commonly less than 20 ML, the two interfaces compose a significant portion of strain contribution in each period and their chemical compositions are usually unknown in simulation and have to be regarded as fitting parameters. We found that the calibration of InAs and GaSb growth rates based on the periodicity of the superlattice diffraction peaks is no longer accurate. RHEED oscillation is recommended for calibration of the GaSb and InAs growth rate.

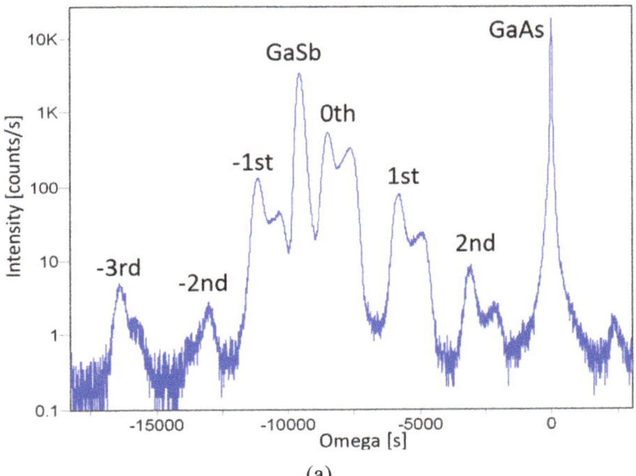

(a)

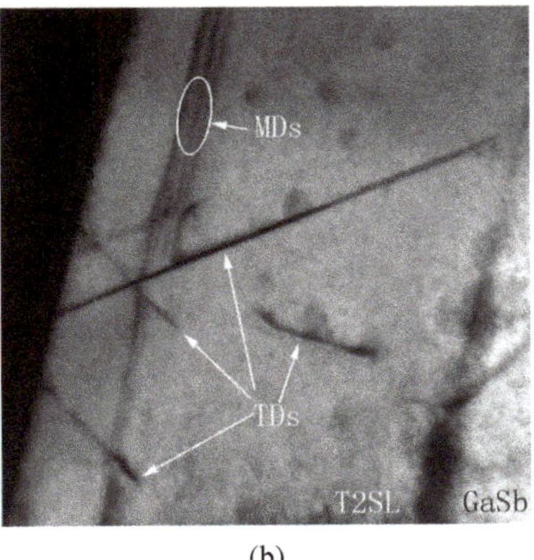

(b)

Figure 1. (a) XRD rocking curve of one partially relaxed T2SL sample (100 periods of 10 ML GaSb/10 ML InAs) grown on a GaAs substrate, and (b) an XTEM image of the same sample. Both MDs and TDs are marked.

2.3. Optical Property Characterization

Photoluminescence (PL) at 77 K was used to examine the bandgap as well as the optical quality of the T2SL material. The PL measurements were based on a Bruker/Fourier transform infrared spectroscopy (FTIR) Equinox 55 system with Hyperion microscope.

3. Optimization of Growth Conditions

Growth temperature is one of the most important parameters for MBE growth of T2SL structures. **Figure 2(a)** compares XRD rocking curves of the two samples grown at 340˚C and 470˚C, respectively. It can be found that the one grown at the low temperature shows better structural quality with narrow diffraction peaks and clear high order peaks, while the one grown at the high T_g has an inferior superlattice structure judged by broadening of the diffraction peaks. High growth temperature will enhance inter-diffusion between different elements and also segregation of large atoms *i.e.* In and Sb, leading to deteriorated interface quality. Therefore it is essential to employ low growth temperatures to obtain sharp interfaces. Due to different heater designs and temperature measurement techniques employed in different MBE systems, the nominal T_g-values reported in the literature are unreliable. We used the temperature and Sb flux dependent (1 × 3) to (2 × 5) surface reconstruction transition on GaSb, described by Bracker *et al.* [34], as the temperature reference. The proper growth temperature is commonly lower than the transition temperature under a certain Sb flux [10]. We optimized the growth temperature in the range of 320˚C - 380˚C and the XRD, atomic force microscopy (AFM) and Hall measurement results are summarized in **Figures 2(b)** and **(c)**, respectively. It can be found from

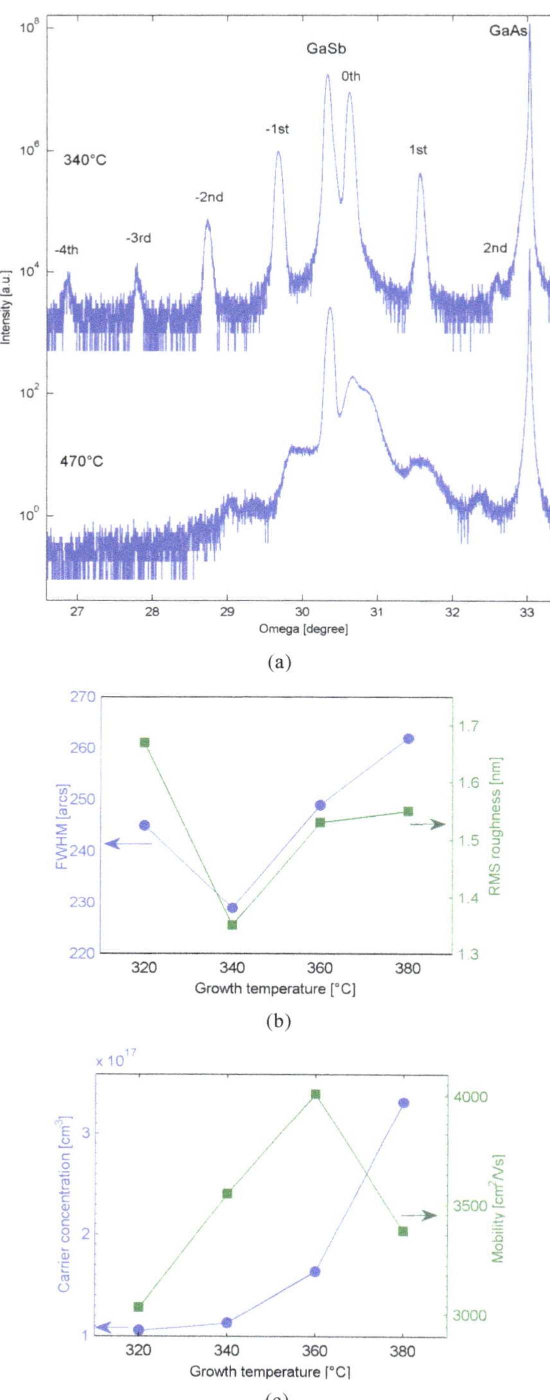

Figure 2. Effects of growth temperature on interface and transport quality of T2SL structures on GaAs comprising 100 periods of 10 ML InAs/10 ML GaSb. (a) shows XRD(004) rocking curves grown at 340°C and 470°C, respectively. (b) shows the FWHM of the -1st order diffraction peaks (blue round) and the RMS roughness values measured in 10 × 10 μm^2 AFM scans (green square) vs. T_g. (c) shows the carrier density (blue round) and mobility (green square) vs. T_g. All measurements were carried out at room temperature.

Figure 2(b) that the sample grown at 340°C has both the best interface quality as indicated by the narrowest −1st order diffraction peak and the smoothest surface. Growth temperature also has an effect on carrier density and mobility, as shown in **Figure 2(c)**. The general trend is that the carrier density increases with T_g, while the electron mobility reaches a maximum at 360°C. We chose T_g = 340°C for the later samples. This temperature is much lower than the optimal temperature for growing GaSb.

The residual doping of an InAs bulk layer is normally *n*-type, while it is *p*-type for a GaSb bulk layer, presumably due to the Fermi level pinning. The residual carrier density in an InAs/GaSb T2SL is determined by the compensation of unintentionally doped carriers in GaSb and InAs layers [29]. The optimal T_g-value for InAs/GaSb T2SLs is far below the optimal T_g-values for GaSb and InAs bulk materials, leading to possibly a higher density of impurities in both layers than those grown at elevated growth temperatures. The measured net *n*-type carriers indicate that there are more electrons contributed by InAs than the holes by GaSb. It's still unclear why the carrier density increases with growth temperature while the mobility has a maximum at 360°C.

Due to the fact that dual group-V elements are involved in growth of T2SL materials, soaking of one element at the interface is commonly used and necessary for tuning the overall strain. Arsenic soaking after growth of GaSb will strengthen GaAs-like interface leading to extra tensile strain in the structure. However, it has been reported that the As soaking can help in obtaining high crystalline and optical quality [27]. We have tested four different As soaking times with otherwise the same growth conditions. The XRD(004) rocking curves shown in **Figure 3** reveal huge differences between samples with and without As soaking. For the sample grown without As soaking, the interfaces have very low quality with

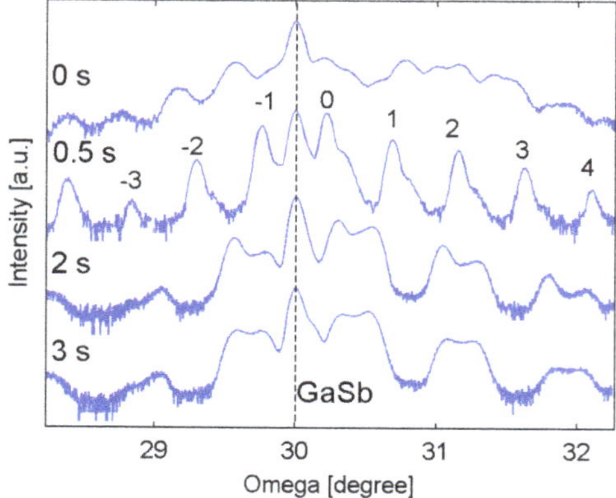

Figure 3. XRD (004) rocking curves of T2SLs composed of 100 periods 10 ML InAs/10ML GaSb grown at 340°C on GaAs substrates with different As soaking time labeled above each curve. The order of the diffraction peaks for the sample with 0.5 s As soaking is marked above the curve.

almost no clear diffraction peaks. However, as short as 0.5 second As soaking makes a significant difference.

This sample has well shaped high order diffraction peaks. Further increasing the As soaking time to 2 s and 3 s leads to peak splitting, indicating strain relaxation possibly caused by As-soaking induced excess tensile strain. This underlines the importance of strain compensation, which will be discussed in detail in the next section. For most of the following samples, 0.5 second As soaking time was employed after growth of GaSb. It should be noted that the proper As soaking time is related to the As flux used as well as the background As during the growth of GaSb. For the optimization shown in **Figure 3**, an Addon As-cracker was used. This cracker has an inefficient cracking efficiency and a severe leakage when the needle valve was closed leading to a high As background. When a Riber As-cracker was used later, the background As was reduced significantly and the optimal As-soaking may shift away from 0.5 second.

Antimony soaking after growth of the InAs layer up to 14 seconds was also tested. No obvious difference was observed. This is probably due to the large amount of accumulated As on the InAs surface, such that the Sb-soaking becomes ineffective. 12-second Sb soaking time was employed for most of the later samples.

4. Strain Compensation

Although InAs and GaSb are commonly regarded as a "lattice matched" material system, the 0.6% lattice mismatch still leads to overall tensile strain in the structure when grown on GaSb substrates and strain relaxation may occur when it is accumulated in micrometer thick devices. Threading dislocations will then strongly degrade the device performance. Therefore, a strain compensation scheme to avoid formation of TDs is essential for T2SL structures.

Different types of interfaces can also bring in excess strain. These, however, can be engineered. There are basically two types of interfaces in InAs/GaSb T2SL structures, GaAs-like and InSb-like. Due to the different bond lengths, GaAs-like interfaces will add tensile strain, while InSb-like interfaces bring in compressive strain. Apparently, an InSb-like interface is preferable to compensate the accumulated tensile strain of InAs. Use of group-V-rich flux is the common condition for MBE growth of III-V compounds, *i.e.* each layer of InAs and GaSb is group V terminated. When growth of an InAs layer is finished, the top atomic layer is As. The next incorporated atomic layer will be the group-III element, Ga, when the growth of a GaSb layer is initiated. Consequently, the InAs/GaSb interface is connected by As-Ga bonds, forming a GaAs-like interface as shown in **Figure 4(a)**. One would expect an InSb-like interface to form in a similar way when an InAs layer is grown on a GaSb layer, as shown in **Figure 4(b)**. However, since the substitution of Sb atoms by As atoms is thermodynamically favored, the result can be a partial or complete GaAs-like interface also in this case; see **Figure 4(c)**. Therefore, the chance to form GaAs-like interfaces is usually higher than that for InSb-like interfaces, leading to increased overall tensile strain, *i.e.* the 0th peak appears at a larger angle than that of the GaSb peak.

For a small lattice mismatched heterostructure grown at low growth temperatures, it is reasonable to assume that there is little intermixing between group-III atoms through diffusion or segregation. For an ordinary InAs/GaSb T2SL structure shown in **Figure 5(a)**, the number of In and Ga atomic layers in each period is denoted by M and N, respectively. We assume that As and Sb intermixing only occurs at the interface to form one atomic layer of AsSb and let x and y to be the As composition at the GaSb/InAs and the InAs/GaSb interfaces, respectively. Subsequently, for the last Ga layer at the GaSb/InAs interface, the bottom (left in the figure) bonds are Ga-Sb bond, while the up (right) bonds become partially Ga-As and Ga-Sb bonds with proportion of x and $(1-x)$, respectively. For the first In layer, the bottom

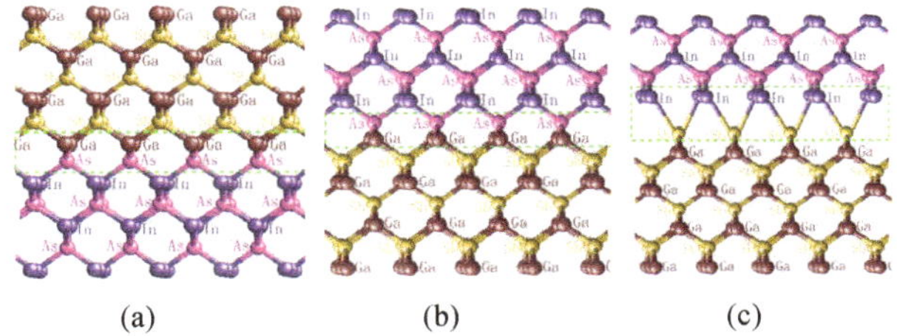

Figure 4. Schematic illustrations of different types of interfaces: (b) InSb-like interface, (a) and (c) GaAs-like interface.

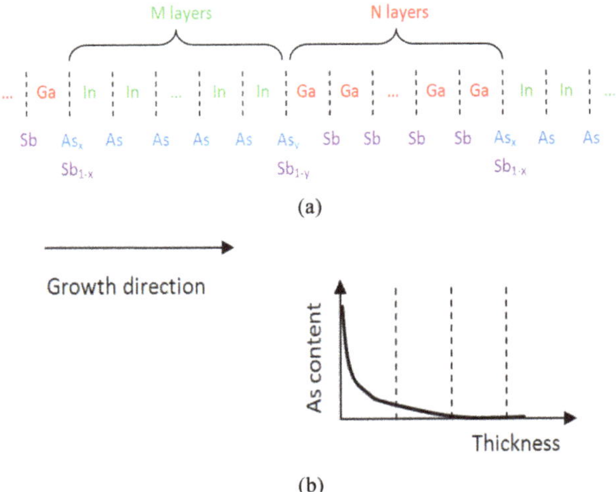

Figure 5. Schematics of (a) an InAs/GaSb T2SL structure, and (b) incorporation of residual As into the following GaSb layer.

bonds are x In-As and $(1-x)$ In-Sb while the up bonds are still In-As. It is a similar situation for the other inter face. The In-As, Ga-As and In-Sb bonds all contribute to the lattice mismatch to the GaSb substrate. The overall lattice mismatch can be expressed as

$$f=\left[\left(M-1+\frac{x}{2}+\frac{y}{2}\right)f_{InAs}+\left(\frac{x}{2}+\frac{y}{2}\right)f_{GaAs}+\left(\frac{1-x}{2}+\frac{1-y}{2}\right)f_{InSb}\right]\Big/(M+N) \quad (1)$$

where f_{InAs}, f_{GaAs} and f_{InSb} are the lattice mismatch of InAs, GaAs and InSb with respect to GaSb, respectively. $(M+N)$ is the total number of atomic monolayers of one period of the T2SL structure. In the above derivation, we neglect the small difference in lattice constants of InAs and GaSb in the denominator. For an MBE chamber used to grow arsenides, a heavy amount of background As is expected to be incorporated during the growth of the GaSb layer, forming $GaSb_{1-z}As_z$, where z is the composition of As. Moreover, at such low growth temperatures, which are comparable to the As bulk temperature, the adsorbed As atoms may not be capable of re-evaporating from the sample surface if not incorporated, leading to a certain amount of residual adsorbed As atoms on the growth front of the InAs layer. Part of these adsorbed As atoms will be incorporated into the following GaSb layer as schematically shown in **Figure 5(b)**. A similar process could occur at the other interface for Sb atoms, but the effect will be much weaker due to the weak bonding of Ga-Sb compared with Ga-As and is negligible here. In addition, the background Sb in InAs is much lower than the background As in GaSb, judged by at least one order of magnitude difference in background pressure when growing GaSb and InAs. As a result, additional tensile strain will be introduced by the adsorbed and background As atoms. Then, Equation (1) becomes

$$f=\left[\left(M-1+\frac{x}{2}+\frac{y}{2}\right)f_{InAs}+\left(\frac{x}{2}+\frac{y}{2}+Nz'\right)f_{GaAs}+\left(\frac{1-x}{2}+\frac{1-y}{2}\right)f_{InSb}\right]\Big/(M+N) \quad (2)$$

where, $z'=\alpha/N+z$ is the average background As composition in GaSb and α is the total amount of incorporated As in GaSb layer due to the excess As atoms adsorbed after the InAs growth.

By inserting the values of f_{InAs}, f_{GaAs} and f_{InSb} with −0.0062, −0.0726 and 0.0629, respectively, we get

$$f=(0.0691-0.0062M-0.0709x-0.0709y-0.0726Nz')/(M+N) \quad (3)$$

Before applying strain compensation, the overall strain is measured to be tensile for $M = 10$, *i.e.* $f<0$. Therefore, the key idea for strain compensation is to reduce the values of x and y, as well as using a small As flux to minimize .

Strengthening the InSb-like interfaces by extensive Sb soaking to reduce y-value is a common strategy for strain compensation in T2SL structures. By using long time Sb soaking after growth of InAs layers, part of the top As atoms would be replaced by Sb atoms, forming partially InSb-like interfaces. **Figure 6(a)** shows the shutter sequence of this method. The overall lattice mismatch can be expressed in the same form as Equations (2) and (3). It is easily seen that the strain can never be fully compensated by only Sb soaking even with both x- and y-values equal to zero if the InAs thickness is more than 11 ML. Sb-As exchange during the Sb soaking has been found to be much less effective than As-Sb exchange during the As soaking [20,27]. Therefore, we expect that the value of y is high, or even close to 1. If we assume that $y=1$, Equation (3) reduces to

$$f=(-0.0018-0.0062M-0.0709x-0.0726Nz')/(M+N) \quad (4)$$

This indicates that f can never reach zero for any thickness of InAs using only soaking. We thus propose new strain compensation methods based on shutter sequences as schematically shown in **Figures 6(b)-(d)**. We introduce an additional In layer to reinforce the InSb-like interfaces. In method (b), the 1 ML In is deposited right after GaSb, and then soaked with Sb to form 1 ML InSb. The short As soak before switching to InAs leads to formation of a monolayer of InAsSb due to As-Sb intermixing. In method (c), a short Sb soaking is employed after the growth of InAs layer to remove residual As and establish an Sb environment on the growth front. Then the 1 ML In is deposited followed by Sb soaking again. Thus, a full monolayer of InSb is formed. In both cases, formation of GaAs-like interfaces can be minimized. In method (d), the 1ML InSb is directly inserted in the middle of GaSb layer, as far as possible from the As-containing interfaces. The overall lattice mismatch in each of these methods can be expressed below in **Table 1**.

Since one additional monolayer of In is inserted for methods (b), (c) and (d), the total thickness in each period is changed to $(M+N+1)$. If we compare the first

Table 1. Expressions of overall lattice mismatch in different strain compensation methods.

Methods	Overall lattice mismatch
(a)	$f=\left[\left(M-1+\frac{x}{2}+\frac{y}{2}\right)f_{InAs}+\left(\frac{x}{2}+\frac{y}{2}+Nz'\right)f_{GaAs}+\left(\frac{1-x}{2}+\frac{1-y}{2}\right)f_{InSb}\right]/(M+N)$ $f=(0.0691-0.0062M-0.0709x-0.0709y-0.0726Nz')/(M+N)$
(b)	$f=\left[\left(M-1+x+\frac{y}{2}\right)f_{InAs}+\left(\frac{y}{2}+Nz'\right)f_{GaAs}+\left(2-x-\frac{y}{2}\right)f_{InSb}\right]/(M+N+1)$ $f=(0.132-0.0062M-0.0691x-0.0709y-0.0726Nz')/(M+N+1)$
(c)	$f=\left[\left(M+\frac{x}{2}\right)f_{InAs}+\left(\frac{x}{2}+Nz'\right)f_{GaAs}+\left(1-\frac{x}{2}\right)f_{InSb}\right]/(M+N+1)$ $f=(0.0629-0.0062M-0.0709x-0.0726Nz')/(M+N+1)$
(d)	$f=\left[\left(M-1+\frac{x}{2}+\frac{y}{2}\right)f_{InAs}+\left(\frac{x}{2}+\frac{y}{2}+Nz'\right)f_{GaAs}+\left(2-\frac{x}{2}-\frac{y}{2}\right)f_{InSb}\right]/(M+N+1)$ $f=(0.132-0.0062M-0.0709x-0.0709y-0.0726Nz')/(M+N+1)$

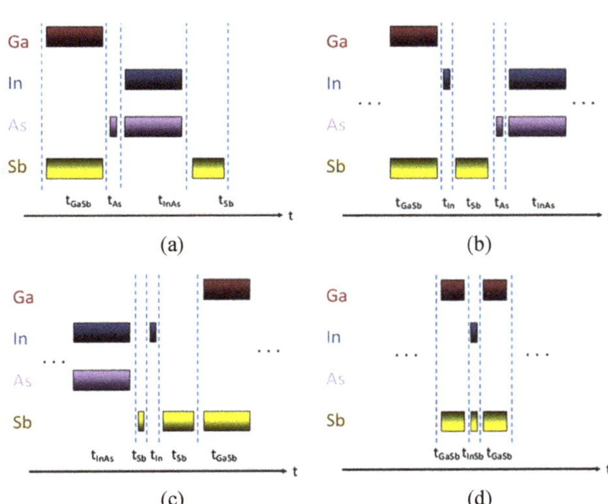

Figure 6. Shutter sequences of different strain compensation methods. (a) Common way with As and Sb soakings at interfaces only. (b) Strengthened In(As)Sb-type interface after a GaSb layer. (c) Strengthened In(As)Sb-type interface after an InAs layer. (d) One additional InSb monolayer inserted in the middle of a GaSb layer.

equation of each method, the first term with f_{InAs} does not differ very much in all the cases. The second term with f_{GaAs} is reduced to roughly half in (b) and (c). This is a direct evidence that insertion of 1 ML InSb at one interface prevents direct contact of As and Ga atoms from forming GaAs-like interface. The last term with f_{InSb} reflects the effect of the one monolayer InSb.

f, M and N are variables, which can be controlled or measured by XRD. So they can be considered as known values. By changing the GaSb thickness, *i.e.* N, but keeping all other growth parameters the same, the average residual As composition in the GaSb layer, z', can be obtained with any methods discussed above. If we assume that the overall lattice mismatch of an original T2SL structure is f_1, and that of another structure with the GaSb layer thickness changed by ΔN, is f_2, we can calculate the z' as

$$z'=\frac{(M+N)f_1-(M+N+\Delta N)f_2}{0.0726\Delta N} \quad (5)$$

It is obvious from **Figure 6** that, under the same growth conditions, the x-value should be the same for methods (a), (c) and (d), while the y-value should be the same for (a), (b) and (d). z' value should be the same for all the cases as long as the same As flux is used. Upon knowing the z'-value, both x- and y-values can be readily obtained by solving the equations. This provides quantitative information of chemical compositions at the two interfaces and indicates the effectiveness for the particular strain compensation technique used. The above arguments also indicate that strain compensation can be implemented in a controlled manner, making it easy to balance strain compensation with other requirements for a detector, such as bandgap, optical and structural qualities.

The strain compensation methods illustrated in **Figure 6** were tested individually on the first batch of GaSb based samples with 100 periods of 10 ML InAs/10 ML GaSb layers. When one method was applied at one interface, the other interface was treated with the normal soaking. The FWHM of the −1st order diffraction peak for the sample with method (a) is 165 arcs, and the residual lattice mismatch is −0.51%. It turns out that the strain compensation method (b), (c) and (d) indeed reduces overall strain compared with (a) by 49%, 78% and 41%, respectively. Method (c) shows the largest strain compensation effect, but the FWHM of the −1st order diffraction peak is 2.3 times compared to that of method (a). Method (b) and (d) show narrower −1st diffraction peaks than that of method (a) by 34% and 76%, respectively, despite being less effective in strain compensation. Using above equations, we find $y = 95\%$ in method (a), close to 1 as we expect, and $x + 1.025Nz' = 26\%$ in method (c). This implies that at most 26% Sb was replaced by As during the 0.5 s As soaking and only 5% As is replaced by Sb when the InAs surface is soaked by Sb for 12 s. A PL comparison for the samples using the scheme (a) and (d) was shown in **Figure 7**. The sample grown with the method (d) has better interface quality and smaller lattice mismatch, and shows a much higher PL intensity than that with the method (a). Since the sample grown with the method (d) has one monolayer of InSb embedded in the GaSb layer, the PL peak wavelength is red-shifted compared to the sample grown with the method (a).

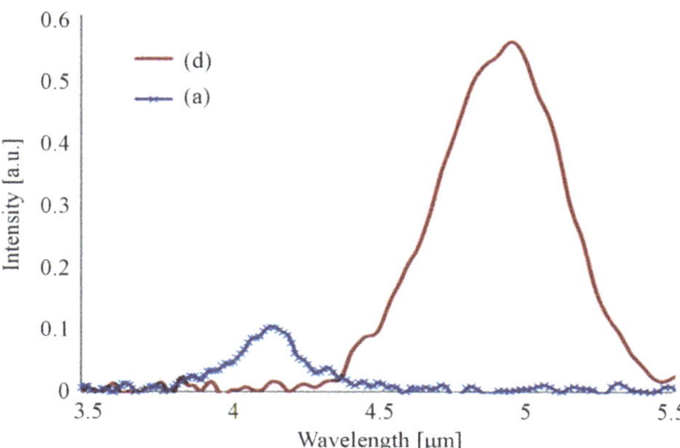

Figure 7. PL results at 77 K of samples with the strain compensation method (a) and (d). Both samples are grown on GaSb substrates with 100 periods of 10 ML InAs and 10 ML GaSb.

Strain compensation schemes with combinations of the above methods were investigated to further compensate the overall strain. The overall lattice mismatch can be expressed in the same way in **Table 2**.

The results on structural properties of this group of samples are summarized in **Table 2**. It should be noted that the thickness of InAs layers in this group of samples is 7 ML in order to adjust the PL wavelength to be shorter than 5 μm. The interface quality is surprisingly improved, compared to the case of employing one strain compensation method (b) or (c). Both the combinations of (b) + (d) and (c) + (d) show very small residual lattice mismatch and narrow −1st order diffraction peaks indicating high interface quality. Combination of (b) + (c) leads to slightly larger residual lattice mismatch and worse interface quality. By combining all the methods (b) + (c) + (d), the strain is tuned from tensile type to compressive type with relative good interface quality. Validation of the equations was tested with these measurement data. For example, for methods (b) + (c) and (b) + (c) + (d), they should have the same interface conditions. The calculated values of $x + 1.05Nz'$ from the equations fall within a deviation of only 6%. The contribution of each strain compensation method alone can be extracted through the above equations. The overall lattice mismatch of (a) was calculated to be 0.51%, while (b), (c) and (d) were found to individually reduce the mismatch by 0.18%, 0.15% and 0.27%, respectively. The extracted overall lattice mismatch of −0.51% for the method (a) is in good agreement with the measured value from the sample using only the method (a) discussed before. It was found that the method (d) is the most effective in strain compensation as it provides In-Sb bonds at both sides of the inserted In layer. The method (b) is slightly more effective than the method (c). PL measurements were carried out and show peak wavelengths around 4.5

Table 2. Expressions of overall lattice mismatch in combined strain compensation schemes.

Schemes	Overall lattice mismatch
(b) + (c)	$f = \left[(M+x)f_{InAs} + (Nz')f_{GaAs} + (2-x)f_{InSb}\right]/(M+N+2)$ $f = (0.1258 - 0.0062M - 0.0691x - 0.0726Nz')/(M+N+2)$
(b) + (d)	$f = \left[\left(M-1+x+\frac{y}{2}\right)f_{InAs} + \left(\frac{y}{2}+Nz'\right)f_{GaAs} + \left(3-x-\frac{y}{2}\right)f_{InSb}\right]/(M+N+2)$ $f = (0.1949 - 0.0062M - 0.0691x - 0.0709y - 0.0726Nz')/(M+N+2)$
(c) + (d)	$f = \left[\left(M+\frac{x}{2}\right)f_{InAs} + \left(\frac{x}{2}+Nz'\right)f_{GaAs} + \left(2-\frac{x}{2}\right)f_{InSb}\right]/(M+N+2)$ $f = (0.1258 - 0.0062M - 0.0709x - 0.0726Nz')/(M+N+2)$
(b) + (c) + (d)	$f = \left[(M-1+x)f_{InAs} + (Nz')f_{GaAs} + (3-x)f_{InSb}\right]/(M+N+3)$ $f = (0.1887 - 0.0062M - 0.0691x - 0.0726Nz')/(M+N+3)$

μm for samples grown with a combination of two methods, and higher intensity than that of the samples which were grown with only one method. To summarize, strain compensation strategies with a combination of inserting one monolayer InSb at one of the interfaces and one monolayer InSb in the GaSb layer are effective for both strain compensation and improvement of interface quality. In this way, the residual strain can be controlled between tensile and compressive type.

5. Band Engineering of the Proposed T2SL Structures

The insertion of 1 ML InSb in the GaSb layer of a T2SL structure has been found to be beneficial for both strain compensation and improving interface quality. The band structure of such T2SL structures has been calculated with a k·p model. **Figures 8(a)** and **(b)** show the alignments of the valence and conduction bands, as well as the envelope wavefunctions of a T2SL structure (10 ML InAs/5 ML GaSb/1 ML InSb/5 ML GaSb). It is found that the lowest lying conduction band state in the superlattice, namely C1, is a mixture of mainly bulk conduction band (C) and light hole (LH) states (78% and 19%, respectively), and very little of spin-orbit (SO) states. The lowest superlattice valence band state, HH1, on the other hand, is pure 100% heavy-hole (HH) in character. The electron wavefunction extends into the GaSb layer, while that of holes in the valence band is confined mostly in the GaSb layer. The corresponding band structure is shown in **Figure 8(c)**. The HH1 band is found to have very little dispersion in the growth direction, which means that the effective mass is large and that it is difficult for the holes to tunnel through the layers. InAs and GaSb thicknesses were varied individually in our calculations. **Figure 8(d)** shows the effective cutoff wavelength of the material as a function of layer thickness. Just as for T2SL structures without an InSb layer in the middle of GaSb layer, the superlattice bandgap is found to be sensitive only to the changes in the InAs thickness. The reason is that, due to the large heavy-hole effective mass of the GaSb valence band, an increased GaSb thickness leads only to a small upward shift of the HH1 energy. With increasing hole confinement layer (GaSb) thickness, the barrier strength for electrons increases, leading to reduced conduction band dispersion in the growth direction. The so created upward movement of the C1 band minimum just happens to almost perfectly match that of the HH1 energy.

The optical matrix element ||<C1|$P(\boldsymbol{k})$|HH1>|| is an indication of optical transition strength. It can be found in **Figure 8(e)** that the strength of optical transition is indeed larger with thinner GaSb and InAs layers. The bandgap and optical transition can thus be optimized individually.

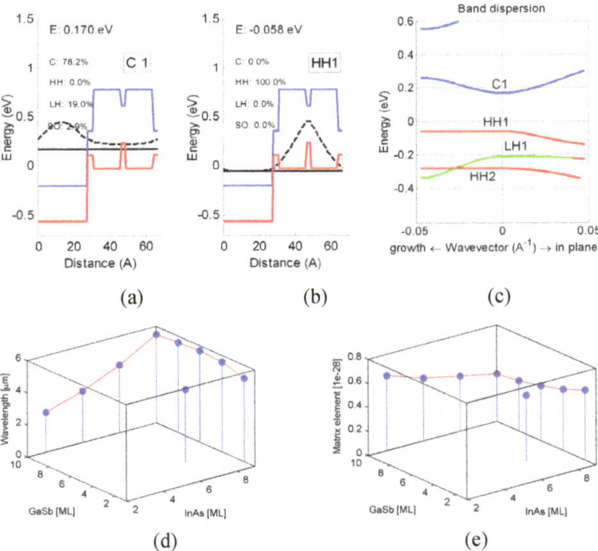

Figure 8. Band structure calculations based on a k·p model. (a) and (b) show band alignments of a T2SL structure with 10 ML InAs/5 ML GaSb/1ML InSb/5 ML GaSb and the calculated energy levels and zone center envelope wavefunctions for the lowest confined levels in the conduction and the valence band, respectively. (c) shows the band dispersion of the same T2SL structure along, and perpendicular to, the growth direction. (d) and (e) show the cutoff wavelength and zone center optical matrix element ||<C1|P(k=0)|HH1>||, respectively, as a function of InAs and GaSb thicknesses.

6. Single Pixel Photodetector

Single pixel *p-i-n* photodetectors are demonstrated. The structure consists of 560 periods of 4 ML InAs/8 ML GaSb T2SLs with strain compensation strategy of (b) + (c) + (d). The process uses a combination of dry- and wet-etching to create pixels. The mesa sidewalls are then passivated with polymer-based passivation and the contacts

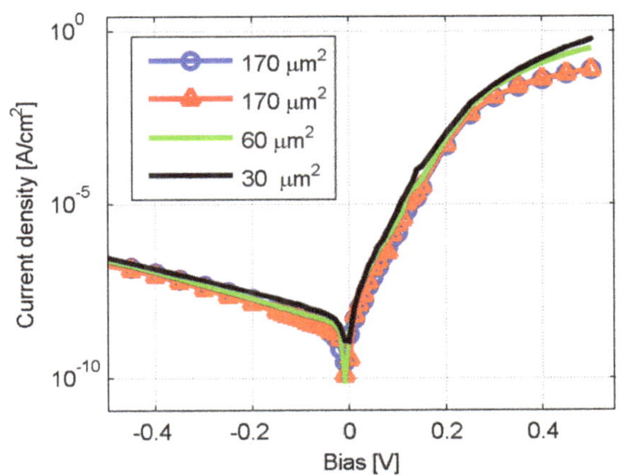

Figure 9. Measured dark current density for different pixel sizes at 80 K.

are deposited. The optical response is measured with a 100% cut off at 3.8 μm. **Figure 9** shows the dark current density for different pixel sizes at 80 K. The dark current density at small reverse biases is in the 10^{-9} A/cm^2 range at 80 K and increases to the 10^{-4} A/cm^2 range at 160 K, and is independent of pixel size as a result of effective device passivation.

7. Summaries

In summary, we have investigated MBE growth conditions of InAs/GaSb T2SL materials aiming for MWIR FPA detectors in an As-dominated growth environment. The growth temperature and the soaking time were optimized in terms of interface, surface and transport quality. The optimal growth temperature was found at around 340˚C - 360˚C. The 12-second Sb soaking on InAs was found to replace only 5% As atoms while the 0.5 second As soaking could replace Sb atoms up to 26%. Different strain compensation strategies by inserting one monolayer of InSb at different positions and their combinations were proposed and tested to be effective. Combinations of the strain compensation methods were found to not only further compensate the tensile strain but also lead to high interface quality and PL intensity. The overall strain can be tuned from tensile type to compressive type without degradation of the interface quality. The effect of the strain compensation is well modeled by analytic functions and the average As compositions at interfaces or in GaSb can be extracted. Optical properties of the samples were examined by PL. The absorption wavelength can be controlled within the MWIR range by adjusting InAs and GaSb thicknesses. Calculations of the band structures indicate that the absorption wavelength depends primarily on the InAs thickness when inserting 1 ML InSb in the middle of the GaSb layer. Single pixel photodetectors with a low dark current were demonstrated.

The Swedish Governmental Agency for Innovation Systems (VINNOVA) is acknowledged for financial support of this project.

8. Acknowledgements

The Swedish Governmental Agency for Innovation Systems (VINNOVA) is acknowledged for financial support of this project.

REFERENCES

[1] G. Hasnain, B. Levine, D. Sivco, A. Cho, "Mid-Infrared Detectors in the 3 - 5 μm Band Using Bound to Continuum State Absorption in InGaAs/InAlAs Multiquantum Well Structures," *Applied Physics Letters*, Vol. 56, No. 8, 1990, pp. 770-772.

[2] A. Rogalski, "Recent Progress in Infrared Detector Technologies," *Infrared Physics & Technology*, Vol. 54, 2011, pp. 136-154.

[3] M. Z. Tidrow, W. A. Beck, W. W. Clark, H. K. Pollehn, J. W. Little, N. K. Dhar, R. P. Leavitt, S. W. Kennerly, D. W. Beekman, A. C. Goldberg and W. R. Dyer, "Device Physics and Focal Plane Array Applications of QWIP and MCT," *Opto-Electronics Review*, Vol. 7, No. 4, 1999, pp. 283-296.

[4] M. Walther, R. Rehm, J. Schmitz, J. Fleissner, F. Rutz, L. Kirste, R. Scheibner, J. Wendler and J. Ziegler,"Quantum Sensing and Nanophotonic Devices VII," SPIE, San Francisco, 2010.

[5] G. Sai-Halasz, R. Tsu and L. Esaki, "A New Semiconductor Superlattice," *Applied Physics Letters*, Vol. 30, No. 12, 1977, p. 651.

[6] D. Smith and C. Mailhiot, "Proposal for Strained Jour II Superlattice Infrared Detectors," *Journal of Applied Physics*, Vol. 62, No. 6, 1987, p. 2545.

[7] G. A. Sai-Halasz, L. Esaki and W. A. Harrison, "InAs-GaSb Superlattice Energy Structure and Its Semiconductor-Semimetal Transition," *Physical Review B*, Vol. 18, No. 6, 1978, pp. 2812-2818.

[8] Y. Wei, A. Gin, M. Razeghi and G. Brown, "Type II InAs/GaSb Superlattice Photovoltaic Detectors with Cutoff Wavelength Approaching 32 μm," *Applied Physics Letters*, Vol. 81, No. 19, 2002, p. 3675.

[9] R. Rehm, M. Walther, J. Schmitz, J. Fleissner, J. Ziegler, W. Cabanski and R. Breiter, "Dual-Colour Thermal Imaging with InAs/GaSb Superlattices in Mid-Wavelength Infrared Spectral Range," *Electronics Letters*, Vol. 42, No. 10, 2006, pp. 577-578.

[10] A. Khoshakhlagh, E. Plis, S. Myers, Y. D. Sharma, L. R. Dawson, S. Krishna, "Optimization of InAs/GaSb Type-II Superlattice Interfaces for Long-Wave (~8 μm) Infrared Detection" *Journal of Crystal Growth*, Vol. 311, No. 7, 2009, pp. 1901-1904.

[11] A. Rogalski, J. Antoszewski, L. Faraone, "Third-Generation Infrared Photodetector Arrays," *Journal of Applied Physics*, Vol. 105, 2009, Article ID: 91101.

[12] M. Walther, J. Schmitz, R. Rehm, S. Kopta, F. Fuchs, J. Fleissner, W. Cabanski and J. Ziegler, "Growth of InAs/GaSb Short-Period Superlattices for High-Resolution Mid-Wavelength Infrared Focal Plane Array Detectors," *Journal of Crystal Growth*, Vol. 278, 2005, pp. 156-161.

[13] M. Walther, R. Rehm, F. Fuchs, J. Schmitz, J. Fleißner, W. Cabanski, D. Eich, M. Finck, W. Rode, J. Wendler, R. Wollrab and J. Ziegler, "256 × 256 Focal Plane Array Midwavelength Infrared Camera Based on InAs/GaSb Short-Period Superlattices," *Journal of Electronic Materials*, Vol. 34, No. 6, 2005, pp. 722-725.

[14] S. A. Pour, E. K. Huang, G. Chen, A. Haddadi, B. M. Nguyen and M. Razeghi, "High Operating Temperature

Midwave Infrared Photodiodes and Focal Plane Arrays Based on Type-II InAs/GaSb Superlattices," *Applied Physics Letters*, Vol. 98, No. 14, 2011, pp. 143501-143503.

[15] R. Rehm, M. Walther, F. Rutz, J. Schmitz, A. Wörl, J. M. Masur, R. Scheibner, J. Wendler and J. Ziegler, "256 × 256 Focal Plane Array Midwavelength Infrared Camera Based on InAs/GaSb Short-Period Superlattices," *Journal of Electronic Materials*, Vol. 40, No. 6, 2011, pp. 1738-1743.

[16] S. D. Gunapala, D. Z. Ting, C. J. Hill, J. Nguyen, A. Soibel, S. B. Rafol, S. A. Keo, J. M. Mumolo, M. C. Lee, J. K. Liu and B. Yang, "Demonstration of a 1024 × 1024 Pixel InAs-GaSb Superlattice Focal Plane Array," *IEEE Photonics Technology Letters*, Vol. 22, No. 24, 2010, pp. 1856-1858.

[17] E. K. W. Huang, P. Y. Delaunay, B. M. Nguyen, S. A. Pour and M. Razeghi, "Photovoltaic MWIR Type-II Superlattice Focal Plane Array on GaAs Substrate," *IEEE Journal of Quantum Electronics*, Vol. 46, No. 12, 2010, pp. 1704-1708.

[18] N. Herres, F. Fuchs, J. Schmitz, K. M. Pavlov, J. Wagner, J. D. Ralston, P. Koidl, C. Gadaleta and G. Scamarcio, "Effect of Interfacial Bonding on the Structural and Vibrational Properties of InAs/GaSb Superlattices," *Physical Review B*, Vol. 53, No. 23, 1996, pp. 15688-15705.

[19] J. H. Li, D. W. Stokes, J. C Wickett, O. Caha, K. E. Bassler and S. C. Moss, "Effect of Strain on the Growth of InAs/GaSb Superlattices: An X-Ray Diffraction Study," *Journal of Applied Physics*, Vol. 107, No. 12, 2010, Article ID: 123504.

[20] E. Plis, S. Annamalai, K. T. Posani, S. Krishna, R. A. Rupani and S. Ghosh, "Midwave Infrared Type-II InAs/GaSb Superlattice Detectors with Mixed Interfaces," *Journal of Applied Physics*, Vol. 100, No. 1, 2006, Article ID: 14510.

[21] J. B. Rodriguez, P. Christol, L. Cerutti, F. Chevrier and A. Joullié, "MBE Growth and Characterization of Type-II InAs/GaSb Superlattices for Mid-Infrared Detection," *Journal of Crystal Growth*, Vol. 274, No. 1-2, 2005, pp. 6-13.

[22] H. J. Haugan, L. Grazulis, G. J. Brown, K. Mahalingam, D. H. Tomich, "Exploring Optimum Growth for High Quality InAs/GaSb Type-II Superlattices," *Journal of Crystal Growth*, Vol. 261, No. 4, 2004, pp. 471-478.

[23] E. Plis, A. Khoshakhlagh, S. Myers, H. S. Kim, N. Gautam, Y. D. Sharma, S. Krishna, S. J. Lee and S. K. Noh, "Molecular Beam Epitaxy Growth and Characterization of Type-II InAs/GaSb Strained Layer Superlattices for Long-Wave Infrared Detection," *Journal of Vacuum Science & Technology B*, Vol. 28, No. 3, 2010, pp. C3-C13.

[24] Y. Wei, J. Bae, A. Gin, A. Hood, M. Razeghi, G. Brown and M. Tidrow, "High Quality Type II InAs/GaSb Superlattices with Cutoff Wavelength ~3.7 μm Using Interface Engineering," *Journal of Applied Physics*, Vol. 94, No. 7, 2003, p. 4720.

[25] B. Satpati, J. Rodriguez, A. Trampert, E. Tournie, A. Joullie and P. Christol, "Interface Analysis of InAs/GaSb Superlattice Grown by MBE," *Journal of Crystal Growth*, Vol. 301-302, 2007, pp. 889-892.

[26] E. Luna, B. Satpati, J. B. Rodriguez, A. N. Baranov, E. Tournié and A. Trampert, "Interfacial Intermixing in InAs/GaSb Short-Period-Superlattices Grown by Molecular Beam Epitaxy," *Applied Physics Letters*, Vol. 96, No. 2, 2010, Article ID: 021904.

[27] R. Kaspi, J. Steinshnider, M. Weimer, C. Moeller and A. Ongstad, "As-Soak Control of the InAs-on-GaSb Interface," *Journal of Crystal Growth*, Vol. 225, No. 2-4, 2001, pp. 544-549.

[28] D. Hoffman, B. Nguyen, P. Delaunay, A. Hood, M. Razeghi and J. Pellegrino, "Beryllium Compensation Doping of InAs/GaSb Infrared Superlattice Photodiodes," *Applied Physics Letters*, Vol. 91, No. 14, 2007, Article ID: 143507.

[29] L. Bürkle, F. Fuchs, J. Schmitz and W. Pletschen, "Control of the Residual Doping of InAs/(GaIn)Sb Infrared Superlattices," *Applied Physics Letters*, Vol. 77, No. 11, 2000, p. 1659.

[30] S. Bandara, P. Maloney, N. Baril, J. Pellegrino and M. Tidrow, "Doping Dependence of Minority Carrier Lifetime in Long-Wave Sb-Based Journal Article II Superlattice Infrared Detector Materials," *Optical Engineering*, Vol. 6, No. 50, 2011, Article ID: 61015.

[31] M. Walther, R. Rehm, J. Schmitz, J. Niemasz, F. Rutz, A. Worl, L. Kirste, R. Scheibner, J. Wendler and J. Ziegler, "Defect Density Reduction in InAs/GaSb Journal Article II Superlattice Focal Plane Array Infrared Detectors," *Proceedings of SPIE*, Vol. 7945, 2011, Article ID: 79451.

[32] B. M. Nguyen, G. Chen, M. A. Hoang and M. Razeghi, "Growth and Characterization of Long-Wavelength Infrared Type-II Superlattice Photodiodes on a 3-in GaSb Wafer," *IEEE Journal of Quantum Electron*, Vol. 47, No. 5, 2011, pp. 686-690.

[33] S. H. Huang, G. Balakrishnan, A. Khoshakhlagh, A. Jallipalli, L. R. Dawson and D. L. Huffaker, "Strain Relief by Periodic Misfit Arrays for Low Defect Density GaSb on GaAs," *Applied Physics Letters*, Vol. 88, No. 13, 2006, Article ID: 131911.

[34] A. S. Bracker, M. J. Yang, B. R. Bennett, J. C. Culbertson and W. J. Moore, "Surface Reconstruction Phase Diagrams for InAs, AlSb, and GaSb," *Journal of Crystal Growth*, Vol. 220, No. 4, 2000, pp. 384-392.

3

Dichroic Electro-Optical Behavior of Rhenium Sulfide Layered Crystal

Ching-Hwa Ho
Graduate Institute of Applied Science and Technology, National Taiwan University of Science and Technology, Taipei, Taiwan

ABSTRACT

Dichroic behaviors of layered ReS_2 have been characterized using angular dependent polarized-absorption and resistivity measurements in the van der Waal plane. The angular dependent optical and electrical measurements are carried out with angles ranging from $\theta = 0°$ ($E \parallel \boldsymbol{b}$) to $\theta = 90°$ ($E \perp \boldsymbol{b}$) with respect to the layer crystal's ***b***-axis. The angular dependence of polarized energy gaps of ReS_2 shows a sinusoidal variation of energies from ~1.341 eV ($E \parallel \boldsymbol{b}$) to ~1.391 eV ($E \perp \boldsymbol{b}$). The experimental evidence of polarized energy gap leaves ReS_2 a potential usage for fabrication of a polarized optical switch suitable for polarized optical communication in near-infra-red (NIR) region. Angular dependence of resistivities of ReS_2 in the van der Waal plane has also been evaluated. The relationship of in-plane resistivities shows a sinusoidal-like variation from $\theta = 0°$ ($E \parallel \boldsymbol{b}$) to 90° ($E \perp \boldsymbol{b}$) and repeated periodically to 360°. The experimental results of optical and electrical measurements indicated that ReS_2 is not only an optical-dichroic layer but also an electrical-dichroism material presented in the layer plane.

Keywords: Triclinic Crystals; Optical Materials; Polarization-Sensitive Devices

1. Introduction

Dichroism and optical-anisotropic materials played an important role for the evolution of light-wave optics in past years. For generation of polarized light, dichroic crystal or birefringent solid is the most commonly used material because the specific axial anisotropy exists in the crystal structure. Among the dichroic or birefringent solids, the mineral tourmaline such as $NaFe_3B_3Al_6Si_6O_{27}(OH)_4$ or the birefringent crystal such as Calcite ($CaCO_3$) are usually the crucial substances for making linear polarizers [1]. Such kinds of polarizers are usually suitable for making visible-to-ultraviolet lasers or for fabrication of linearly polarized devices.

Dichroic semiconductor—rhenium disulfide (ReS_2) has been synthesized and proposed to possess the potential capability for fabrication of polarization sensitive photodetector applied in multi-channel optical communications [2]. The polarization sensitive behavior of ReS_2 is arisen from its specific in-plane optical anisotropy in the layer [2]. The crystallography of ReS_2 is usually crystallized in a distorted $CdCl_2$ layer-type structure with triclinic symmetry (space group $P\bar{1}$) [3-5]. ReS_2 is an indirect semiconductor with an unpolarized gap of about 1.37 eV [5]. From the observation of polarization-dependent spectral quantum efficiency, the layered ReS_2 presents optically biaxial behavior with the linearly polarized lights incident normal to the basal plane [*i.e.* $\mathbf{E} \parallel (001)$ and $\mathbf{k} \perp (001)$] [6]. Polarized photocurrent measurements of an ITO/ReS_2 device with different azimuthal angles revealed that angular dependence of photoresponses in ReS_2 follows the Malus law [1,6]. The polarization-dependent spectral quantum efficiency also verified ReS_2 an effective polarization sensitive photodetector applied in visible range [6].

In this paper, the optical and electrical dichroic properties of ReS_2 are shown to simultaneously present in the van der Waal plane. The dichroic behaviors of ReS_2 are characterized by angular dependent polarized-absorption and electrical-resistivity measurements in the layer plane. The measurements were done in the angular range from $\theta = 0°$ ($E \parallel \boldsymbol{b}$) to $\theta = 90°$ ($E \perp \boldsymbol{b}$) with respect to the ***b***-axis of the layer crystal. The polarized energy gaps of ReS_2 were obtained from the analysis of polarization-dependent absorption spectra. The relation of the polarized gaps of ReS_2 was determined to be $E_g(\theta) = 1.366 - 0.025 \cdot \cos(2\theta)$ eV. Angular dependence of resistivities of

ReS_2 in the van der Waal plane showed a sinusoidal-like variation from $\theta = 0^\circ$ ($E \parallel \boldsymbol{b}$) to $\theta = 90^\circ$ ($E \perp \boldsymbol{b}$). The angular dependency of in-plane resistivities of ReS_2 is $\rho(\theta) = 21.054 - 14.535 \cdot \cos(2\theta)$ Ω-cm. The optical and electrical evidences showed that ReS_2 is not only an optical-dichroic layer but also an electrical-dichroism crystal in the van der Waal plane.

2. Experimental Procedure

2.1. Growth and Structural Property

Single crystals of ReS_2 were grown by chemical vapor transport (CVT) method [7] using ICl_3 as a transport agent. The growing process of ReS_2 using CVT had been described elsewhere [8]. The as-grown crystals of ReS_2 were essentially layer type. X-ray diffraction measurement confirmed triclinic structure of the crystal [8]. Weak van der Waals bonding between the layers means that they display good cleavage property parallel to the layers. The strong bonding force along the Re cluster chains also implies that we can easily tear the layer along ***b***-axis via the auxiliary of weak bonding force between the individual chains. Hall effect measurement revealed p-type semiconducting behavior of the layer. The carrier density is $\sim 1 \times 10^{16}$ cm^{-3}.

2.2. Optical Measurement

Measurements of reflectance and transmittance at near-normal incidence were made on a scanning monochromatic measurement system. An 150 W tungsten-halogen lamp filtered by a PTI 0.2 m monochromator provided the monochromatic light. Transmission intensity was closely monitored to obtain an incidence as close to 90° as possible. Single crystals with a thickness of about 100 μm were used in the transmission measurements. The reflected light of the sample was detected by an EG & G type HUV-2000B silicon photodiode and the signal was recorded from an EG & G model 7265 dual phase lock-in amplifier. A pair of OPTOSIGMA near-infrared-dichroic-sheet polarizers with the measurement range of 760 - 2000 nm was employed in the polarization dependent optical measurements.

2.3. Electrical Measurement

Angular dependent resistivity measurements of layered ReS_2 were made on ten different samples cut in bar type. The cutting edge of each sample was varied from $\theta = 0^\circ$ ($\parallel \boldsymbol{b}$) to $\theta = 90^\circ$ ($\perp \boldsymbol{b}$) with an angle increment of 10° starting from the layer crystal's ***b***-axis. Displayed in **Figure 1(a)** is a scanning-electron microscope (SEM) photograph of ReS_2 in the layered plane. The crystal image is 50 × amplification and the crystal orientation of ***b***-axis is shown. The ***b***-axis is parallel to the metal Re cluster

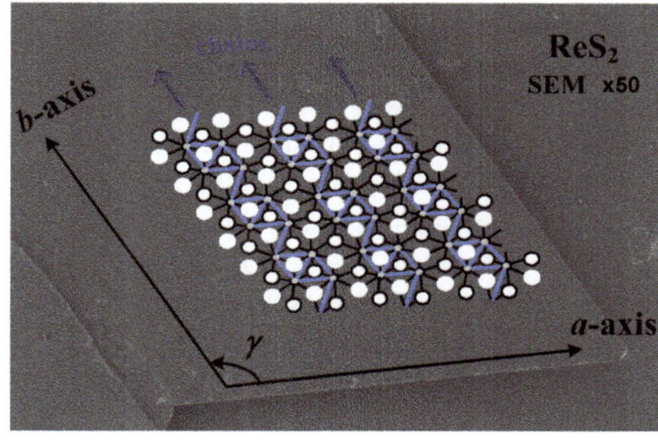

(a)

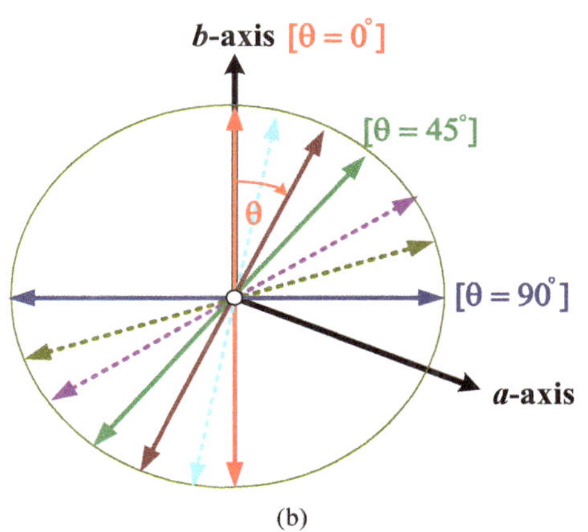

(b)

Figure 1. (a) SEM image of layered ReS_2 on the van der Waal plane, where the Re_4 clustering chains that corresponding to the crystal orientation of *b*-axis are illustrated. (b) Schematic illustrations for the angular dependent optical-absorption and electrical-resistivity measurements. The angle θ is defined as the angular difference between the orientation of electric field and that of the crystal's *b*-axis.

chains, which corresponds to the longest edge of the plate [6,9]. The angular dependence of electrical-resistivity and polarized-absorption measurements of ReS_2 was evaluated by the rotation indication as shown in **Figure 1(b)**. All the measurements were carried out with the angular range starting from $\theta = 0^\circ$ to $\theta = 90^\circ$. The angular increment of the polarized-absorption measurement is 5° while the increment is 10° for the electrical-resistivity measurement.

3. Results and Discussion

3.1. Dichroic Optical Property

Polarized optical-absorption measurements were implemented with the polarization angles from $\theta = 0^\circ$ to $\theta = 90^\circ$. Absorption coefficient α of ReS_2 can be determined from the transmittance T_r by taking into account the spectral dependence of the reflectance R using the relation $T_r = \left[(1-R)^2 \times e^{-\alpha d}\right] / \left[1 - R^2 \times e^{-2\alpha d}\right]$ [10]. This equation assumes that there are multiple reflections within the sample, but that they add incoherently due to sample inhomogeneity or a sufficiently large spread of the incident angles. Because αd is large for the sample crystals, the second term in the denominator of the T_r relation can be neglected. **Figure 2(a)** displays the polarized absorption coefficients of ReS_2 as a function of photon energy from $\theta = 0^\circ$ to $\theta = 90^\circ$ at 300 K. The absorption edges of ReS_2 clearly indicate an energy blue-shift behavior with the increase of polarization angles from $\theta = 0^\circ$ to 90°. The absorption-edge anisotropy in rhenium disulfide provides relevant evidence that the energy gap of ReS_2 is polarization dependent. This property also lends a potential ability for ReS_2 to fabricate a polarized optical switch suitable for polarized optical communication in the NIR region. The analysis of the experimental results in **Figure 2(a)** also shows that the absorption coefficient α for ReS_2

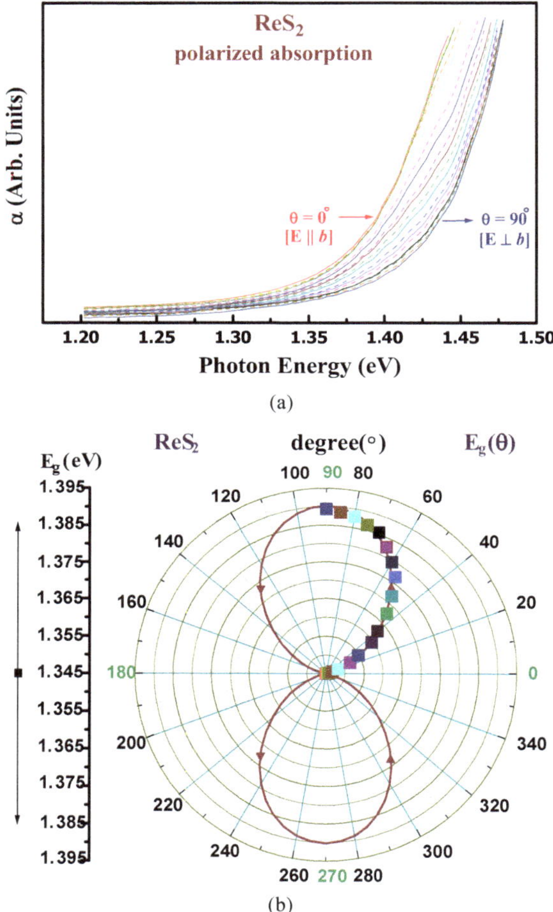

Figure 2. (a) Angular dependent polarized-absorption spectra of ReS_2 from $\theta = 0°$ to $\theta = 90°$ with an angular increment of 5° at 300 K. (b) The angular dependency of polarized energy gaps of ReS_2 in polar-coordination plot. The solid squares are the experimental data points and the solid lines are least-square fits to Equation (1). A double circle line that moving from $\theta = 0°$ to 360° depicts the varying trace of the polarized energy gaps for layered ReS_2.

is proportional to $(h\nu - E_g)^n$ with n = 2.0 ± 0.1. This suggests an indirect allowed transition. A more complete analysis by taking into account both absorption and emission phonons [5] could be utilized to determine the polarized band gaps and average phonon energies from the polarized absorption spectra. Shown in **Figure 2(b)** by the solid squares (in polar-coordination plot) are the values of polarized energy gaps of ReS_2 ranging from $\theta = 0^\circ$ to $\theta = 90^\circ$ with an increment of 5°. The average phonon energies for the indirect ReS_2 from the polarized absorption spectra are determined to be about 25 ± 5 meV. The angular dependence of polarized energy gaps of ReS_2 shows a sinusoidal-like variation of energies from 1.341 eV ($\theta = 0^\circ$) to 1.391 eV ($\theta = 90^\circ$). The angular dependence of the polarized gaps of ReS_2 can be analyzed by fitting the experimental data to a sinusoidal functional form of

$$E_g(\theta) = E_{g0} - E_\Delta \cdot \cos(2 \cdot \theta) \quad (1)$$

where E_{g0} is related to the unpolarized band gap, and E_Δ is the energy amplitude of the sinusoidal variation. The energy-variation relation of Equation (1) is similar to the generalized Malus law [11] that describing the dependence of the linearly polarized lights. The solid curve displayed in **Figure 2(b)** is the results by fitting the polarized energy gaps to Equation (1). From the least-square fits using Equation (1), the energy values of E_{g0} and E_Δ are determined to be 1.366 eV and 0.025 eV, respectively. As shown in **Figure 2(b)**, a polar-coordination plot of a double circle line just depicts the curve trace of the polarized gaps moving from $\theta = 0^\circ$ to 360° for the layered ReS_2. This result concludes that the energy variation of polarized energy gaps for a dichroic semiconductor also follows the generalized Malus rule.

3.2. Axial Dependence of Electrical Carrier Transport

The dichroic electrical property of the p-type layered ReS_2 in van der Waal plane was studied using a regular four-point method [12]. The measurement configuration is shown in **Figure 3(a)**. A dc current source supplied the current and a dc voltmeter used for detection unit. The in-plane resistivities were measured on ten different rectangular shape samples with different orientations. The direction of cutting-edge for each specimen was varied from $\theta = 0^\circ$ (∥ ***b***) to $\theta = 90^\circ$ (⊥ ***b***) with an angular increment of 10°. Displayed in **Figure 3(b)** by the solid circles are the experimental data points for the angular dependent in-plane resistivities of layered ReS_2. The in-plane electrical anisotropy of ReS_2 in **Figure 3(b)** clearly indicates the lowest resistivity is occurred with the electric field parallel to the ***b***-axis while the maximum resistivity is arisen along the direction perpendicular to ***b***-axis. The

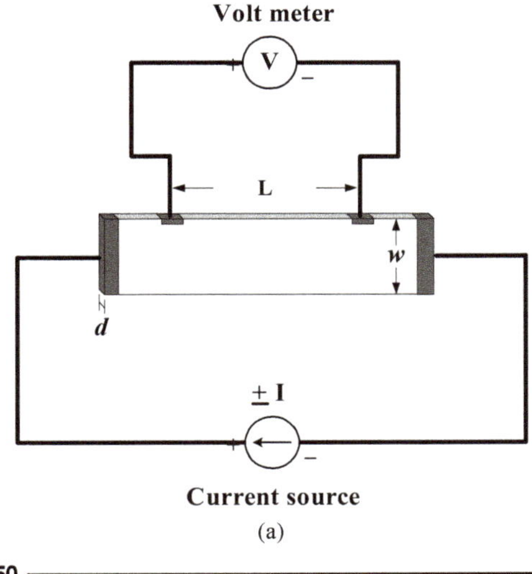

(a)

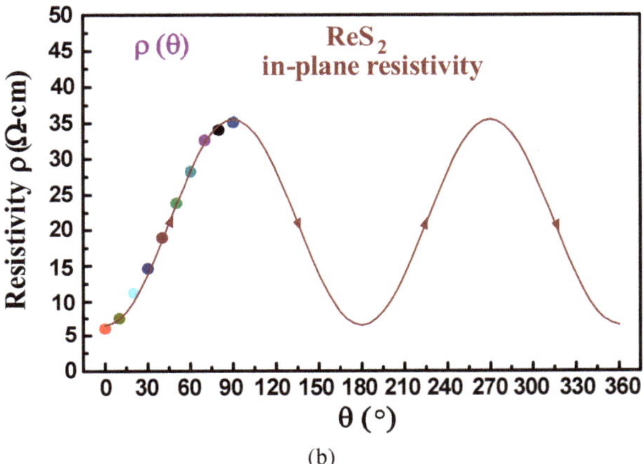

(b)

Figure 3. (a) Measurement configuration of electrical resistivity using regular four-point method. (b) Angular dependent in-plane resistivities of ReS_2. The solid circles are the data points measured from $\theta = 0°$ to $\theta = 90°$. The solid curve is the fitting result extended from 0° to 360°.

lowest value of resistivity along ***b***-axis is related to the strongest bonding force existed along the crystal orientation of the Re cluster chains. The angular-dependent in-plane resistivities of ReS_2 in **Figure 3(b)** also reveal a sinusoidal-like variation from $\theta = 0°$ (E || ***b***) to $\theta = 90°$ (E ⊥ ***b***). The relationship of angular dependent resistivities of ReS_2 can also be analyzed by using the generalized Malus formula of $\rho(\theta) = \rho_0 - \rho_\Delta \cdot \cos(2\cdot\theta)$. The solid curve in **Figure 3(b)** is the fitting result of angular-dependent in-plane resistivities of ReS_2 extended from 0° to 360°. The angular dependence of in-plane resistivities of ReS_2 is determined to be $\rho(\theta) = 21.054 - 14.535\cdot\cos(2\theta)$ Ω-cm. The electric anisotropy in **Figure 3(b)** confirms that ReS_2 is an electrical dichroism which presents a special axial selectivity of electrical conduction along the layer plane. This property is attributed to the in-plane axial anisotropy of mobilities that existed in the triclinic-layered ReS_2.

4. Conclusion

In conclusion, the dichroic optical and electrical behaveiors of layered ReS_2 were characterized using angular dependent polarized-absorption as well as electrical-resistivity measurements in the van der Waal plane. The polarized energy gaps of ReS_2 show a sinusoidal angular dependence of $E_g(\theta) = 1.366 - 0.025\cdot\cos(2\theta)$ eV. The in-plane resistivities of ReS_2 were measured on ten different rectangular-shape samples with dissimilar orientations. The angular dependent in-plane resistivities of ReS_2 also show a sinusoidal-like variation from $\theta = 0°$ (E || ***b***) to $\theta = 90°$ (E ⊥ ***b***). The relationship of angular dependent resistivities of ReS_2 was determined to be $\rho(\theta) = 21.054 - 14.535\cdot\cos(2\theta)$ Ω-cm. The angular dependence of resistivity is due to the in-plane axial anisotropy of mobilities in the triclinic-layered ReS_2. The experimental observations verified that ReS_2 is not only an optical-dichroic layer but also an electrical dichroism in the van der Waal plane.

5. Acknowledgements

The author would like to acknowledge the research funding supported by the National Science Council of Taiwan under the Project No. NSC 101-2221-E-011-052-MY3.

REFERENCES

[1] E. Hecht, "Optics," 4th Editon, Addison Wesley, San Francisco, 2002, p. 333.

[2] C. H. Liang, K. K. Tiong, Y. S. Huang, D. Dumcenco and C. H. Ho, "In-Plane Anisotropic Electrical and Optical Properties of Gold-Doped Rhenium Disulphide," *Journal of Materials Science*: *Materials in Electronics*, Vol. 20, No. 1, 2009, pp. S476-S479.

[3] J. C. Wildervanck and F. Jellinek, "The Dichalcogenides of Technetium and Rhenium," *Journal of the Less Common Metals*, Vol. 24, No. 1, 1971, pp. 73-81.

[4] H. -J. Lamfers, A. Meetsma, G. A. Wiegers, and J. L. de Boer, "The Crystal Structure of Some Rhenium and Technetium Dichalcogenides," *Journal of Alloys and Compounds*, Vol. 241, 1996, pp. 34-39.

[5] C. H. Ho, P. C. Liao, Y. S. Huang, T. R. Yang and K. K. Tiong, "Optical Absorption of ReS_2 and $ReSe_2$ Single Crystals," *Journal of Applied Physics*, Vol. 81, No. 9, 1997, pp. 6380-6383.

[6] K. Friemelt, M.-Ch. Lux-Steiner and E. Bucher, "Optical Properties of the Layered Transition-Metal-Dichalcogenide ReS_2: Anisotropy in the van der Waals Plane," *Journal of Applied Physics*, Vol. 74, No. 8, 1993, pp. 5266-5268.

[7] C. H. Ho, "Single Crystal Growth and Characterization of Copper Aluminum Indium Disulfide Chalcopyrites,"

Journal of Crystal Growth, Vol. 317, No. 1, 2011, pp. 52-59.

[8] C. H. Ho, Y. S. Huang, P. C. Liao and K. K. Tiong, "Crystal Structure and Band-Edge Transitions of $ReS_{2-x}Se_x$ Layered Compounds," *Journal of Physics and Chemistry of Solids*, Vol. 60, No. 11, 1999, pp. 1797-1804.

[9] B. A. Parkinson, J. Ren and M.-H. Whangbo, "Relationship of STM and AFM Images to Local Density of States in Valence and Conduction Bands of ReS_2," *Journal of the American Chemical Society*, Vol. 113, No. 21, 1991, pp. 7833-7837.

[10] J. I. Pankove, "Optical Processes in Semiconductors," Dover, New York, 1975.

[11] G. A. Medvedkin, Yu. V. Rud and M. A. Tairov, "Photoelectric Anisotropy of II-IV-V_2 Ternary Semiconductors," *Physica Status Solidi* (*a*), Vol. 115, No. 1, 1989, pp. 11-50.

[12] H. H. Wieder, "Laboratory Notes on Electrical and Galvanomagnetic Measurements," Elsevier, Amsterdam, 1979.

Synthesis, Spectroscopic and Crystal Structure Studies on Ethyl 5,7-Dimethyl Coumarin-4-Acetate

Kiran K. Pujar[1], G. N. Anil kumar[2], Manohar V. Kulkarni[1*]
[1]Department of Chemistry, Karnatak University, Dharwad, India
[2]M S Ramaiah Institute of Technology, Department of Physics, Bangalore, India

ABSTRACT

Ethyl ester of 5,7-dimethyl coumarin-4-acetic acid has been synthesized from 3,5-Xylenol in a two step sequence of reaction involving Pechmann cyclisation and acid catalyzed esterification. The title compound 2 crystallizes in Monoclinic form, space group P 1 21/c 1, with a = 8.6248(4) Å, b = 18.9103(8) Å, c = 8.4204(4) Å, β = 101.241(2), V = 1347.00(11) Å, D cal = 1.283 Mg/cm3, Z = 4. The molecule is stabilized by intermolecular C-H … O bonds.

Keywords: Coumarin-4-Acetic acid Ethyl Ester; X-Ray Diffraction

1. Introduction

Coumarins are a group of naturally occurring lactones with wide ranging biological activities [1]. Varieties of 4-substituted coumarins possessing hydroxyl [2], aminoalkyl [3] arylaminomethyl [4] sulphonamido [5] and aryloxymethyl [6] groups have exhibited anti coagulant, anti microbial and anti inflammatory activities. Coumarin 4-acetic acids have been found to exhibit good inflammation inhibiting activity in animal models [7]. They have been employed as key intermediates in the design and synthesis of polycyclic coumarins related to Protoberberine alkaloids [8]. The reactivity of C_4-methylene group has been employed for the construction of many 4-substituted bi-heterocyclic coumarins. Solid state conformational studies on coumarin 4-acetic acid [9], ester [10,11] and dithionate [12] have been reported recently. It needs to be emphasized that the crystallographic study on esters is of considerable biological interest [13] since it determines the preferred orientation of the rotamer (**Figure 1**), in turn the direction of the electron pair on alkyl oxygen. In view of the biological significance of 4-substituted coumarins and interest associated with the geometry of esters the title compound has been synthesized, charac terized and studied for its solid state conformation by X-ray studies.

Figure 1. General structure for Rotamers of Esters.

*Corresponding author.

2. Experimental

2.1. Synthesis of the Title Compound

Compound **1** has been prepared by the reaction of 3,5-xylenol and citric acid by using sulphuric acid as condensing agent according to general procedure reported for coumarin 4-acetic acids [14]. Compound **1** was further converted into its ester (compound **2**) by refluxing in dry ethanol with catalytic quantity of sulphuric acid by Fischer esterification. The resulting solution was poured into crushed ice, the precipitate was filtered off, washed with 5% $NaHCO_3$, and with water and recrystallized from ethanol. (**Figure 2**) M.P. 123-125.C, yield 95%. The authenticity of the compound has been established by UV, IR, ^1H, ^{13}C NMR, and Mass spectra. Crystals suitable for diffraction study were grown by a slow evaporation technique using 1:1 mixture of ethanol and dioxan at room temperature.

2.2. Characterization and Physical Measurements

The melting point was determined in open capillaries and is uncorrected. The IR absorption spectrum was obtained by KBr pellet using a FT IR spectrometer. UV-vis spectrum was recorded in the range of 200 nm to 800 nm using a Lambda 35 Perkin-Elmer spectrophotometer. The ^1H and ^{13}C NMR spectra were determined on a Bruker Avance 400 model at 400 and 200 MHz respectively. Solvents were dried by refluxing with the appropriate drying agents and distilled before use.

2.3. X-Ray Crystallography

For the crystal structure determination, the single-crystal of the compound $C_{15}H_{16}O_4$ was used for data collection. The structure was solved and refined using the Bruker SHELXTL Software Package, using the space group P 1 21/c 1, with Z = 4 for the formula unit $C_{15}H_{16}O_4$. The final anisotropic full-matrix least-squares refinement on F2 with 176 variables converged at R1 = 4.03%, for the observed data and w R2 = 12.44% for all data. The goodness-of-fit was 1.008. The largest peak in the final difference electron density synthesis was 0.148 $e^-/Å$ and the largest hole was −0.136 $e^-/Å$ with an RMS deviation of 0.029 $e^-/Å$. On the basis of the final model, the calculated density was 1.283 g/cm3 and F(000), 552 $e^-/Å$. Molecular structure of the compound showing the atomic numbering scheme is shown in **Figure 3**. The crystallographic parameters of the compound were presented in **Table 1**. Selected bond distances and bond angles are listed in **Table 2** and selected Hydrogen-bonding parameters listed in **Table 3**.

3. Results and Discussion

Title compound crystallizes in the monoclinic centrosymmetric space group P 1 21/c 1with Z = 4. The structure of the compound consists of ethyl ester group. Selected bond lengths and bond distances for the title com ding parameters are listed in **Table 3**.

3.1 Molecular Skeleton

Coumarin ring is planar and carbethoxy group attached to the C_4-CH_2 is out of plane by an angle of 11.26. The methyl group oriented at an angle of 90° to the mean plane of the bicyclic system. The ester carbonyl is oriented perpendicular to the C_3-C_4 double bond. The molecule adopts the S-cis arrangement across the alkyl-oxygen bond as revealed in the ORTEP diagram (**Figure 3**). The exocyclic Oxygen is involved in intermolecular hydrogen bonding ($O_4 \cdots H$-C_1) with the CH_3 group of ester. A weak hydrogen bonding ($O_2 \cdots H$-C_4) is also observed between the C_4-CH_2 and the carbonyl oxygen of ester. The molecular packing diagram (**Figure 4**) shows hydrogen bonding and the crystalline state is stabilized by vander waals forces of attraction. The S-cis arrangement is also supported by the dihedral angle of 1.8 between C_2-O_1-C_3-O_2 indicating an eclipsed arrangement (**Table 3**).

3.2. Bond Lengths

Bond lengths have been discussed in **Table 2**. The aromatic C-C bond lengths are in the range of 1.30 - 1.40 **Å**. The ester C_{13}-O_4 bond shows a bond length of 1.20 **Å** which is similar to the lactone distance of 1.19 **Å** between O_2-C_3 .The alkyl oxygen O_1-C_3 bond is 1.32 **Å** whereas the corresponding bond of O_3-C_{13} bond is slightly longer by 1.35 **Å**.

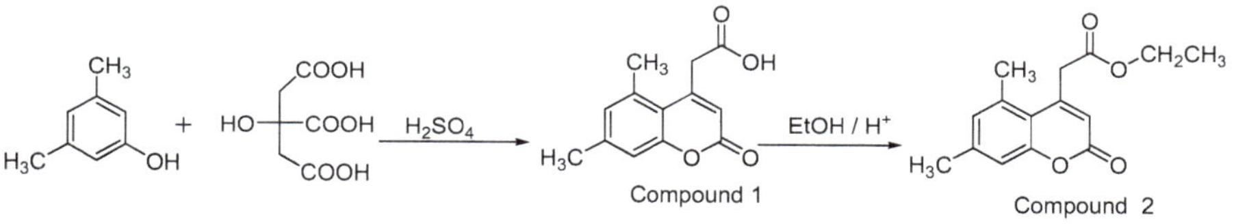

Figure 2. Chemical structure and synthetic pathway of the title compound.

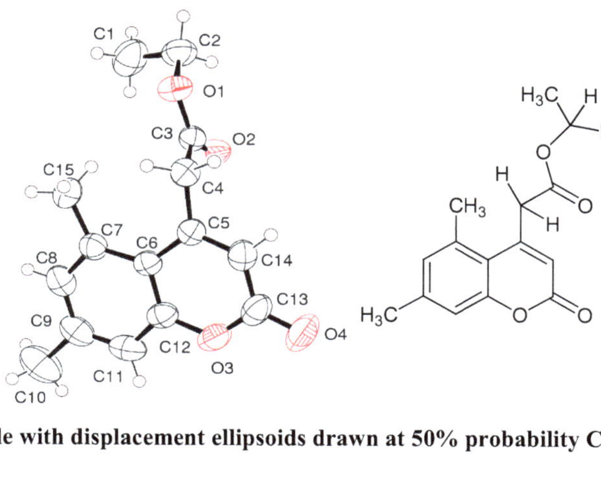

Figure 3. ORTEP plot of molecule with displacement ellipsoids drawn at 50% probability Covalent structure of compound 2.

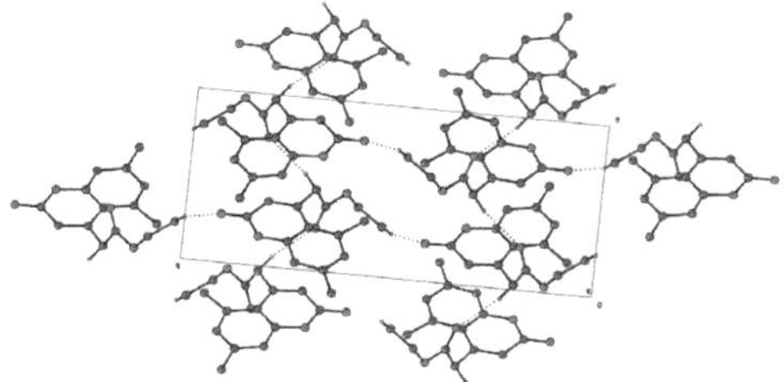

Figure 4. Packing diagram dotted lines indicate C-H····O hydrogen bonds of the molecule viewed down a-axis.

Table 1. Crystallographic data and structure refinement parameters.

Empirical formula	$C_{15}H_{16}O_4$
Formula weight	260.28 g/mol
Crystal colour Temperature Wavelength	Colourless 296 K 0.71073 Å
Crystal system	*Monoclinic*
Space group	*P*2$_1$/*c*
Unit cell dimensions *a*, *b*, *c* (Å) *α*, *β*, (°)	8.6248 (4), 18.9103 (8), 8.4204 (4) 90, 101.241 (2), 90
Volume	1347.00 (11) (Å^3)
Z, Calculated density	4 Mg m^{-3}
Absorption coefficient	0.09 mm^{-1}
F(000)	552
Crystal size	0.3 × 0.25 × 0.2 mm
Theta range for data collection	2.41 to 26.07^0
Limiting indices	$-10 \le h \le 10, -23 \le k \le 22, 10 \le l \le 10$
Refinement method	Full-matrix least-squares on F^2
Absorption correction	Multi-scan
No. of measured, independent12192, 2675, 2113 and observed [*I* > 2(*I*)] reflections *R*$_{int}$0.021 (sin *θ*/*λ*)$_{max}$ (Å^{-1})$_{max}$ (Å^{-1}) 0.618	
Data/restraints/parameters	2675/0/176
Goodness-of-fit on F^2	1.008
H-atom treatment	H atoms constrained
$\Delta\rho_{max}$, $\Delta\rho_{min}$ (e Å^{-3})	0.15, −0.14
Final R indices	R1 = 0.0403, wR2 = 0.1138
R indices(all data)	R1 = 0.0533, wR2 = 0.1244
Large diff. peak and hole	0.148 and −0.136e Å^{-3}

3.3. Bond Angles

Bond angles have been discussed in **Table 2**. All the aromatic C-C-C bond angles are in the expected range

Table 2. Selected bond lengths and bond angles (Å, °).

O_1-C_2	1.458 (2)	C_5-C_{14}	1.346 (2)
O_1-C_3	1.3240 (17)	C_6-C_7	1.421 (2)
O_2-C_3	1.1949 (18)	C_6-C_{12}	1.402 (2)
O_3-C_{12}	1.3777 (19)	C_7-C_8	1.384 (2)
O_3-C_{13}	1.359 (2)	C_7-C_{15}	1.507 (2)
O_4-C_{13}	1.207 (2)	C_8-C_9	1.385 (2)
C_1-C_2	1.482 (3)	C_9-C_10	1.511 (3)
C_3-C_4	1.504 (2)	C_9-C_{11}	1.363 (3)
C_4-C_5	1.504 (2)	C_{11}-C_{12}	1.380 (2)
C_5-C_6	1.4586 (19)	C_{13}-C_{14}	1.437 (2)
C_2-O_1-C_3	117.29 (11)	C_6-C_7-C_{15}	124.66 (14)
C_{12}-O_3-C_{13}	122.35 (13)	C_8-C_7-C_{15}	116.36 (15)
O_1-C_2-C_1	110.46 (16)	C_7-C_8-C_9	123.46 (16)
O_1-C_3-O_2	123.71 (13)	C_8-C_9-C_10	120.61 (19)
O_1-C_3-C_4	110.87 (11)	C_8-C_9-C_{11}	117.94 (15)
O_2-C_3-C_4	125.41 (13)	C_10-C_9-C_{11}	121.45 (18)
C_3-C_4-C_5	112.83 (11)	C_9-C_{11}-C_{12}	120.23 (16)
C_4-C_5-C_6	124.09 (12)	O_3-C_{12}-C_6	122.36 (13)
C_4-C_5-C_{14}	116.56 (13)	O_3-C_{12}-C_{11}	114.29 (14)
C_6-C_5-C_{14}	119.33 (13)	C_6-C_{12}-C_{11}	123.34 (14)
C_5-C_6-C_7	127.90 (12)	O_3-C_{13}-O_4	117.41 (16)
C_5-C_6-C_{12}	116.07 (12)	O_3-C_{13}-C_{14}	116.33 (14)
C_7-C_6-C_{12}	116.02 (12)	O_4-C_{13}-C_{14}	126.25 (17)
C_6-C_7-C_8	118.97 (14)	C_5-C_{14}-C_{13}	123.52 (14)
C_3-O_1-C_2-C_1	−86.04 (18)	C_5-C_6-C_7-C_{15}	4.1 (2)
C_2-O_1-C_3-O_2	−1.8 (2)	C_{12}-C_6-C_7-C_8	1.8 (2)
C_2-O_1-C_3-C_4	179.39 (12)	C_{12}-C_6-C_7-C_{15}	−176.54 (15)
C_{13}-O_3-C_{12}-C_6	−1.2 (2)	C_5-C_6-C_{12}-O_3	0.02 (19)
C_{13}-O_3-C_{12}-C_{11}	−180.00 (14)	C_5-C_6-C_{12}-C_{11}	178.72 (14)
C_{12}-O_3-C_{13}-O_4	178.99 (15)	C_7-C_6-C_{12}-O_3	−179.43 (13)
C_{12}-O_3-C_{13}-C_{14}	0.5 (2)	C_7-C_6-C_{12}-C_{11}	−0.7 (2)
O_1-C_3-C_4-C_5	−166.99 (12)	C_6-C_7-C_8-C_9	−1.8 (2)
O_2-C_3-C_4-C_5	14.3 (2)	C_{15}-C_7-C_8-C_9	176.68 (16)
C_3-C_4-C_5-C_6	76.53 (17)	C_7-C_8-C_9-C_10	−179.68 (17)
C_3-C_4-C_5-C_{14}	−101.80 (15)	C_7-C_8-C_9-C_{11}	0.6 (2)
C_4-C_5-C_6-C_7	2.9 (2)	C_8-C_9-C_{11}-C_{12}	0.6 (2)
C_4-C_5-C_6-C_{12}	−176.50 (13)	C_10-C_9-C_{11}-C_{12}	−179.14 (17)
C_{14}-C_5-C_6-C_7	−178.84 (14)	C_9-C_{11}-C_{12}-O_3	178.28 (14)
C_{14}-C_5-C_6-C_{12}	1.79 (19)	C_9-C_{11}-C_{12}-C_6	−0.5 (2)
C_4-C_5-C_{14}-C_{13}	175.85 (14)	O_3-C_{13}-C_{14}-C_5	1.4 (2)
C_6-C_5-C_{14}-C_{13}	−2.6 (2)	O_4-C_{13}-C_{14}-C_5	−176.91 (18)
C_5-C_6-C_7-C_8	−177.55 (14)		

Table 3. Selected hydrogen-bond parameters.

D-H···*A*	*D*-H (Å)	H···*A* (Å)	*D*···*A* (Å)	*D*-H···*A* (°)
C_1-H_1B···O4^{i}	0.9600	2.5100	3.448 (3)	164.00
C_4-H_4B···O2^{ii}	0.9700	2.4500	3.3963 (19)	166.00

Symmetry code(s): (i) $-x + 1, y - 1/2, -z + 1/2$; (ii) $x, -y + 1/2, z - 1/2$

carbon. The angles O_2-C_3-O_1 and O_2-C_3-C_4 are 123.7° and 125.4° whereas the O_1-C_3-C_4 angle is significantly reduced to 110.8°. The values supporting the fractional (sp) hybridized orbitals are involved in bonding.

3.4. Spectral Studies

UV—Visible spectrum—The UV spectrum of the title compound 2 in DMSO shows a strong band at 297 nm whereas fluorescence band appears at 387 nm which shows emission over a long range.

FT IR spec trum—The FT IR spectrum of the title compound 2 shows the strong characteristic absorption band at 1725 cm^{-1} region due to lactone carbonyl stretching frequency and band at 1615 cm^{-1} observed due to C = C stretching.

^{1}H NMR spectrum—The structure of the title compound 2 was further characterized by NMR absorption. Important structural information can be obtained from its NMR Spectra. ^{1}H NMR(400 MHz, DMSO-d_6) δ ppm: 2.34 (s, 3H, C_5-CH_3), 2.54 (s, 3H, C_7-CH_3), 1.15 (t, 3H, CH_3-of ester, J = 7.2Hz), 4.15 (q, 4H, C4-CH_2 and CH_2 of ester J = 7.2Hz) 6.39 (s, 1H, C_3-H), 7.0 (s, 1H, C_6-H), 7.1 (s, 1H, C_8-H)

^{13}C NMR spectrum: ^{13}C NMR spectrum (100 MHz DMSO-d_6) (δ ppm) 14-CH_3, 20 C_7-CH_3, 22 C_5-CH_3, 40 C_4-CH_2, 61 O-CH_2, 115, 118, 130, 136, 142, 150, 155 these shifts correspond to aromatic carbons. Low intensity signals at 159, 170 are for carbonyl carbons of coumarin and ester respectively.

3.5. Crystal Structure

A specimen of $C_{15}H_{16}O_4$ was used for the X-ray crystallographic analysis. The X-ray intensity data were measured. The total exposure time was 2.50 hours. The frames were integrated with the Bruker SAINT software package using a narrow-frame algorithm. The integration of the data using a monoclinic unit cell yielded a total of 12192 reflections to a maximum θ angle of 26.07° (0.81 Å resolution), of which 2675 were independent (average redundancy 4.558, completeness = 99.9%, R_{int} = 2.13%, R_{sig} = 1.71%) and 2113 (78.99%) were greater than $2\sigma(F^2)$. The final cell constants of a = 8.6248 (4) Å, b = 18.9103 (8) Å, c = 8.4204 (4) Å, β = 101.241 (2)°, volume = 1347.00 (11) $Å^3$, are based upon the refinement of the XYZ-centroids of 3786 reflections above 20 σ(I) with 5.382° < 2θ < 51.85°. Data were corrected for absorption effects using the multi-scan method (SADABS). The ratio of minimum to maximum apparent transmission was 0.920. The structure was solved and refined using the Bruker SHELXTL Software Package, using the space group P 1 21/c 1, with Z = 4 for the formula unit, $C_{15}H_{16}O_4$. The final anisotropic full-matrix least-squares refinement on F^2 with 176 variables converged at R1 = 4.03%, for the observed data and wR2 = 12.44% for all data. The goodness-of-fit was 1.008. The largest peak in the final difference electron density synthesis was 0.148 $e^-/Å^3$ and the largest hole was −0.136 $e^-/Å^3$ with an RMS deviation of 0.029 $e^-/Å^3$. On the basis of the final model, the calculated density was 1.283 g/cm^3 and F(000), 552 e^-.

4. Conclusion

It can be concluded that the newly synthesized 5, 7-dimethyl coumarin-4-acetic ethyl ester exists mainly in the S-cis arrangement in the solid state. The packing diagram revealed the existence of only S-cis isomer which is stabilized by C-H····O bonds.

5. Acknowledgements

Fellowship from SC-ST cell Karnatak University, Dharwad is gratefully acknowledged.

REFERENCES

[1] M. V. Kulkarni, G. M. Kulkarni, C. H. Lin and C. M. Sun, "Coumarins and 1-Azacoumarins as Biodynamic Agents," *Current Medicinal Chemistry*, Vol. 13, No. 23, 2006, pp. 2795-2818.

[2] S. Roseman, C. F. Heubuer, R. Pankratz and K. P. Link, "Studies on 4-Hydroxy Coumarins. XVI. The Metabolism of 4-Hydroxy Coumarins in the Dog," *Journal of the American Chemical Society*, Vol. 76, No. 6, 1954, pp. 1650-1652.

[3] P. Laurin, D. Ferroud, L. Schiol, M. Klich, C. Dupuis-Hamelin, P. Mauvais, P. Lassaigne, A. Bonnefoy and B. Musicki, "Structure-Activity Relationship in Two Series of Aminoalkyl Substituted Coumarin Inhibitors of Gyrase B," *Bioorganic and Medicinal Chemistry Letters*, Vol. 9, No. 19, 1999, pp. 2875-2880.

[4] M. V. Kulkarni and V. D. Patil, "Studies on Coumarins (Part 1)," *Archiv der Pharmazie*, Vol. 314, No. 8, 1981, pp. 708-711.

[5] S. S. Hanmantgad, M. V. Kulkarni and V. D. Patil, "Syn-

thesis and Biological Activity of Some 4-(Sulphonamido-Methyl)-Coumarins," *Indian Journal of Chemistry*, Vol. 24B, No. 4, 1985, pp. 459-461.

[6] M. Ghate and M. V. Kulkarni, "Synthesis of Vanillin Ethers from 4-Bromomethyl Coumarins as Anti-Inflammatory Agents," *European Journal of Medicinal Chemistry*, Vol. 38, No. 3, 2003, pp. 297-302.

[7] M. Basanagouda, V. B. Jadhav, M. V. Kulkarni and R. Nagendra Rao, "Computer Aided Prediction of Biological Activity Spectra: Study of Correlation between Predicted and Observed Activities for Coumarin-4-Acetic Acids," *Indian Journal of Pharmaceutical Sciences*, Vol. 73, No. 1, 2011, pp. 88-92.

[8] V. B. Jadhav, S. K. Nayak, T. N. Row and M. V. Kulkarni, "Synthesis, Structure and DNA Cleavage Studies of Coumarin Analogues of Tetrahydroisoquinoline and Protoberberine Alkaloids," *European Journal of Medicinal Chemistry*, Vol. 45, No. 9, 2010, pp. 3575-3580.

[9] K. Sivakumar, F. Jesurethinam, K. Subramanian and S. Natarajan, "7-Hydroxy-4-Coumarin Acetic Acid Monohydrate," *Acta Crystallographica Section C*, Vol. 46, 1990, pp. 1663-1665.

[10] S. Naveen, S. M. Anandwalkar, *et al*., "7,8-Dimethyl Coumarin 4-Acetic Acid Ethyl Esters," *Analytical Sciences*, Vol. 22, 2006, pp. X173-X174.

[11] I. Fujii, Y. Mano and N. Hirayama, "Crystal of 4-Ethoxycoumarin" *Analytical Science*, Vol. 21, 2005, pp. x7-x8

[12] N. M. Mahabaleshwaraiah, K. M. Kumar, O. Kotresh, W. F. A. Al-Eryani and H. C. Devarajgowda, "(6-Methoxy-2-Oxo-2H-Chromen-4-yl)Methylpyrrolidine-1-Carbodithiate," *Acta Crystallographica*, Vol. E68, No. 5, 2012, p. 1566.

[13] P. Deslongchamps, U. O. Cheriyan, J. P. Pradere, P. Soucy and R. J. Taillefer, "Hydrolysis and Isomerization of syn Unsymmetrical N,N-Dialkylated Immidate Salts. Experimental Evidence for Conformational Changes and for Stereo Electronically Controlled Cleaves in Hemi-Orthoamide Tetrahedral Intermediates," *Nouveau Journal de Chimie*, Vol. 3, No. 2, 1979, pp. 343-350.

[14] S. C. Laskowski and R. O. Clinton, "Coumarins. II. Derivative of Coumarins-3- and -4-Acetic Acids," *Journal of the American Chemical Society*, Vol. 72, No. 9, 1950, pp. 3987-3991.

5

ZnO Heteroepitaxy on Sapphire Using a Novel Buffer Layer of Titanium Oxide: Optoelectronic Behavior

Satoshi Yamauchi[1*], Yoh Imai[2]
[1]Department of Biomolecular Functional Engineering, Ibaraki University, Hitachi, Japan
[2]Department of Electric and Electronic Engineering, Ibaraki University, Hitachi, Japan

ABSTRACT

Optoelectronic property of ZnO epitaxial layer grown by plasma-assisted epitaxy at temperature as low as 340°C using Ti_2O_3 buffer layer on a-sapphire were studied by low temperature photoluminescence at 10 K comparing to the layers on c-sapphire and a-sapphire without the buffer layer. The near band-edge emission consisting of free-exciton emissions and neutral-donor bound exciton emissions was significantly dependent on the buffer thickness and dominated by the free-exciton emissions in the layer grown on the very thin buffer layer about 0.8 nm, whereas the intense emissions by neutral-donor bound excitons were observed in the ZnO layer on c-sapphire. The structural behavior indicated the donor was originated from the three-dimensional growth of ZnO layer and details of the optoelectronic feature suggested the residual donors were Al and interstitial-Zn.

Keywords: ZnO; Ti_2O_3; Plasma-Assisted Epitaxy; Photoluminescence; Free-Exciton; Bound-Exciton

1. Introduction

ZnO is candidate for highly efficient blue or ultraviolet light-emitting devices using the wide direct band gap of 3.37 eV and the large exciton binding energy of 60 meV at room temperature. For such device applications, advanced processes are required for the high-quality growth and highly efficient impurity doping. To date, MBE using oxygen-plasma cell [1], MOMBE using H_2O vapor [2], PLD [3] etc. have been attractively studied to improve the structural and optoelectronic properties. At the first stage for ZnO growth, reduction of oxygen deficiency which generates strong green-emission due to oxygen vacancies in ZnO had been required to improve the optoelectronic property. For the purpose, plasma-assisted epitaxy (PAE) using oxygen gas plasma was also an useful process as demonstrated for undoped-ZnO growth at the temperature as low as 400°C on c-sapphire [4,5], in which the green-emission was sufficiently decreased comparing to the band-edge emissions. However, the near band-edge emissions were dominated by neutral donor bound emissions as same as the other processes, which indicated unexpected shallow donors were included with relatively high-density in the layer. It is important to prevent the donors during ZnO growth, especially for p-type ZnO growth. Therefore, the origin and the removal process have been studied at the second stage, in addition to the effective acceptor doping into the layer [6,7]. As well recognized, low-temperature photoluminescence (PL) is so useful for evaluation of the impurities. In the case of ZnO, the bound exciton emissions concerned with shallow donor impurities such as Ga, Al, In and H have been established with the free-exciton emissions [8]. In addition to such impurity donors, interstitial-Zn should also be taken into account for the shallow donor defect in ZnO as suggested by electron paramagnetic resonance [9]. In the growth process, the defect-donor can be easily introduced during three-dimensional growth of ZnO by high-sticking coefficient of Zn-adatom at steps or kinks on the growth surface. Further, the unfavorable feature is probably enhanced in non-equilibrium growth at low temperatures required for highly efficient doping. Therefore, suitable two-dimensional ZnO growth at low temperatures with reducing impurity-donors and the intrinsic donor-defects is required for the optoelectronic device applications. For the purpose, we previously demonstrated drastic improvement of ZnO growth by Ti_2O_3 buffer layer on a-sapphire [10], where two-dimensional epitaxial growth of ZnO layer was successfully achieved without the rotational domain.

*Corresponding author.

In this paper, optoelectronic property of PAE-ZnO layer grown on the Ti_2O_3 buffer layer is examined by low-temperature photoluminescence and shows significant reduction of the residual donor. In addition, details of the near band-edge emissions suggest the donor species.

2. Experimental

2.1. Layer Growth

Titanium oxide layer was grown by LPCVD using titanium tetra-iso-propoxide (TTIP: Ti-$(OC_3H_7)_4$) and oxygen gas. Details of the apparatus and the condition was shown elsewhere [10]. In the case of buffer layer growth for ZnO layer, the thickness was controlled by the growth period estimated from the growth rate.

ZnO layer was grown at 340°C or 400°C in 3 mtorr by plasma-assisted epitaxy (PAE) using oxygen gas plasma generated by 10 W radio-frequency (rf) power at 13.56 MHz through a capacitively coupled rf-electrode. Details of the PAE-apparatus and the growth process were described elsewhere [4,10]. ZnO growth rate was different by change of the growth mode on c-, a- and the buffer layer as shown elsewhere [10], but effective Zn/O supply ratio during the growth was kept at 1.0 by control of Zn-flux determined by dependence of the growth rate on the Zn-flux.

Single crystalline a- and c-sapphire with mirror surface and 300 μm-thick were used as substrates after cleaning by organic-solvents and hot $H_2SO_4 + H_2O_2$. In the case of etching, the substrates were treated in a hot $3H_2SO_4 + H_3PO_4$ solution at 130°C for 15 min.

2.2. Evaluation

Thickness of ZnO layers and Ti_2O_3 layers were checked by a contact-type surface profiler (Veeco, DEKTAK150). Surface morphology of the ZnO films was observed by Nomarski differential interference microscope (OLYMPUS, BX60). Optoelectronic property was examined by photoluminescence (PL) at 10 K using a cryogenic system (Janis Research, CCS-150). UV-light around 313 nm radiated from a deep UV-lamp (USHIO, UXM-501MA) was selected by a band-pass filter and irradiated on the ZnO layer through a sapphire window equipped on the cryostat chamber with the light power density about 2 mW/cm^2. The luminescent light modified to 20 Hz-AC by an optical chopper was introduced into a monocrometor (JASCO, SS-50), then detected and amplified by a photomultiplier (Hamamatsu R374) driven by 1 kV. The current signal was amplified and converted to voltage signal by an amplifier (FEMTO, DLPCA-200), and then the noise components was removed in lock-in-amplifier (NF, LI5640) synchronous to the frequency by the chopper. Then, the signal was recorded in a PC after A/D conversion. The relative sensitivity of the system for photon energy was corrected by black-body radiation spectrum from a standard lamp.

3. Results and Discussions

3.1. ZnO Layer on C-Sapphire

Growth rate of ZnO layers at 400°C (closed circles) and 340°C (open circles) on c-sapphire for various Zn-flux in oxygen plasma excited by 10 W rf-power are shown in **Figure 1**. The growth rates same at both temperatures were increased with the Zn-flux then saturated by the flux above 80 μmol/min. The results indicated the effective supply flux rate of Zn/O could be determined at both temperatures since the growth rate was limited by smaller flux of Zn or O. **Figure 2** shows typical near band-edge PL spectrum of PAE-ZnO layer with the thickness of 1.0 m grown at 400°C on c- sapphire by the Zn/O supply ratio of 1.0 (Zn-flux: 80 μmol/min). The inset shows surface morphology of the layer observed by Nomarski microscope. The layer consisting of three-dimensionally grown columnar grains was preferentially oriented along c-axis but including 30°-rotational domains. In the PL spectrum, prominent emission peak at 3.3623 eV originated from neutral donor-bound exciton emission (DBE) was observed with a weak DBE emission peak 3.3666 eV and free-exciton emissions of A-E_x and B-E_x at 3.3774 eV and around 3.385 eV respectively [8]. FWHM of the DBE emission of the layer grown by the supply ratio around 1.0 was below 2 meV and the intensity was 4-orders magnitude larger than the green-emission due to deep-level. It is noted that the PL-spectrum was obviously degraded in the ZnO layer grown at temperatures lower than 400°C and the sharp near band-edge emissions could not be observed.

3.2. ZnO Layer on Ti_2O_3 Buffer Layer

3.2.1. PL Spectrum on A-Sapphire and Buffer Layer

Figure 3 shows near band-edge spectra of PAE-ZnO layers grown on (a) a-sapphire and (b) Ti_2O_3 buffer layer grown at 320°C with the thickness of 0.8 nm, where the ZnO layers were grown at 340°C in oxygen plasma excited by 10 W rf-power. It is noted here the near band-edge emissions could be observed in ZnO layers grown at the temperature around 340°C but obviously degraded by the high temperature growth around 400°C. The insets show the surface morphologies observed by Nomarski microscope. The layer was epitaxially grown on the buffer layer with the epitaxial relationship of [1-100] ZnO//[1-100] Ti_2O_3, whereas the layer was polycrystallized on a-sapphire. The hexagonal pyramid-like grains with the facets according to the epitaxial relationship as shown in the inset of (b) were observed in ZnO

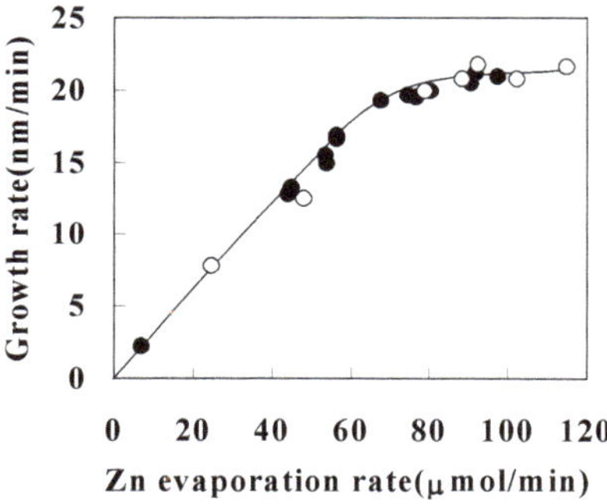

Figure 1. Dependence of ZnO growth rate on Zn-flux, where the ZnO layers were grown on c-sapphire at 400°C (solid circle) and 340°C (open circle) in oxygen plasma excited by 10W rf-power.

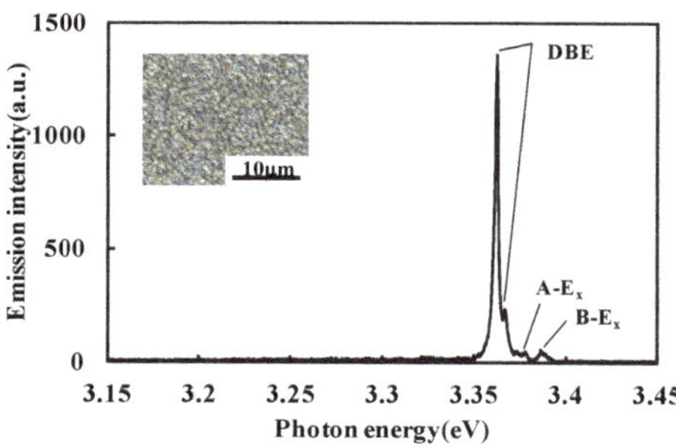

Figure 2. Photoluminescence spectrum of PAE-ZnO layer on c-sapphire.

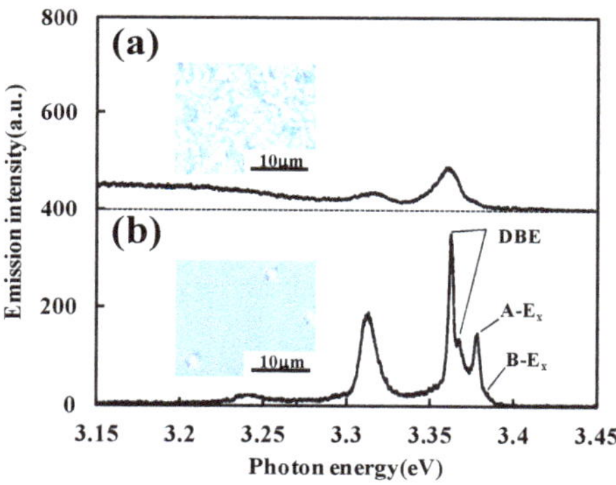

Figure 3. Photoluminescence spectra of PAE-ZnO layer on (a) c-sapphire and (b) 0.8 nm-thick Ti_2O_3/a-sapphire.

layer on the buffer layer, which indicated the two-dimensional growth was enhanced on the buffer layer as described elsewhere [10]. In the near band-edge PL-feature, strong and sharp exciton emissions could be observed from ZnO layer on the buffer layer, in contrast to board and weak emissions from poly-crystallized ZnO layer directly grown on a-sapphire. The spectrum of **Figure 3(b)** was dominated by DBE emissions with the same peak energies in **Figure 2** (grown on c-sapphire), however, free-exciton emissions of A-E_x and B-E_x were clearly observed with decreasing the DBE intensities. The results indicate that although the ZnO layer was grown at low temperature comparing to the growth on c-sapphire, the crystallinity was improved on the buffer layer. In addition, the residual shallow donor could also be reduced in the ZnO on the buffer layer.

3.2.2. Dependence on the Buffer-Layer Growth Condition

The PL feature was significantly dependent on the growth condition and the thickness of buffer layer. **Figure 4** shows variation of the intensity ratio of A-E_x to DBE emission peak at 3.3623 eV on the thickness of buffer layer grown at 320°C and 340°C (closed circles) with the intensity of A-E_x emission (open circles). Crystal quality should not be discussed only by the intensity ratio because PL-spectrum is also influenced by non-radiative defects, but it can be considered the free-exciton emission intensity suggests the quality. In the case of 340°C growth of the buffer layer, A-E_x emission in ZnO layer on the 1.8 nm-thick buffer layer was most intense, but optimized the thickness of buffer layer for the intensity ratio of A-E_x/DBE was around 2.8 nm since not only the intense A-E_x but also strong donor bound exciton emission were observed from ZnO layer on the 1.8 nm-thick buffer layer. The result indicated non-radiative defects were significantly reduced in ZnO layer on the buffer layer with the thickness around 1.8 nm comparing to the ZnO layers on c-sapphire but the donor could not be reduced in the ZnO layer. In contrast, when the buffer layer was grown at 320°C, the PL-property was more sensitive to the buffer layer thickness, but the thickness could be optimized at 0.8 nm for both of A-E_x intensity and the intensity ratio. The layer grown on the optimized buffer layer grown at 320°C showed smooth surface partially including hexagonal pyramid-like grains as shown in **Figure 3(b)**, where the number of pyramid-like grain was most reduced in the ZnO layer on the optimized buffer layer [10]. It was found that the pyramid-like grains were originated from surface roughness and dislocations on a-sapphire and could be prevented by chemical etching of the substrate using a hot $3H_2SO_4 + H_3PO_4$ solution. It is noted that density of the pyramid-like grains were decreased but columnar grains without the fine facets observed in the pyramid-like grains were increased with the buffer thickness above 2 and 4 nm for the growth temperature of 320°C and 340°C, respectively.

3.2.3. PL Spectrum of ZnO Layer on Optimized Buffer-Layer

Figure 5 shows near band-edge PL-spectrum of ZnO layer on 0.8 nm-thick buffer layer grown at 320°C on chemically etched a-sapphire. As shown in the inset, hexagonal pyramid-like grains were almost disappeared in the ZnO layer. In comparison to **Figure 3(b)**, donor

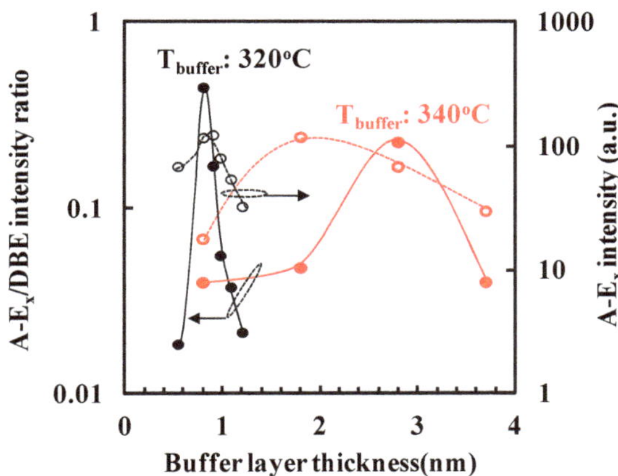

Figure 4. Dependence of A-Ex/DBE intensity ratio (closed circles) and A-E_x intensity (open circles) on the thickness of Ti_2O_3 buffer layer grown at 320°C (black) and 340°C (red).

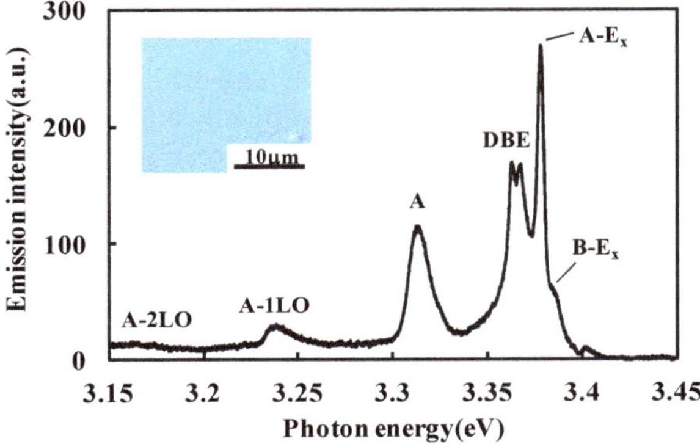

Figure 5. Photoluminescence spectra of PAE-ZnO layer on 0.8 nm-thick Ti_2O_3/a-sapphire, where the a-sapphire was chemically etched by a hot $3H_2SO_4$ + H_3PO_4 solution at 130°C for 15 min before the Ti_2O_3 growth.

bound exciton emissions were drastically decreased and dominated by intense and sharp A-E_x with the FWHM below 5 meV. The fine feature clearly indicated residual donor was successfully reduced during the ZnO growth and the three-dimensional growth increased the donor density.

Broad-line around 3.313 eV denoted "A", which was shifted between 3.313 - 3.315 eV by the ZnO growth condition, with the photon-replicas (LO-phonon energy: 73 meV) was also observed in the ZnO layers. The emission was seemed to be similar to emission discussed as A-line [11], which was concluded as surface defects-related emission. The thermal activation energy of the broad-line obtained by Arrhenius-relationship on the intensity was about 60 meV, which indicated the emission was not due to electron recombination in deep-level but originated from recombination of exciton bound to defects. Further, the emission was not observed from the ZnO layer on c-sapphire as shown in **Figure 2**, which indicated the emission was not derived from the grain boundaries. The origin of defect bounding exction is not identified yet, but may be related to surface structural defects [11,12].

3.2.4. Residual Donor

It has been well recognized a lot of sharp donor bound exciton emission lines in the narrow energy range from 3.348 to 3.373 eV [8], in which the lines are due to neutral or ionized donor defect pair [8,13]. **Figure 6** shows PL spectra in log-scale of ZnO layers on (a) buffer layer and (b) c-sapphire. The bound exciton emission peak at 3.3623 eV could be assigned to be due to neutral donor bound exciton with the localization energy of 15.1 meV from the A-E_x with peak at 3.3774 eV. Impurities corresponding to neutral donor bound exciton emissions have been identified in bulk-ZnO for IIIB-elements such as Al, Ga and In [8]. Although peak energy of the neutral donor bound exciton emission was different from the impurity related bound exciton emission as previously reported, the localization energy of 15.1 meV was coincident to the binding-energy of Al. The exciton line bound to hydrogen at 3.363 eV [14] was in agreement to the bound exciton in PAE-ZnO layer, but the impurity was rule out because the emission peak at 3.3623 eV was not decreased by annealing at 600°C for 30 min whereas the emission bound to hydrogen-donor was drastically reduced by the annealing [14]. It is considered that the impurity can be identified by the localization energy from free-exciton emission because the free-exciton line is shifted by bi-axial stress [15]. The A-E_x emission of PAE-ZnO was shifted toward higher energy side than that with peak 3.3759 eV (transversal free A-exciton state) dominated in A-exciton emissions bulk-ZnO [8]. As a result, it could be suggested that a residual donor in the PAE-ZnO layer was Al. Meyer, *et al.* showed Haynes'

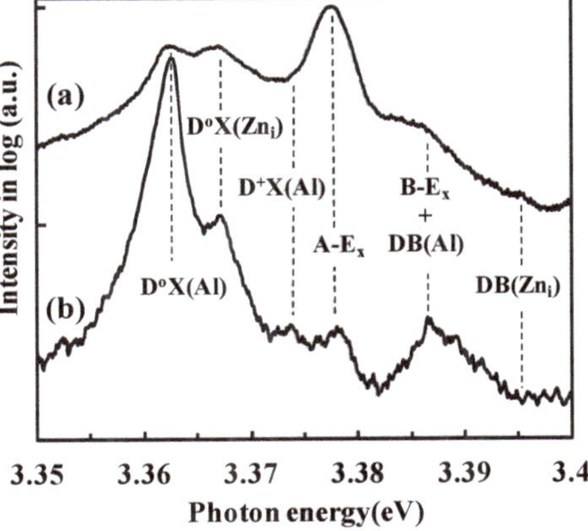

Figure 6. Photoluminescence spectra in excitonic emission region of PAE-ZnO layers on (a) 0.8 nm-thick Ti_2O_3/a-sapphire and (b) c-sapphire, where the intensity was shown in log scale and the ground-line of (a) was shifted in vertical.

rule in ZnO as $E_{loc} = 0.37E_D - 4.2$ meV for neutral donor bound exiton and $E_{loc} = 0.50E_D - 23$ meV for ionized donor bound exciton, where E_{loc} and E_D are activation energy of donor and localization energy of the bound excitons from free A-exciton, respectively [16]. According to the localization energy of 15.1 meV for the neutral donor bound exciton (D^oX(Al)), binding energy of the donor was obtained as 52.2 meV. In addition, localization energy the ionized donor bound exciton (D^+X(Al)) was expected to be at 3.1 meV by the above relationship. The exciton line could be observed as the weak emission around 3.374 eV on the low-energy side shoulder of the A-E_x in (a) and at 3.3736 eV in (b). Further, the donor to valence-band (DB(Al)) emission was expected to be observed at 3.3852 eV, in which the band-gap was obtained by the A-E_x line and the binding energy of 60 meV. It was difficult to identify the emission in PL spectrum because the emission line was overlapped on the B-E_x emission, however, the intensity around 3.385 eV in the spectrum with intense D^oX(Al) emission was larger than that of A-E_x emission as shown in **Figure 6**(b).

On the other, the donor concerned to the emission line at 3.3666 eV with the localization energy of 10.8 meV could not be identified by the IIIB-impurities. Previously, Shibata, *et al.* indicated the emission at 3.366 eV is neutral donor bound exciton emission due to interstitial-Zn [17], where the localization energy was obtained as 11 meV because free-exciton emission was observed at 3.377 eV. The localization energy was in good agreement to that of the emission at 3.3666 eV in PAE-ZnO. The spectral feature broader than the D^oX(Al) also suggested the emission at 3.3666 eV was due to exciton emission bound to interstitial-Zn complex ($D^oX(Zn_i)$). Binding energy of the donor was obtained to 40.5 meV and exciton will not be bound to the ionized donor with such shallow level (E_D < 47 meV) as recognized by the previous relationship [16]. In contrast, it is considered that the weak emission peak at 3.3955 eV in **Figure 6** (a) was the donor to valence-band emission (DB(Zn_i)), in which the donor binding energy was obtained as 41.9 meV. It is noted that the emission of (DB(Zn_i) was not observed from ZnO layer including Al with relatively high density as shown in **Figure 6** (b).

Both of the D^oX(Al) and $D^oX(Zn_i)$ were drastically reduced in the ZnO layer grown at 340˚C on the buffer layer comparing to that on c-sapphire grown at 400˚C. The Al was probably diffused from sapphire substrate because the D^oX(Al) was drastically reduced in the ZnO layer grown at 340˚C on the buffer layer comparing to the layer on c-sapphire grown at 400˚C although Al on the growth surface was expected to be with higher sticking coefficient at lower temperature. Therefore, two possible effects can be considered for reduction of the donor in ZnO on the buffer layer. The one is decrease of the growth temperature, which brings decrease of the diffusion coefficient. As expected increasing diffusion coefficient in grain boundaries, the other is concerned to structural behavior of ZnO layer, which can be expected by variation of A-E_x/D^oX(Al) intensity ratio for the buffer layer thickness as shown in comparison to the growth behavior. That is, it is considered that the Al diffusion was suppressed in ZnO layer on the optimum buffer layer by significantly decreased grain boundaries. In contrast, interstitial-Zn is promised to be introduced on the surface during ZnO growth. Also in this case, it is expected that interstitial-Zn is easily introduced into the layer at low temperature, however, the $D^oX(Zn_i)$ was decreased by the buffer layer even at low temperature of 340˚C comparing to 400˚C for the growth on c-sapphire, which suggest that the surface structure is more important to prevent interstitial-Zn than the growth temperature. Further, the intensity ratio of A-E_x/$D^oX(Zn_i)$ of ZnO layer was drastically decreased on the buffer layer grown at 320˚C with thickness above 1.2 nm, where the three-dimensional ZnO growth was drastically enhanced and a lot of columnar-grains without facets were observed, however, the increased ratio of ZnO layer including pyramid-like grains on the buffer layer with optimized thickness as shown in **Figure 3(b)** was similar to that of the layer successfully removed the grains by sapphire-etching as shown in **Figure 4**. The results indicate interstitial-Zn is increased on the three-dimensional growth surface but prevented on hexagonal pyramid surface.

4. Conclusion

Optoelectronic property of PAE-ZnO layers grown on c-sapphire, a-sapphire and Ti_2O_3 buffer layer was examined by PL at low temperature of 10K. Neutral donor bound exciton emissions were dominated in the near band-edge region of ZnO layer on c-sapphire. In contrast, although the PL feature was significantly dependent on the buffer layer thickness and the growth condition, successful reduction of donor bound exciton emissions by the optimized buffer layer were resulted in the spectrum dominated by free-exciton emissions. The residual donors in the ZnO layer could be identified as Al and interstitial-Zn by the localization energy of the neutral bound exciton emissions from free-exciton emission. The obvious decrease of grain boundaries in ZnO layer was so effective to the Al-reduction in addition to low temperature ZnO layer was nominally decreased by the low temperature growth of 340˚C on the optimized buffer layer.

REFERENCES

[1] M. A. L. Johnson, S. Fujita, W. H. Rowland, Jr., W. C. Hughes, J. W. Cook, Jr. and J. F. Schetzina, "MBE Growth and Properties of ZnO on Sapphire and SiC Sub-

strates," *Journal of Electronic Materials*, Vol. 25, No. 5, 1996, pp. 855-862.

[2] A. B. M. A. Ashrafi, I. Suemune, H. Kumano and S. Tanaka, "Nitrogen-Doped p-Type ZnO Layers Prepared with H_2O Vapor-Assisted Metalorganic Molecular-Beam Epitaxy," *Japanese Journal of Applied Physics*, Vol. 41, 2002, pp. L1281-L1284.

[3] R. D. Vispute, V. Talyansky, Z. Trajanovic, S. Choopun, M. Downes, R. P. Sharma, T. Venkatesan, M. C. Woods, R. T. Lareau and K. A. Jones, "High Quality Crystalline ZnO Buffer Layers on Sapphire (001) by Pulsed Laser Deposition for III-V Nitrides," *Applied Physics Letters*, Vol. 70, No. 20, 1997, pp. 2735-2737.

[4] S. Yamauchi, H. Handa, A. Nagayama and T. Hariu, "Low Temperature Epitaxial Growth of ZnO Layer by Plasma-Assisted Epitaxy," *Thin Solid Films*, Vol. 345, No. 1, 1999, pp. 12-17.

[5] S. Yamauchi, T. Ashiga, A. Nagayama and T. Hariu, "Plasma-Assisted Epitaxial Growth of ZnO Layer on Sapphire," *Journal of Crystal Growth*, Vol. 214-215, 2000, pp. 63-67.

[6] S. Yamauchi, Y. Goto and T. Hariu, "Photoluminescence Studies of Undoped and Nitrogen-Doped ZnO Layers Grown by Plasma-Assisted Epitaxt," *Journal of Crystal Growth*, Vol. 260, No. 1-2, 2004, pp.1-6.

[7] D. C. Look, D. C. Reynolds, C. W. Litton, R. L. Jones, D. B. Eason and G. Gantwell, "Characterization of Homoepitaxial p-Type ZnO Grown by Molecular Beam Epitaxy," *Applied Physics Letters*, Vol. 81, No. 10, 2002, pp. 1830-1832.

[8] B. K. Meyer, H. Alves, D. M. Hofmann, W. Kriegseis, D. Forster, F. Bertram, J. Christen, A. Hoffmann, M. Straßburg, M. Dworzak, U. Haboeck and A. V. Rodina, "Bound Exciton and Donor-Acceptor Pair Recombinations in ZnO," *Physica Status Solidi (B)*, Vol. 241, No. 2, 2004, pp. 231-260.

[9] L. S. Vlasenko and G. D. Watkins, "Intrinsic Defects in ZnO: A Study Using Optical Detection of Electron Paramagnetic Resonance," *Physica B*, Vol. 376-377, 2006, pp. 677-681.

[10] S. Yamauchi and Y. Imai, "ZnO Heteroepitaxy on Sapphire Using a Novel Buffer Layer of Titanium Oxide: Crystallographic Behavior," *Crystal Structure Theory and Applications*, Vol. 2, No. 2, 2013, pp. 39-45.

[11] S. S. Kurbanov and T. W. Kang, "Spectral Behavior of the Emission Around 3.31 eV (A-Line) from ZnO Nanocrystals," *Journal of Luminescence*, Vol. 130, No. 5, 2010, pp. 767-770.

[12] A. Teke Ü. Özgür, S. Dogan, X. Gu, H. Morkoç, B. Nemeth, J. Nause, H. O. Everitt, "Excitonic fine Structure and Recombination Dynamics in Single-Crystalline ZnO," *Physical Review B*, Vol. 70, No. 19, 2004, Article ID. 195207-1-10.

[13] D. C. Reynolds, D. C. Look, B. Jogai, C. W. Litton, T. C. Collins, W. Harsch and G. Cantwell, "Neutral-Donor-Bound-Exciton Complexes in ZnO Crystals," *Physical Review B*, Vol. 57, 1998, pp. 12151-12155.

[14] H. Alves, D. Pfisterer, A. Zeuner, T. Riemann, J. Christen, D. M. Hofmann and B. K. Meyer, "Optical Investigations on Excitons Bound to Impurities and Dislocations in ZnO," *Optical Materials*, Vol. 23, No. 1-2, 2003, pp. 33-37.

[15] B. Gil, "Oscillator Strengths of A, B, and C Excitons in ZnO Films," *Physical Review B*, Vol. 64, 2001, Article ID. 201310-1-3.

[16] B. K. Meyer, J. Sann, S. Lautenschläger, M. R. Wagner and A. Hoffmann, "Ionized and Neutral Donor-Bound Excitons in ZnO," *Physical Review B*, Vol. 76, 2007, Article ID. 184120-1-10.

[17] H. Shibata, M. Watanabe, M. Sakai, K. Oka, P. Fons, K. Iwata, A. Yamada, K. Matsubara, K. Sakurai, H. Tampo, K. Nakahara and S. Niki, "Characterization of ZnO Crystals by Photoluminescence Spectroscopy," *Physica Status Solidi (C)*, Vol. 1, No. 4, 2004, pp. 872-875.

Investigation of Some Structural and Mechanical Properties of $Ba_{0.5}Ca_xSr_{0.5-x}TiO_3$ Ceramics

Lubna M. Sharaf El-Deen, Mohamed H. Badr*, Abdel-Mageed H. Khafagy, Dalia U. Nassar
Physics Department, Faculty of Science, Minufiya University, Shibin El Kom, Egypt

ABSTRACT

$Ba_{0.5}Ca_xSr_{0.5-x}TiO_3$ (BCST) ceramics, where x = 0, 0.1, 0.2, 0.3 and 0.4, were prepared by the conventional solid state reaction technique. X-ray diffraction (XRD) analysis confirmed the formation of BST perovskite phase structure besides some calcium oxide peaks for samples with high Ca content, x. Scanning electron microscopy (SEM) results confirmed the XRD results, *i.e.*, as x increased, the average grain size decreased. Energy dispersive X-ray (EDX) analysis verified the increase of the amount of Ca element with increasing of its content. Mechanical properties such as ultrasonic attenuation, longitudinal wave velocity, and longitudinal elastic modulus were studied by an ultrasonic pulse echo technique at 2 MHz frequency. Investigations of ceramic microstructures and mechanical properties showed their dependence on composition. Increasing of Ca content resulted in a decrease in bulk density and ultrasonic attenuation and an increase in porosity, velocity, and modulus. High temperature ultrasonic studies showed, in addition to Curie phase transition, three or more relaxation peaks and its origin was investigated.

Keywords: BST Ceramics; BCST Ceramics; Phase Transition; Ultrasonic Attenuation; SEM

1. Introduction

$BaTiO_3$ and $SrTiO_3$ are representatives for ABO_3-type perovskite materials. $BaTiO_3$ is usually a ferroelectric material with Curie temperature of 120°C. $SrTiO_3$ is a paraelectric one with no ferroelectric phase transition [1]. Nevertheless, the combined production of $BaTiO_3$ and $SrTiO_3$ (*i.e.*, $Ba_{1-x}Sr_xTiO_3$) is a solid solution system between $BaTiO_3$ and $SrTiO_3$. Therefore, $Ba_{1-x}Sr_xTiO_3$ (BST) has the simultaneous advantage of high dielectric constant of $BaTiO_3$ and the structural stability of $SrTiO_3$ [1,2]. These ferroelectric materials have attracted considerable attention owing to their unique properties such as chemical stability, high permittivity, high tunability, and low dielectric losses. Furthermore, BST has shown a great promise in applications, such as phase shifting elements in phased array antennas and as tuning elements in devices operating at microwave frequencies [1,3-9]. In view of their merits, the investigation on BST solid solution is therefore significantly important [1,2,9,10]. Furthermore, the Curie temperature of BST can be controlled by adjusting the Ba/Sr ratio and/or doping ions to substitute for A or B sites in the ABO_3 perovskite systems [11,12].

In $BaTiO_3$ materials, an elastic modulus anomaly and a mechanical loss peak are induced at three phase transitions: cubic-tetragonal (ferroelectric-paraelectric), tetragonal-orthorhombic, and orthorhombic-rhombohedra. Some losses due to relaxation processes have been observed in materials having coarse grains in ferroelectric phase [13]. These relaxations were ascribed to the interaction between domain walls and oxygen vacancies diffusion.

The $Ba_{0.5}Sr_{0.5}TiO_3$ (BST) material exhibits the transition from the ferroelectric state to the paraelectric state below room temperature [14]. In a recent work [15], we have investigated the impact of changing barium content on the mechanical properties (such as elastic modulus, attenuation, and velocity of ultrasonic waves) and Curie transition of $Ba_xSr_{1-x}TiO_3$ ceramics. In this work, we aim to investigate thoroughly the effect of Ca-doping on the structural and mechanical properties of $Ba_{0.5}Ca_xSr_{0.5-x}TiO_3$ (BCST) ceramics with x = 0.0, 0.1, 0.2, 0.3 and 0.4. X-ray diffraction, SEM, EDX, and ultrasonic techniques (at a frequency of 2 MHz) were also used to characterize the structure and phase transitions of these ceramics.

2. Materials and Methods

Calcium-doped ceramics with the chemical formula

*Corresponding author.

$Ba_{0.5}Ca_xSr_{0.5-x}TiO_3$ (BCST), where x = 0.0, 0.1, 0.2, 0.3, and 0.4, were prepared by the conventional solid state reaction technique according to the following reaction:

$$(0.5)BaCO_3 + (0.5-x)SrCO_3 + (x)CaO + TiO_2$$
$$\rightarrow Ba_{0.5}Ca_xSr_{0.5-x}TiO_3 + (1-x)CO_2 \uparrow$$

For all prepared samples, the reagent grade chemicals of high purity (99.99%) $BaCO_3$, $SrCO_3$, CaO and TiO_2 powders were used as the raw materials and weighed according to the above indicated compositions.

The raw materials were weighed and mixed in the appropriate ratios. Mixtures of required ceramics were first ground thoroughly by a ball milling for 4 h to insure homogeneity. Then, they were calcined at 1100˚C for 11 h in alumina crucibles opened to the air. The calcined compositions were again ground for 6 h. The produced fine calcinated powders were pressed into disc-shaped pellets with 10mm in diameter and 0.6 ~ 1.5 mm in thickness at an iso-static pressure of 5 tons with polyethylene glycol $[(C_2H_4O)_n.H_2O]$ as an organic binder with 2.0% of the weight of the sample. The pelletized samples were finally sintered at 1250˚C for 6h.

The bulk density of ceramics was measured by the conventional Archimedean method. X-ray diffraction (XRD) patterns were recorded with Bruker AXS X-ray diffractometer (D8 Advance) using Cu Kα radiation. The two SEM and EDX measurements were made on JEOL scanning electron microscope (JXA-840A Electron Probe Microanalyzer, INCA x-sight, Oxford Instruments) for the elemental analysis and chemical characterization of the samples. Attenuation of ultrasonic waves (α), longitudinal velocity of wave propagation (V_L), and the longitudinal elastic modulus (L) of tested ceramics were determined by employing the conventional pulse-echo technique at room temperature and during heating as has been reported elsewhere [15].

3. Results

The Archimedean bulk density (ρ_{exp}) and the theoretical density (ρ_x) from the X-ray diffraction patterns of BCST prepared samples were used to calculate the percentage porosity (**Table 1**) according to the equation % porosity = $(\rho_x - \rho_{exp})/\rho_x$ [16]. **Figure 1** shows variations of both the bulk density and porosity of prepared BCST ceramics with the Ca content x where $0.0 \le x \le 0.4$. This figure revealed a linear decease in the density from 4803 to 4339 kg/m^3 as x increased from 0 to 0.4, in harmony with previously reported results [17-19]. Whereas, the percent porosity has increased linearly from 13.4% to 16.5%, for the same variation in Ca content.

Figure 2 shows the room temperature X-ray diffract-graphs for all tested BCST ceramics between 20˚ and 80˚. The intensity of peaks for different values of Ca content, x, were normalized and shifted for clarity pur-

Table 1. The variation of the density (bulk density, ρ_{exp}, and theoretical density, ρ_x) the percent porosity, the ultrasonic attenuation, longitudinal velocity, modulus, and lattice parameter of BCST ceramics with different values of Ca content, x.

Ca cont. x	Density ρ_{exp} (kg/m^3)	ρ_x (kg/m^3)	Porosity (%) (dB/cm)	Attenuation α V_L (m/s)	Long. Velocity L(GPa)	Long. Elastic Modulus	Lattice Parameter a (Å)
0.0	4803	5509	13.423	24.261	1316.1	8.314	3.9510
0.1	4748	5601	13.650	19.983	1642.5	12.815	3.9461
0.2	4521	5491	15.157	16.945	1873.9	15.872	3.9415
0.3	4483	5481	16.168	15.678	2012.9	18.152	3.9368
0.4	4339	5375	16.535	13.482	2167.2	20.384	3.9325

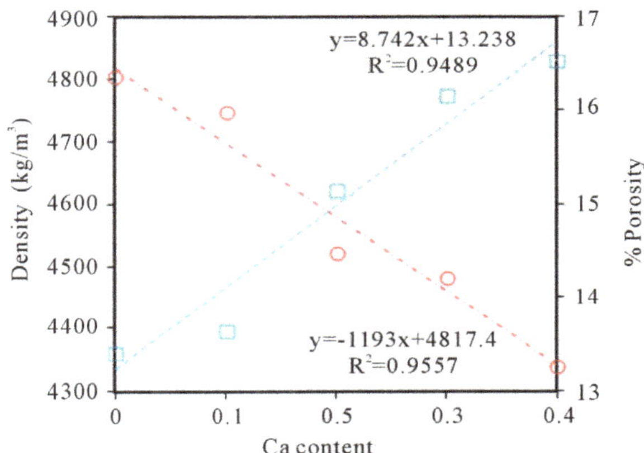

Figure 1. Variation of the density and porosity with the Ca content of $Ba_{0.5}Ca_xSr_{0.5-x}TiO_3$ ceramics sintered at 1250˚C for 6 h.

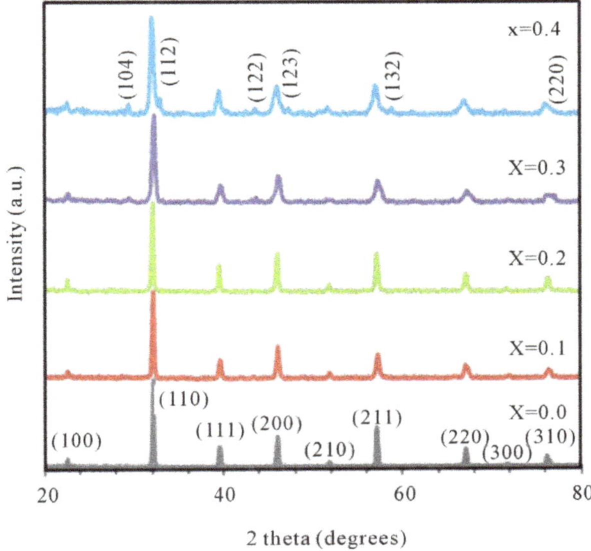

Figure 2. XRD patterns of $B_{0.5}Ca_xSr_{0.5-x}TiO_3$ (BCST) ceramics with Ca content x = 0.0, 0.1, 0.2, 0.3 and 0.4.

poses. All peaks were indexed in the cubic structure due to their observed reflections of different polycrystalline orientations [20-22] as indicated by the (110) index; the major peak of highest intensity. Variation of the lattice parameter a (Å) with different Ca content was listed in **Table 1**.

The chemical compositions of tested BCST ceramics were determined from EDX spectra and listed in **Table 2**. The scanning electron microscopy (SEM) is a powerful experimental technique to determine the particle size, pore concentration, and inclusions in a material. **Figures 3(a)-(d)** show the scanning electron micrographs taken at room temperature of BCST ceramics with Ca content in the range $0.0 \le x \le 0.4$. The SEM for x = 0.2 was not included in this figure for brevity reasons. These micrographs describe the surface property of samples, microstructure, size, and distribution of particles. Whereas, the average grain size dependence (as determined from XRD and SEM investigations) on the Ca content was illustrated in **Figure 4**.

Figure 5 shows the regression line which illustrates the variation of ultrasonic attenuation (α) of ultrasonic waves, measured at room temperature, with the Ca content of the prepared $Ba_{0.5}CaSr_{0.5-x}TiO_3$ ceramics sintered at 1250°C for 6 h. Inspection of the figure reveals that α is dependent on the composition of the tested ceramic, *i.e.*, it is exponentially decreased with increasing of Ca content over the above mentioned investigated range (see also **Table 1**).

Figure 6 shows the dependences of both longitudinal ultrasonic velocity and longitudinal elastic modulus on composition of $Ba_{0.5}Ca_xSr_{0.5-x}TiO_3$ (BCST) ceramics with x= 0.0, 0.1, 0.2, 0.3 and 0.4 were sintered at 1250°C for 6 h. As could be deduced from this figure, both of the ultrasonic velocity and modulus have increased nonlinearly with the increase in Ca content from 0 to 0.4.

Figures 7(a) and **(b)**, show the variations of ultrasonic attenuation (α) of ultrasonic waves, at 2 MHz frequency, with temperature for $Ba_{0.5}Ca_xSr_{0.5-x}TiO_3$ ceramics sintered at 1250°C for 6h with x = 0.0, 0.1, 0.4 and x = 0.0, 0.2, 0.3, respectively. The figures reveal well-defined damping peaks at 266°C, 313°C, 319°C, 325°C and

Table 2. Initial Ca contents and EDX analyses results for the investigated BCST ceramics.

Ca cont., x	EDX analyses results (wt.%)	EDX Sr/Ca
0.0	$Ba_{0.551}Sr_{0.500}Ti_{1.121}O_3$	--
0.1	$Ba_{0.577}Ca_{0.015}Sr_{0.407}Ti_{1.173}O_3$	27.97
0.2	$Ba_{0.593}Ca_{0.151}Sr_{0.370}Ti_{1.058}O_3$	2.50
0.3	$Ba_{0.588}Ca_{0.256}Sr_{0.370}Ti_{0.979}O_3$	1.42
0.4	$Ba_{0.590}Ca_{0.327}Sr_{0.198}Ti_{1.057}O_3$	0.61

Figure 3. SEM micrographs of $Ba_{0.5}Ca_xSr_{0.5-x}TiO_3$ ceramics. The bars represent 10 μm in length.

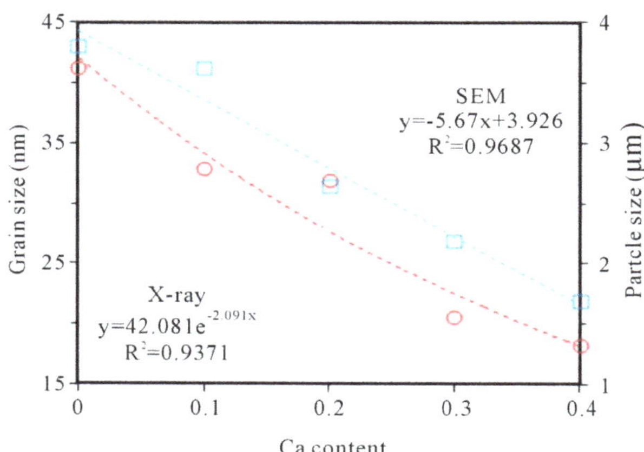

Figure 4. The dependence of average grain size and particle size on the Ca content, x, for BCST ceramics.

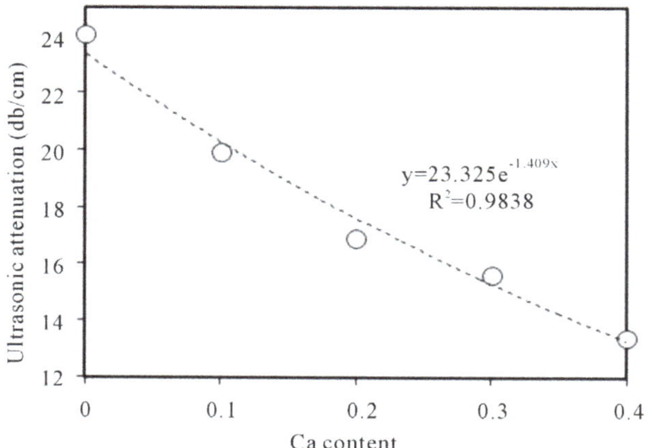

Figure 5. Variation of the ultrasonic attenuation with Ca content in BCST ceramics sintered at 1250°C for 6 h.

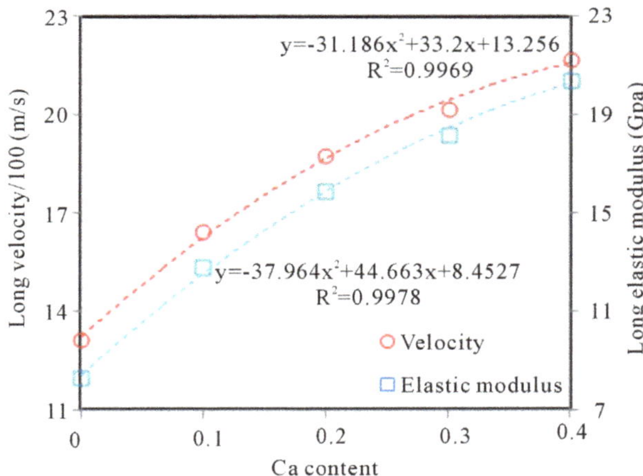

Figure 6. Variation of longitudinal ultrasonic velocity and longitudinal elastic modulus with the Ca content of BCST ceramics sintered at 1250˚C for 6 hr.

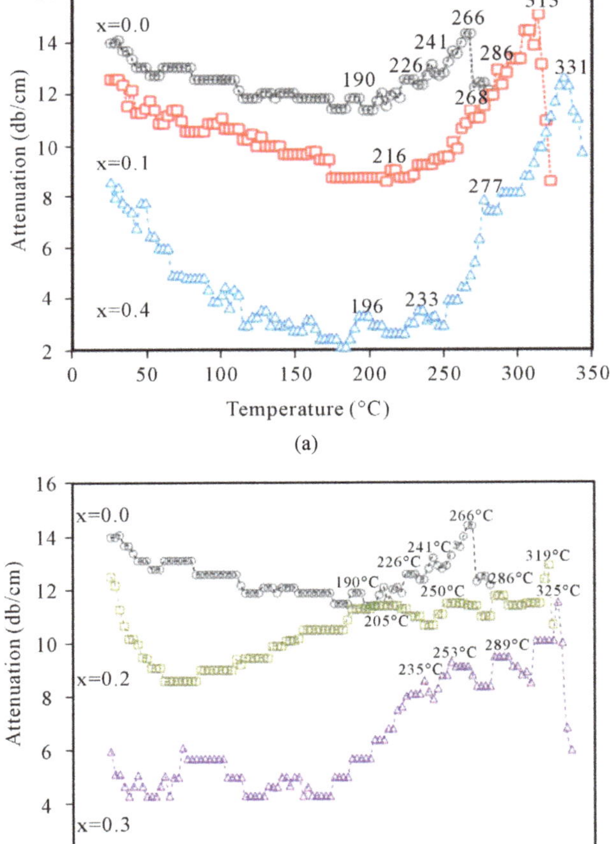

Figure 7. (a): Variation of ultrasonic attenuation with temperature at 2 MHz for x = 0, 0.1 and 0.4; (b): Variation of ultrasonic attenuation with temperature at 2 MHz for with x = 0, 0.2 and 0.3.

331˚C for all tested specimens of x = 0.0, 0.1, 0.2, 0.3 and 0.4, respectively. Also, a shoulder was observed for each ceramic near the room temperature. However, locations and shapes of all observed peaks and shoulders depend on the composition of ceramics.

4. Discussion

4.1. Density and Porosity

The observed decrease of the density of tested BCST ceramics (**Figure 1**) is attributed to: 1) distortion occurred due to substitution of Ca atoms of lower atomic weight (40.08 a.m.u.) with Sr ones of higher atomic weight (87.62 a.m.u.) in the (BCST) ceramics [17-19]; and 2) the observed increase in the percent porosity which is attributed to substitution of Ca atoms of lower ionic radius (0.99 Å) with Sr ones of higher ionic radius (1.12 Å) [18] (see **Table 1**).

4.2. XRD, SEM and EDX

XRD patterns, illustrated in **Figure 2**, showed all major X-ray peaks of diffraction for BCST ceramics. The patterns confirm the formation of perovskite phase of BST ceramics obtained previously in [16,20-22]. Also, these XRD patterns showed a cubic perovskite phase which confirmed that Ca^{+2} ions substitution for Sr^{+2} can be mostly incorporated into the perovskite structure of BST material. Moreover, as the Ca content increased (x = 0.3 and 0.4) very small assigned peaks were observed and indexed to their corresponding orientations as shown on x = 0.4 pattern, **Figure 2**. The intensity of these subsidiary peaks slightly increased for x = 0.4 and were attributed to the secondary phase of calcium oxide in these ceramics. The lattice parameter a for the polycrystalline cubic phase was calculated from the $d_{(hkl)}$ and 2θ degrees values, and typical values were listed in **Table 1**. As could be seen from **Figure 2**, the samples with $x \leq 0.2$ showed no apparent changes in the relative intensities and broadening of peaks. This suggested x = 0.2 sets the limit of the salability range when Ca substitutes Sr in BST ceramics sintered at 1250˚C for 6 h for the studied values of x. As Ca content increases (x = 0.3 and 0.4) the peaks become broadened to some extent in addition to calcium oxide peaks. The previous observations indicate that increasing the content of Ca in the BCST materials decreases the particle size, lattice parameter a (Å), and volume of unit cell a^3 ($Å^3$) when Sr^{2+} is partly substituted with Ca^{2+}, which is consistent with the fact that the radius of Ca^{2+} ion is smaller than that of Sr^{2+} ions [19], see also **Table 1**.

Referring to EDX analysis results tabulated in **Table 2**, the x = 0.0 sample consists of Ti, Ba and Sr, *i.e.*, it is Ca-free sample. Whereas for x > 0.0 samples, EDX results confirm the presence of Ca in addition to the Ti, Ba, and Sr elements. Also, it is clear to note the decrease in intensity of strontium in the final products with increasing Ca content than that of the initial reactants. For the above tested specimens, the wt.% of each of these elements, as deduced from EDX analysis given in **Table 2**, indicates that the amount of Ca element in the sample increases depending on increasing Ca incorporation in the $B_{0.5}Ca_xSr_{0.5-x}TiO_3$ ceramics.

It is evident that the Sr/Ca ratio of BCST samples resulted from EDX analysis are different from that in the initial reactants compositions. Regardless of these differences the trends of Sr/Ca ratios in the two cases are the same, *i.e.*, the Sr/Ca ratio decreases as the x value increases, and the Sr/Ca ratios from EDX analyses are consistently smaller than that in the initial reaction system. Consequently, Ca incorporation has its effect on the mechanical and electrical properties of tested ceramics as will be discussed later on in the following sections.

Figure 3(a) reveals that the average particle size is about 6.77 µm for $Ba_{0.5}Sr_{0.5}TiO_3$ (BST) ceramic which is bigger than that of BCST ceramics in **Figures 3(b)-(e)** doped with Ca. This is in agreement with the work of [23] in which BST with x = 0.5 sample sintered at 1260˚C showed a large distribution of grain dimensions between 5 and 25 µm. Sub-micronic grains ≈ 0.90 µm, are located at the boundary of the largest ones as observed by [24]. Also, a less homogeneous microstructure, with an evident bi-modal grain size distribution was pointed out by the SEM images (**Figures 3(b)-(e)**). Besides, large particles, exaggerated smaller grains resulted by abnormal grain growth, were also noticed. The average values of particle size of studied ceramics, shown in **Figure 4**, were calculated from the micrographs of **Figure 3** according to the selected intersected lines method [25]. However, the average particle size values which found from ESM micrographs were decreased from 3.81 µm to 1.69 µm with increasing of Ca content in the sample from x = 0.0 to x = 0.4, which are larger than those values obtained for these grains as calculated by using Scherrer's equation (see **Figure 4**). The dependence of grain size on Ca content, **Figure 4**, seems to be decreased for both values determined from the X-ray and the SEM measurements. In other words, as the calcium content, x, increases in the ceramic from 0 to 0.4, the grain size is decreased in consistent with the A-site cation size effect reported in previous literature [18,19].

4.3. Mechanical Properties Investigations

4.3.1. Room Temperature Characterization

The observed decrease in the ultrasonic attenuation (**Figure 5**) may be attributed to the observed decrease in the crystallites or grain sizes associated with the replacement of Sr by Ca in BCST ceramics, as seen in **Figure 4**. This view agrees with our XRD and SEM investigations and in harmony with previous results [18,19,26].

4.3.2. Ultrasonic Velocity and Elastic Modulus

Variations of both the ultrasonic velocity and longitudinal elastic modulus, **Figure 6**, could be understood based on the effect of Ca content on microstructures of these ceramics. We have shown that the increase in Ca content in the ceramics decreases the grain or particles sizes due to its insertion in A-sites with its smaller radius as observed from X-ray results and SEM micrographs (see **Figure 4**). This decrease in grain size is known to raise the velocity of wave propagation [15,26]. Therefore, the observed increase in velocity is attributed to the decrease in grain sizes in the ceramics' microstructures with the increase in Ca content inside the samples. Also, it is known that the elastic modulus is proportional to the velocity squared according to the equation $L = V^2\rho$. Even though the density decreases with increasing calcium content, the increase in L with the increase in the wave velocity (V) overwhelms the density effect due to the velocity squared character.

4.3.3. Temperature Dependence of Ultrasonic Attenuation

Variation of ultrasonic attenuation (α) with temperature, shown in **Figures 7(a)** and **(b)**, could be interpreted in terms of the effect of Ca doping in tested BCST ceramics. It has been noticed that the observed damping peaks were found to be dependent on the composition. Explicitly, each peak position shifts to higher temperatures with a reduction in its maximum and the peak shape becomes broadened and diffused, especially in the ferroelectric side, as Ca replaces Sr in BCST ceramics. These observations are in good agreement with previously reported investigations of dielectric constant in $Ba_{1-2x}Sr_xCa_xTiO_3$ ceramics [18]. The sample with x = 0.0 (calcium-free sample, **Figure 7**) has the lowest Curie temperature (T_C) and occurred at 266˚C. As the calcium content increases at the expense of strontium content, T_C increases, namely, for BCST with x = 0.1, 0.2, and 0.3 the Curie temperature T_C = 313, 319, and 325˚C, respectively. Furthermore, for the highest calcium content with x = 0.4, T_C occurred at the highest Curie temperature of 331˚C. Therefore, it can be said that, on replacing Sr by Ca the peak temperature of the Curie transition shifts to higher temperatures which is in good agreement with previous results [18,19,26]. However, their occurrence in this range of higher temperatures, may be due to the high operated ultrasonic frequency (2 MHz) and perhaps also to the effect of electric field produced on the ceramic surfaces during these measurements (since tested samples were spring-loaded in the holder as well as these ferroelectric materials are pyroelectric and piezoelectric). This induced electric field can preserve the ordered states in ferroelectric domains for more time, in the ferroelectric phase, in the manner that it needs—in addition to the applied ultrasonic energy—more thermal heat which in turn raises the temperature of order-disorder temperature phase transition or the Curie temperature. However, this needs more investigations which will be considered in future work on these ferroelectric materials.

The above conclusion may be summed in another way as follows: when the Ca content increases at the expenses of the Sr in the investigated range, the average grains size decreases (as seen from X-ray measurements, and SEM images) and seen in **Figure 4**. Therefore, the ceramic becomes to some extent more rigid, in addition to the effect of the electric field induced due to both the piezoelectric (direct and indirect) and the pyroelectric effects, which allows the ceramic to require more thermal energy for its disorder state to be attained through the Curie transition, thus the increase in the transition temperature from 266˚C to 331˚C as obtained. Also, the observed shoulders on the plots of all ceramics in **Figures 7(a)** and **(b)** refer to the high temperature wings of the orthorhombic to tetragonal phase transition in each tested BCST ceramic which seems to occur around 0˚C temperature [26-28].

Regarding the relaxation peaks observed below the structural phase transition (T_C) in the ferroelectric regions, there are three (or more) relaxation peaks on each plot in the ultrasonic attenuation-temperature spectra BCST ceramics (see **Figures 7(a)** and **(b)**) whose heights and positions are composition dependent. All of these relaxation peaks can be described by a thermal activated Arrhenius relationship [29] and their origin could be understood according to their values of activation energy as has been done in the case of BST ceramics in published work [15,26]. Anyhow, the calculated values of activation energy (W) of these relaxation processes were listed in **Table 3**. Referring to the table, it can be observed that the calculated values for activation energies of relaxation peaks corresponding to each tested ceramic are dependent on their origin. Namely, it could be attributed to diffusion of point defects in the ferroelectric phase [26, 27-29] as occurred in the first relaxation (W is larger than 1.1 eV) and to diffusion of oxygen vacancies associated to domain walls and domain wall motions as occurred in the second and third relaxations when W is less than 1.1 eV, respectively [26-28]. Besides, it is also observed from **Table 3** that, for each relaxation, both of the activation energy value and position increased and decreased with the gradual increase in Ca content overall the tested range from x = 0.0 to x = 0.4.

Table 3. Temperatures and activation energies of relaxation peaks occurred in BCST ceramics, sintered at 1250˚C for 6 h, with different Ca contents, x.

Ca	1st Relax.		2nd Relax.		3rd Relax.	
content,	T	W	T	W	T	W
x	(˚C)	(eV)	(˚C)	(eV)	(˚C)	(eV)
0.0	190	1.209	226	1.051	241	1.002
0.1	216	1.088	268	0.929	286	0.890
0.2	205	1.135	250	0.976	286	0.890
0.3	235	1.020	253	0.967	289	0.884
0.4	196	1.175	233	1.027	277	0.909

5. Conclusion

In this work, $Ba_{0.5}Ca_xSr_{1-x}TiO_3$ ceramics with x = 0.0, 0.1, 0.2, 0.3 and 0.4 were prepared by solid state reaction technique and studied by XRD, SEM, EDX, and ultrasonic techniques. XRD data confirmed the formation of the perovskite phase structure in addition to peaks observed for some tested ceramics which have been related to excess Ca contents in doped samples. SEM images and EDX analysis confirmed the appearance of the major perovskite phase and an increase of Ca ions upon increasing its content. The effect of increasing of Ca content in BCST samples resulted in a decrease of particle size, as confirmed by XRD and SEM images. Also, BCST ceramics showed composition dependence for their density, longitudinal velocity, elastic modulus, and attenuation upon increasing Ca content (from x = 0 to 0.4) and resulted in a decrease of bulk density, and ultrasonic attenuation, and an increase in velocity, and longitudinal modulus. The temperature dependence of ultrasonic attenuation investigations for BCST samples have revealed, the occurrence of not only, the Curie (or tetragonal to cubic) transition and the right shoulder of (orthorhombic to tetragonal) transition, but also some relaxation peaks associated with point defects and diffusion of oxygen vacancies. However, the curie transition temperature T_C showed to be dependent on the content of Ca; its increasing resulted in an increase in T_C of the tested BCST samples.

REFERENCES

[1] K. Abe and S. Komatsu, "Ferroelectric Properties in Epitaxially Grown $Ba_xSr_{1-x}TiO_3$ Thin Films," *Journal of Applied Physics*, Vol. 77, No. 12, 1995, pp. 6461-6465.

[2] Z. L. Wang and Z. C. Kang, "Functional and Smart Materials—Structural Evolution and Structural Analysis," Science Press, Beijing, 2002.

[3] L. C. Sengupta and S. Sengupta, "Novel Ferroelectric Materials for Phased Array Antennas," *IEEE Transactions on Ultrasonics, Ferroelectrics, and Frequency Control*, Vol. 44, No. 4, 1997, pp. 792-797.

[4] S. S. Gevorgian and E. L. Kollberg, "Do We Really Need Ferroelectrics in Paraelectric Phase Only in Electrically Controlled Microwave Devices?" *IEEE Transactions on Microwave Theory and Techniques*, Vol. 49, No. 11, 2001, pp. 2117-2124.

[5] P. C. Joshi and M. W. Cole, "Mg-doped $Ba_{0.6}Sr_{0.4}TiO_3$ Thin Films for Tunable Microwave Applications," *Ap-*

plied Physics Letters, Vol. 77, No. 2, 2000, pp. 289-291.

[6] W. Chang and L. Sengupta, "MgO-mixed Ba0.6Sr0.4TiO3 Bulk Ceramics and Thin Films for Tunable Microwave Applications," *Journal of Applied Physics*, Vol. 92, No. 7, 2002, pp. 3941-3946.

[7] M. Kuwabara, H. Matsuda and Y. Ohba, "Varistor Characteristics in PTCR-Type (Ba,Sr)TiO_3 Ceramics Prepared by Single-Step Firing in Air," *Journal of Materials Science*, Vol. 34, No. 11, 1999, pp. 2635-2639.

[8] J. F. Scott, M. Azuma, E. Fujii, T. Otsuki, G. Kano, M. C. Scott, C. A. Paz de Araujo, L. D. McMillan and T. Roberts, "Microstructure-Induced Schottky Barrier Effects in Barium Strontium Titanate (BST) Thin Films for 16 and 64 Mbit (DRAM cells)," *Proceedings of International Symposium on Integrated Ferroelectrics*, New York, 1992, p. 356.

[9] X. Weidong, L. Yanrong and Y. Chun, "First Principle Studies on Fine Structure for $Ba_xSr_{1-x}TiO_3$," *Chinese Journal of Chemical Physics*, Vol. 18, No. 2, 2005, pp. 179-182.

[10] T. Hu, H. Jantunen, A. Uusimaki and S. J. Leppavuori, "BST Powder with Sol-Gel Process in Tape Casting and Firing," *Journal of the European Ceramic Society*, Vol. 24, No. 6, 2004, pp. 1111-1116.

[11] V. V. Lemanov, "Concentration Dependence of Phonon Mode Frequencies and the GrüNeisen Coefficients in BaxSr$_{1-x}$TiO3 Solid Solutions," *Physics of the Solid State*, Vol. 39, No. 2, 1997, pp. 318-322.

[12] B. Jaffe, W. R. Cook and H. Jaffe, "Piezoelectric Ceramics," Academic Press, London, 1971.

[13] B. L. Cheng, M. Gabbay, M. Maglione and G. Fantozzi, "Relaxation Motion and Possible Memory of Domain Structures in Barium Titanate Ceramics Studied by Mechanical and Dielectric Losses," *Journal of Electroceramics*, Vol. 10, No. 1, 2003, pp. 5-18.

[14] A. Ioachim, M. I. Toacsan, M. G. Banciu, L. Nedelcu, C. Plapcianu, H. V. Alexandru, C. Berbecaru, D. Ghetu, G. Stoica and R. Ramer, "Frequency Agile BST Materials for Microwave Applications," *Journal of Optoelectronics and Advanced Materials*, Vol. 5, No. 5, 2003, pp. 1389-1393.

[15] M. H. Badr, L. M. Sharaf El-Deen, A. H. Khafagy and D. U. Nassar, "Structural and Mechanical Properties Characterization of Barium Strontium Titanate (BST) Ceramics," *Journal of Electroceramics*, Vol. 27, No. 3-4, 2011, pp. 189-196.

[16] O. P. Thakur, C. Prakash and D. K. Agrawal, "Dielectric Behavior of $Ba_{0.95}Sr_{0.05}TiO_3$ Ceramics Sintered by Microwave," *Materials Science and Engineering: B*, Vol. 96, No. 3, 2002, pp. 221-225.

[17] J. F. Scott, "High-Dielectric Constant Thin Films for Dynamic Random Access Memories (DRAM)," *Annual Review of Materials Research*, Vol. 28, No. 1, 1998, pp. 79-100.

[18] C. Berbecaru, H. V. Alexandru, C. Porosnicu, A. Velea, , A. Ioachim, L.Nedelcu and M. Toacsan, "Ceramic Materials $Ba_{(1-x)}Sr_xTiO_3$ for Electronics—Synthesis and Characterization," *Thin Solid Films*, Vol. 516, No. 22, 2008, pp. 8210-8214.

[19] S. Yun, X. Wang, B. Li and D. Xu, "Dielectric Properties Ca-Substituted Barium Strontium Titanate Ferroelectric Ceramics," *Solid State Communications*, Vol. 143, No. 10, 2007, pp. 461-465.

[20] V. V. Lemanov, A. V. Sotnikov, E. P. Smirnova, P. P. Syrnikov and E. A. Tarakanov, "Phase Transitions and Glasslike Behavior in $Sr_{(1-x)}Ba_xTiO_3$," *Physical Review B*, Vol. 54, No. 5, 1996, pp. 3151-3157.

[21] A. K. Singh, Subrat K. Barik, R. N. P. Choudhary and P. K. Mahapatra, "Ac Conductivity and Relaxation Mechanism in $Ba_{0.9}Sr_{0.1}TiO_3$," *Journal of Alloys and Compounds*, Vol. 479, No. 1-2, 2009, pp. 39-42.

[22] Y.-C. Liou and C.-T. Wu, "Synthesis and Diffused Phase Transition of $Ba_{0.7}Sr_{0.3}TiO_3$ Ceramics by a Reaction-Sintering Process," *Ceramics International*, Vol. 34, No. 3, 2008, pp. 517-522.

[23] C. Fu, C. Yang, H. Chen, W. Wang, and L. Hu, "Microstructure and Dielectric Properties of $Ba_xSr_{1-x}TiO_3$ Ceramics," *Materials Science and Engineering: B*, Vol. 119, No. 2, 2005, pp. 185-188.

[24] A. Ioachim, R. Ramer, M. I. Toacsan, M. G. Banciu, L. Nedelcu, C. A. Dutu, F. Vasiliu, H. V. Alexandru, C. Berbecaru, G. Stoica and P. Nita, "Effect of the Sintering Temperature on the Ba(Zn 1/3Ta 2/3)O3 Dielectric Properties," *Journal of the European Ceramic Society*, Vol. 27, No. 2-3, 2007, pp. 1117-1122.

[25] A. Ioachim, H. V. Alexandru, C. Berbecaru, S. Antohe, F. Stanculescu, M. G. Banciu, M. I. Toacsan, L. Nedelcu, D. Ghetu, A. Dutu and G. Stoica, "Dopant Influence on BST Ferroelectric Solid Solutions Family," *Materials Science and Engineering: C*, Vol. 26, No. 5-7, 2006, pp. 1156-1161.

[26] G. Cigna, "Dynamic Mechanical Properties, Struc- ture, and Composition of Impact Polystyrene," *Journal of Applied Polymer Science*, Vol. 14, No. 7, 1970, pp. 1781-1793.

[27] H. Frayssignes, B. L. Cheng, G. Fantozzi and T. W. Button, "Phase Transformation in BST Ceramics Investigated by Internal Friction Measurements," *Journal of the European Ceramic Society*, Vol. 25, No. 13, 2005, pp. 3203-3206.

[28] B. L. Cheng, B. Su, J. E. Holmes, T. W. Button, M. Gabbay and G. Fantozzi, "Dielectric and Mechanical Losses in (Ba,Sr)TiO_3 Systems," *Journal of Electroceramics*, Vol. 9, No. 1, 2002, pp. 17-23.

[29] A. S. Nowick and B. S. Berry, "Anelastic Relaxation in Crystalline Solids," Academic Press, New York, 1972.

7

ZnO Heteroepitaxy on Sapphire Using a Novel Buffer Layer of Titanium Oxide: Crystallographic Behavior

Satoshi Yamauchi[1*], Yoh Imai[2]

[1]Department of Biomolecular Functional Engineering, Ibaraki University, Hitachi, Japan

[2]Department of Electric and Electronic Engineering, Ibaraki University, Hitachi, Japan

ABSTRACT

A novel buffer layer consists of titanium oxide grown on a-sapphire by low-pressure chemical vapor deposition using titanum-tetra-iso-propoxide and oxygen gas was used for ZnO epitaxial growth at temperature as low as 340°C by plasma-assisted epitaxy using radio-frequency oxygen-gas plasma. XRD and RHEED indicated (0001)Ti_2O_3 layer in corundum crystal system was epitaxially grown on the substrate in an in-plane relationship of [1-100]Ti_2O_3// [0001]Al_2O_3 by uniaxial phase-lock system. Growth behavior of ZnO layer was significantly dependent on the Ti_2O_3 buffer-layer thickness, for example, dense columnar ZnO-grains were grown on the buffer layer thinner than 10 nm but the hexagonal pyramid-like grains were formed on the thin buffer layers below 2 nm. RHEED observations showed ZnO layer including the pyramid-like grains was epitaxially grown with single-domain on the thin buffer layer of 0.8 nm in the in-plane relationship of [1-100]ZnO//[1-100]Ti_2O_3//[0001]Al_2O_3, whereas the multi-domain was included in ZnO layer on the buffer layer above 10 nm.

Keywords: ZnO; Ti_2O_3; Plasma-Assisted Epitaxy; Hexagonal Pyramid Grain

1. Introduction

ZnO has been candidate of interesting material for highly efficient blue or ultraviolet light-emitting devices because of the wide direct band gap of 3.37 eV at room temperature and the large exciton binding energy of 60 meV. However, it has been recognized that undoped-ZnO shows highly n-type conductivity by native-defects such as interstitial-Zn and strong green emission due to oxygen deficiency, which are introduced by the high temperature growth and/or poor reactivity of supplied oxygen for the growth. Several advanced processes such as Moleculer-Beam Epitaxy (MBE) using oxygen-plasma cell [1], Metal-Organic MBE (MOMBE) using H_2O vapor [2], Pulsed Laser Deposition (PLD) [3] etc. have been developed to reduce the native defects. Plasma-assisted epitaxy (PAE) using oxygen plasma excited by radio-frequency (rf) power is also useful to achieve ZnO-epitaxial growth at low temperatures supplying reactive oxygen as demonstrated for undoped-ZnO growth at 400°C on c-sapphire [4,5] and Si [6]. In addition, acceptor-doping with sufficient dopant preventing the self-compensation should be required to fabricate p-type ZnO and the pn-junction. Nitrogen has been used for acceptor-doping into ZnO layer [2,7] including the co-doping [8], however, the heavily doped nitrogen-acceptor showed the large activation energy such as 224 meV [9]. In contrast, relatively shallow nitrogen acceptor with the activation energy of 132 meV was doped into ZnO layer by PAE using $O_2 + N_2$ gas plasma with precise control of O/Zn supply ratio during the growth [10]. Such features have been observed in the growth of other II-VI compound semiconductor such as ZnSe [11] and are origin-nated from self-compensation in the heavy acceptor-doping by native donor-defects. For ZnO growth, the growth mode should be taken into account to control the doping since the layer has been grown on sapphire substrate with large lattice mismatching. Indeed, it was reported that the electron concentration is high in ZnO layers including three-dimensional grains and/or multi-domains [12], where the donor defects were probably come from interstitial-Zn formed on the three-dimensional surface with higher sticking coefficient of Zn on the steps and kinks. In general, suitable buffer layer is required for two-dimensional epitaxial growth in lattice mismatching

*Corresponding author.

system such as ZnO/sapphire to satisfy the relationship of $\sigma_{sv} > \sigma_{sc} + \sigma_{cv}$ by reducing the interface energy, where σ_{sv}, σ_{sc} and σ_{cv} are surface energy on the substrate, interface energy and surface energy of film respectively. For the purpose, Y. Chen *et al.* reported interest results to relax the lattice-mismatching by using an initial buffer layer of thin MgO layer with rocksalt crystal structure on c-sapphire, where the mismatch could be reduced to 9% comparing the large value of 18% on c-sapphire [13]. In the hetero-epitaxial system, initial ZnO layer grown at 500°C on the buffer layer was annealed at 750°C to reduce the surface roughness, and then single domain ZnO layer was epitaxially grown with smooth surface at 620°C. However, it was also reported that the low temperature photoluminescence spectrum was improved but dominated by neutral donor-bound exciton emission at 3.369 eV, which indicates native donor was still included in the ZnO layer. In contrast, while Ti_2O_3 with corundum crystal structure is also an interesting material as the initial buffer layer, which is expected to reduce the lattice mismatch in the system of (0001)ZnO/(0001)Ti_2O_3 as low as 5% with large surface energy on the (0001)Ti_2O_3, the heteroepitaxial system has not been studied because Ti_2O_3 single has not been established.

In this paper, we demonstrate epitaxial growth of Ti_2O_3 layer on a-sapphire by low-pressure chemical vapor deposition and then enhanced two-dimensional epitaxial growth of ZnO layer at the temperature as low as 340°C by PAE on the Ti_2O_3 buffer layer.

2. Experimental

Titanium oxide layer was grown in bell-jar type chamber at low temperature of 320°C in 3 mtorr on a-sapphire by LPCVD using titanium tetra-iso-propoxide (TTIP: $Ti(OC_3H_7)_4$) and oxygen gas. The liquid TTIP (97% purity) was vaporized in a quartz-cell at 70°C after purification in vacuum at 50°C for 3 hr, then the TTIP-gas was introduced into the growth chamber through a hot alumina nozzle at 85°C monitoring and controlling the pressure in the chamber using Schultz gage and a variable valve, where high-purity oxygen gas (6N-purity) was simultaneously supplied with the TTIP-gas but through the individual gas lines for precise control O_2/TTIP supply ratio. It is mentioned here that detection sensitivity in such ion-gage should be corrected for each gas species to determine the supply rate but the directly monitored pressure was used without the correction in this study because it is not easy to evaluate the sensitivity of the metal-organic gas catalyzed by the oxidant.

ZnO layer was grown at 340°C in 3 mtorr by plasma-assisted epitaxy (PAE) using oxygen gas plasma which was excited by 10 W rf-power at 13.56 MHz using a capacitively coupled rf-electrode. Details of the PAE apparatus were described elsewhere [9]. High-purity elemental Zn (6N-up purity) and oxygen gas (6N purity) were used for ZnO growth. Elemental Zn (6N-up purity) was evaporated from a Knudsen cell with a nozzle to prevent oxidation of the surface and supplied to the growth surface through the oxygen gas plasma, where high-purity oxygen gas with 6N-purity was used. Effective Zn/O supply ratio was determined by the dependence of ZnO growth rate on Zn-supply rate [9].

300 m-thick single crystalline a- and c-sapphire with mirror surface used as substrates were cleaned in organic-solvents and hot $H_2SO_4 + H_2O_2$ at 60°C for 5 min, then, loaded into the chamber after rinse in ultra pure-water with the resistivity above 18.2 M cm and blow by nitrogen-gas. In the case of substrate etching, a hot $3H_2SO_4 + H_3PO_4$ solution at 130°C was used as the etchant and the substrate treated by $H_2SO_4 + H_2O_2$ was etched for 15 min.

Thickness of the titanium-oxide layer and the ZnO layer were checked by a contact-type surface profiler, where the thickness of the thin titanium-oxide layer below 10 nm was estimated by the growth rate evaluated for the layer above 20 nm-thickness and the period rate. Crystal-system and crystal-orientation was examined by a-2XRD measurement using $Cuk_{1,2}$. Surface morphology of the ZnO films was observed by Nomarski differential interference microscope. Reflection high-energy electron diffraction (RHEED) system was also used on the surfaces of the layers and the substrate to disclose the surface structures and the in-plane epitaxial relationship.

3. Results and Discussions

3.1. Ti_2O_3 Epitaxial Growth on A-Sapphire

It is recognized that titanium-oxide growth using TTIP and O_2 is owing to thermal catalysis of TTIP enhanced by the oxidation, therefore, the growth behavior includeing the crystallinity is significantly dependent on not only the growth temperature but also the supply ratio of O_2/TTIP [14]. As a typical result, FTIR results indicated that some types of impurities such as hydrocarbons and hydroxyls derived from the TTIP dissociation sequences were slightly remained in the layer by the low supply ratio below 1 but not detected in the layer grown by O_2-rich conditions above 1.5, therefore, the layer was grown in the large supply ratio of 8 to remove such impurities completely. On the other, it has also been recognized that titanium oxide compound can be crystallized in diverse compositions and crystal systems due to the hybrid-orbital of titanium including the d-orbital, where the valence of Ti is changed from 0.33 to 5.0. Tokita *et al.* reported (112) anatase-TiO_2 with tetragonal crystal system can be epitaxially grown on c-sapphire at relatively low temperature about 500°C by atmospheric chemical vapor deposition using TTIP and O_2 [15]. In contrast, we found out (0001) Ti_2O_3 with corundum in hexagonal system was epitaxially grown on a-sapphire by LPCVD as shown below. **Figure 1** shows XRD profile of 400 nm-thick titanium-oxide layer grown on a-sapphire by LPCVD, where the diffraction corresponding to (0006)Ti_2O_3 was only appeared at 39.5° [16] from the layer with some peaks from the substrate, which indicates preferentially (0001) oriented Ti_2O_3 layer was grown on the substrate. **Figures 2(a)** and **(b)** show RHEED patterns of the layer with the thickness of 10 nm, where the electron beam was incident along [0001]Al_2O_3 and [1-100]Al_2O_3, respectively. It is noted that the patterns corresponding to (a) [1-100] and (b) [-1-120]Ti_2O_3 showed sixfold-symmetry because each pattern was repeated every 60° and alternated every 30° when the film was rotated with respect to the electron beam incident. The RHEED patterns showed the layer was three-dimensional growth but clearly indicated (0001)Ti_2O_3 layer was epitaxially grown on a-sapphire with an inplane relationship of [1-100]Ti_2O_3//[0001]Al_2O_3. Representation of oxygen in the two surfaces can be shown as **Figure 3**, where the oxygen on Al_2O_3 and Ti_2O_3 surface is shown by open and closed circles, respectively. The symmetry of the oxygen arrangement between the two surfaces was different in each other, however, the distance of oxygen on Ti_2O_3 is periodically matched along [1-100]Al_2O_3, in which the threefold oxygen on Ti_2O_3 surface along [1100]Ti_2O_3 is just matched to the fivefold oxygen in Al_2O_3 along [1-100]Al_2O_3 with the mismatch less

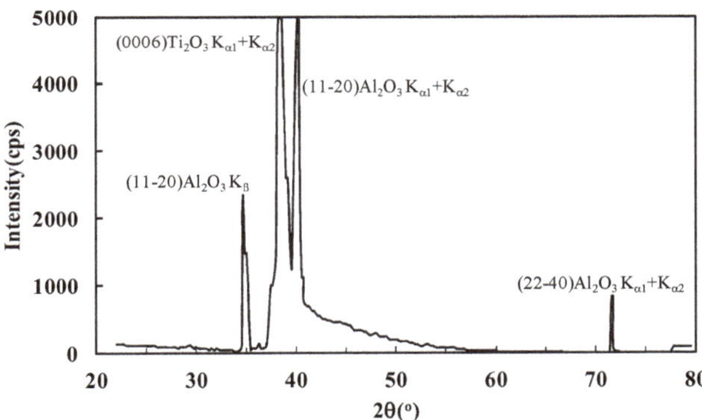

Figure 1. θ **-** 2θ **XRD pattern of 400 nm-thick Ti_2O_3 layer grown on a-sapphire by LPCVD.**

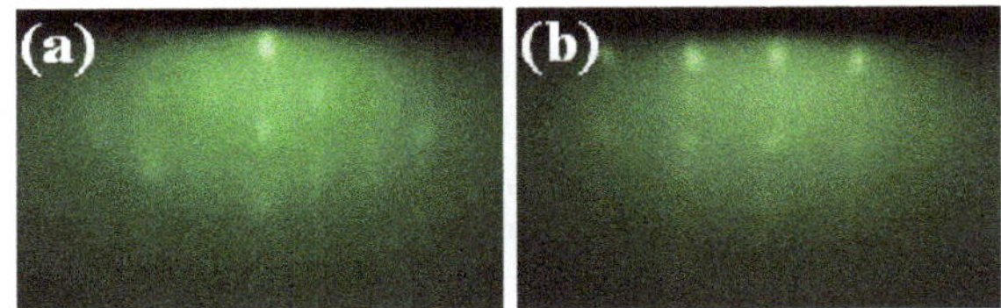

Figure 2. RHEED patterns on Ti_2O_3 layer grown on a-sapphire, where the electron beam was incident along (a) [0001] and (b) [1-100]Al_2O_3.

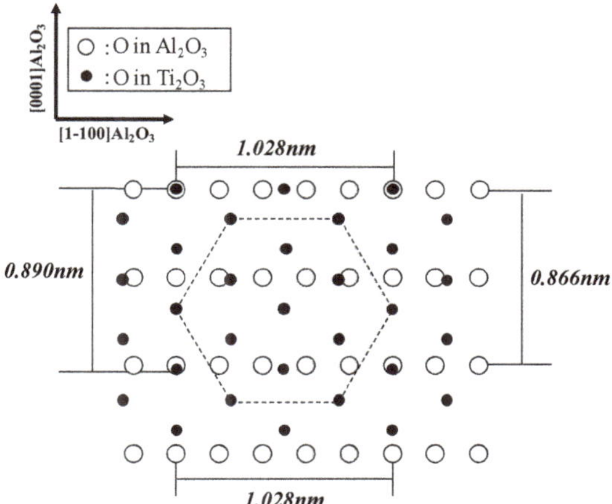

Figure 3. Representation of oxygen on a-sapphire (open-circles) and (00 01)Ti_2O_3 (closed-circles) in the in-plane relationship of [1-100]Ti_2O_3//[0001]Al_2O_3.

than 0.06% in addition to the relatively low mismatching of 2.8% along [0001]Al_2O_3. Previously, it was demonstrated that (0001)ZnO with sixfold-symmetry is locked by uniaxially matched oxygen along [0001]Al_2O_3 with twofold symmetric oxygen on a-sapphire surface [17]. Also in the Ti_2O_3/a-sapphire system, it can be considered that (0001)Ti_2O_3 was locked by the asymmetry on a-sapphire surface along [1-100]Al_2O_3. The spotty RHEED patterns as shown in **Figure 2** were significantly dependent on the layer thickness as shown in **Figure 4**, which showed RHEED patterns of (a) 2 nm- and (b) 0.8 nm-thick Ti_2O_3 layers, where the electron beam was incident along [1-100]Al_2O_3 as same as **Figure 2(b)**. The spotty pattern from the relatively thick layer above 10 nm (**Figure 2(b)**) was changed to streak with decreasing the thickness below 2 nm and showed the fine Kikuchi-line on the 0.8 nm-thick layer.

The sixfold symmetry observed for the 10 nm-thick layer was same as the very thin layers, however, the streak-line distance for the 0.8 nm-thick layer was slightly large (about 2%) comparing to that for the thick layer, which suggested the layer was stressed by the lattice mismatch along [1-100]Al_2O_3. These results indicated the layer was two-dimensionally grown with slightly shrunk hexagonal unit at initial stage but three-dimensionally above the critical thickness probably within the lattice constant along c-axis (1.364 nm). The growth features can be considered that the layer was grown by Stranski-Krastanov mode [18], in which the energy relationship between the substrate and the layer is satisfied to be $\sigma_{sv} > \sigma_{sc} + \sigma_{cv}$ for the thickness below the critical thickness but changed to be $\sigma_{sv} < \sigma_{sc} + \sigma_{cv}$ for the layer thicker than the critical thickness, or secondary nucleation caused by dislocation in the layer grown on the lattice mismatched substrate. Of course, the growth behaveior of Ti_2O_3 layer including the critical thickness was

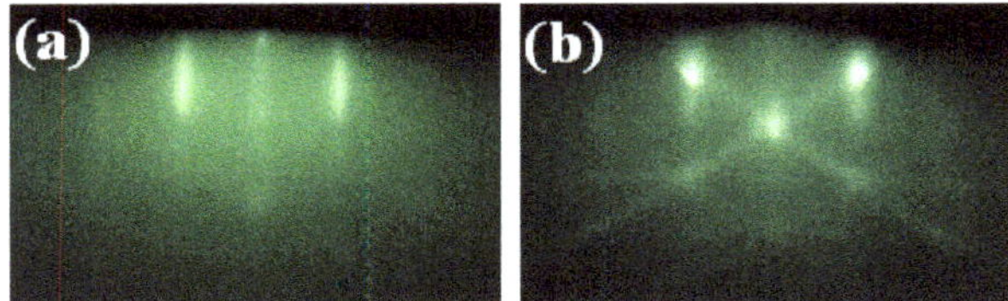

Figure 4. RHEED patterns of Ti_2O_3 layer with the thickness of (a) 2 nm and (b) 0.8 nm on a-sapphire, where the electron beam was incident along [1-100]Al_2O_3.

significantly influenced by the growth conditions such as the introduced gas rate, growth temperature and so on, and the thin layer with smooth surface as shown in **Figure 4(b)** could be grown in oxygen rich condition (O_2/TTIP supply ratio above 8) with the low growth rate about 0.3 nm/min at 320°C.

3.2. ZnO Growth on Ti_2O_3 Buffer Layer

3.2.1. Surface Morphology

Figure 5 shows Nomarski-photographs of PAE-ZnO layers with the thickness about 600 nm grown at 340°C. Preferentially (0001) oriented ZnO layer consists of densely three-dimensional columnar grains was grown on c-sapphire (**Figure 5(a)**), however, the 30°-rotational domains with in-plane relationships of [11-20]ZnO//[11-20]Al_2O_3 and [10-10]ZnO//[11-20]Al_2O_3 due to large lattice mismatch between (0001)ZnO and c-sapphire were included in the layer as reported elsewhere [19]. In contrast, polycrystalline ZnO layer with rough surface including (10-11) and (0001) oriented grains was grown on a-sapphire at the temperature as low as 340°C by PAE, while single-crystalline (0001)ZnO layer was grown on a-sapphire at 500°C by radical source MBE using two step process demonstrated by the initial step for nucleation at 400°C and the in-situ annealing for the coalescence at 500°C [17]. It is considered that the growth behavior on a-sapphire by PAE was come from three-dimensional nucleation with the low density at the initial stage, which was resulted in textured three-dimensional grain growth and secondary nucleation in deep grain boundaries on the ZnO surface during the growth. In contrast, the feature on the Ti_2O_3 layer was quite differed as shown in **Figures 5(c)-(e)**, where ZnO layer with the average thickness about 600 nm was grown at 340°C. ZnO layer consists of highly dense columnar grains coming from high density nucleation on the Ti_2O_3 layer comparing to that on sapphires was grown on the buffer layer above 10 nm (**Figure 5(c)**), which speculated that the surface reactivity of the layer was higher than sapphire. Further, pyramid-shape grains with the hexagonal facets could be clearly observed in the ZnO layers on the thinner Ti_2O_3 layer than 2 nm and the density was decreased with decreasing the buffer layer thickness (**Figures 5(d)** and **(e)**). It is interesting that

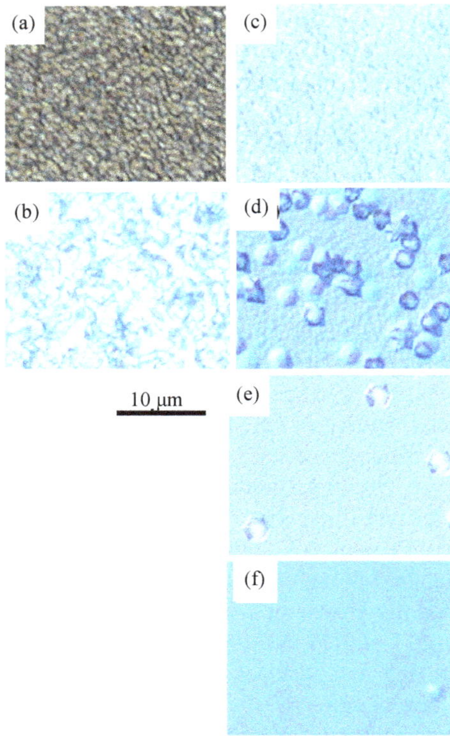

Figure 5. Surface morphologies observed by Nomarski microscope of ZnO layers on (a) c-sapphire, (b) a-sapphire, (c) 10 nm-thick buffer layer, (d) 2 nm-thick buffer layer, (e) 0.8 nm-thick buffer layer on a-sapphire and (f) 0.8 nm-thick buffer layer on a-sapphire etched by $3H_2SO_4 + H_3PO_4$ solution.

the small columnar-like grain was observed at center of the pyramid in the layer on 2 nm-thick buffer layer (**Figure 5(d)**), which was expected the three-dimensional nuclei was firstly created at the initial stage and then the pyramid grain grew around the columnar-like grain, that is, the columnar-like grain was grown with significantly higher rate comparing to the smooth surface. The pyramid was also formed in ZnO layer on the 0.8 nm-thick Ti_2O_3 layer but the columnar-like grain in the center could not be observed, which indicates growth rate of the columnar-like grain was lower than that on the 2 nm-thick Ti_2O_3 layer. **Figure 6** shows growth rates of ZnO layer on c-sapphire and the 0.8 nm-thick buffer layer as a function of Zn evaporated rate from the cell during the growth in constant O_2-supply rate and induced rf-power, where the growth rate was evaluated by the thickness of ZnO layer grown for 1 hour and the thickness on the buffer layer was checked on the smooth surface without the pyramids. In both cases, the growth rate was limited by the smaller supply rate of Zn or O, which indicated the sticking coefficients of Zn on Zn and O on O were negligible in this condition, but obviously larger on c-sapphire than on the buffer layer. On the other, both of the saturated growth rate in the height and the Zn-flux for

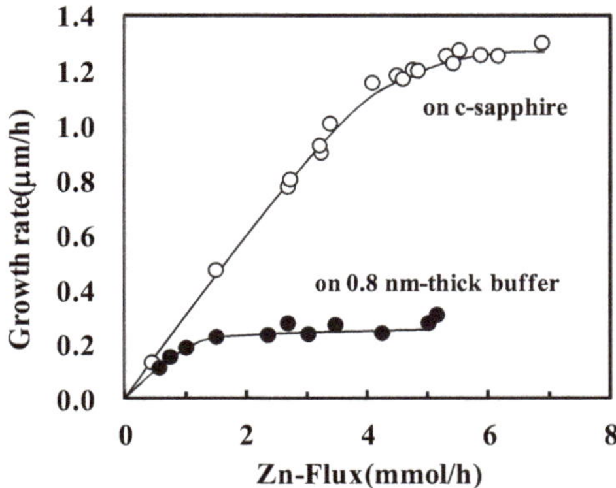

Figure 6. Growth rate of ZnO layer on c-sapphire and 0.8 nm-thick buffer layer.

the effective supply rate of Zn/O = 1 were increased for the pyramids. In addition, the values were increased with the buffer thickness, for example, the pyramids were grown with the rate about 0.7 m/h by the Zn-flux above 2.5 mm/h on the 0.8 nm-thick layer but about 1 m/h by the flux above 3.5 mmol/h on the 2 nm-thick layer. It is noted here that the ZnO layer consists of columnar grains on the thick buffer layer above 10 nm was grown with the high rate comparable to that on c-sapphire. Previously, it was reported that growth rate of ZnO layer on Zn-polar surface is higher than that on O-polar surface [20] because of the higher sticking coefficient of O-adatom on the Zn-face than that of Zn-adatom on the O-face. If the high growth rate of the partially formed pyramid grain comparing to the other smooth area was originated from the growth faces, it is expected the grain could not be formed with the fine hexagonal facets and the in-plane size should be determined at initial growth stage of ZnO, however, the grains showed the fine facets and the in-plane size were increased with the growth period. Therefore, it should be concluded that the high rate growth of the pyramid grains was not caused by difference of the polar on the growth surface but by the higher sticking coefficient of adatoms at the steps or the kinks than that on the terraces. The growth feature of the pyramid grain was also dependent on the surface condition of sapphire substrate, for example, the pyramid grain was almost disappeared by chemical etching of a-sapphire in a hot $3H_2SO_4$ + H_3PO_4 solution at 130°C for 15 min. as shown in **Figure 5(f)**, where the growth conditions and the thickness for ZnO and Ti_2O_3 layers were same as the sample in **Figure 5(e)**. The result clearly indicated that some types of defect including surface roughness on the buffer layer could be removed by the etching.

3.2.2. Epitaxial Relationship

Figure 7 show RHEED patterns corresponding to (a) [1-210] and (b) [-1010]ZnO on the ZnO layer shown in **Figure 5(e)** when the electron beam was incident along

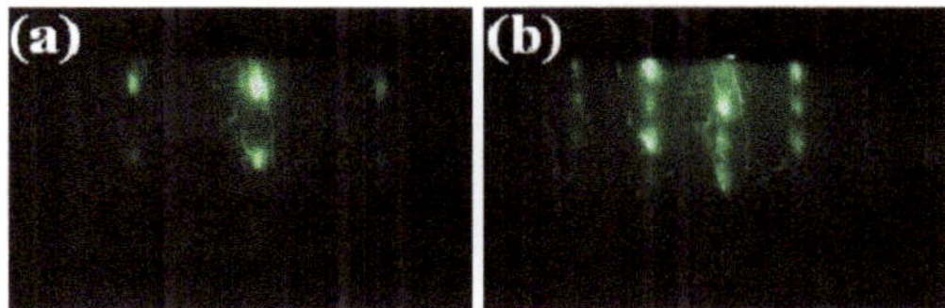

Figure 7. RHEED patterns of ZnO layer on 0.8 nm-thick buffer layer when the electron beam was incident along (a) [0001] and (b) [1-100]Al_2O_3.

[0001] and [1-100]Al_2O_3, respectively. The each pattern was repeated every 60° and alternated every 30° as the film was rotated with respect to the electron beam incident. The patterns clearly indicated that the ZnO layer was epitaxially grown without rotational domains on the buffer layer, whereas the layer was including hexagonal pyramids. [1-210] of the hexagonal pyramids as shown in **Figure 5(e)** was also coincident to [0001]Al_2O_3. These results investigate the three-dimensional nuclei were partially formed with the same in-plane epitaxial relationship of the two-dimensional growth surface. As a result, the in-plane pyramid size could be increased with the growth period. The growth behavior is seemed that the ZnO layer was grown by two growth modes of Frank-van der Merwe [17] (two-dimensional growth) and Volmer-Weber [18] (three-dimensional pyramid grain growth). It is considered that the three-dimensional nuclei might be originated from the surface defects of buffer layer, which was come from sapphire-substrate, and prevented by the surface treatment of the substrate such as the high-temperature annealing.

Figures 8(a) and **(b)** show schematic representations of oxygen on (0001)Ti_2O_3 and (0001)ZnO surfaces, where the oxygen-atoms on Ti_2O_3 and ZnO were shown by closed- and open-circle with the unit cell by dot-line and solid-line hexagon based on the a-axis lattice constant of 0.324 nm and 0.514 nm for ZnO and Ti_2O_3, respectively. In the case of [1-210]ZnO//[1-100]Ti_2O_3 (**Figure 8(a)**), hexagon unit of the ZnO is larger than subunit hexagon of the Ti_2O_3 (about +9.2%), which is much smaller than that between (0001)ZnO/(0001)Al_2O_3 in the relationship of [1-210]ZnO//[1-100]Al_2O_3 (more than +18%). In contrast, the mismatch between the ZnO and the Ti_2O_3 with in-plane relationship of [1-100]ZnO// [1-100]Ti_2O_3 as shown in **Figure 8(b)** is reduced to −5.4% which is better than that in [1-210]ZnO//[1-100]Ti_2O_3, therefore, the in-plane epitaxial relationship resulted in [1-100]ZnO//[1-100]Ti_2O_3. It is noted here that the mismatches between the ZnO and the 0.8 nm-thick Ti_2O_3 layer were expected to be smaller (about 2%) than the values because the hexagon unit at the 0.8 nm-thick Ti_2O_3 surface was slightly smaller than that of the bulk but the mismatch is estimated to be above 3%, however, the mismatch was still larger than that of (0001)ZnO/(0001)Al_2O_3 in the relationship of [1-100]ZnO

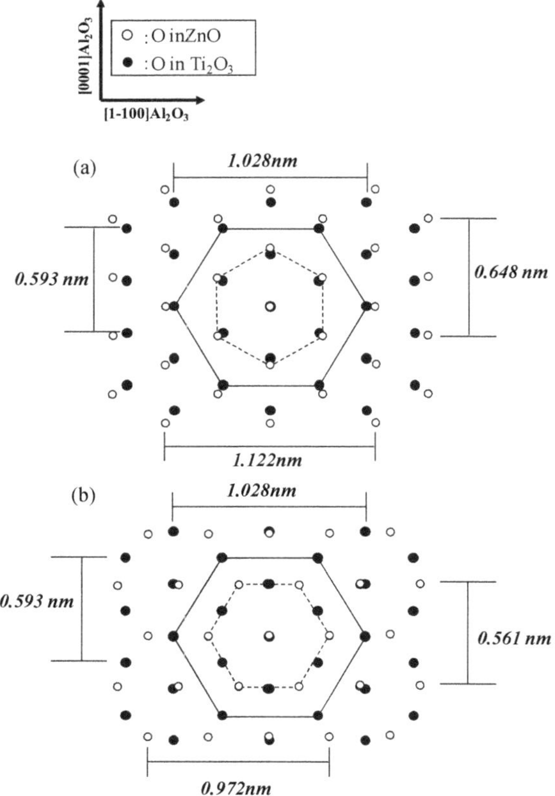

Figure 8. Representation of oxygen on (0001)Ti_2O_3 (closed-circles) and (0001) ZnO (open-circles) in the in-plane relationship of (a) [1-210]ZnO//[1-100]Ti_2O_3 and (b) [1-100]ZnO //[1-100]Ti_2O_3.

//[1-100]Al_2O_3 (about −2.3%). It has been commonly recognized that the phase-locking without rotational domain is difficult on c-sapphire with the isotropic hexagonal surface, but the results in this study indicated the phase could be successfully locked on the isotropic hexagonal surface of Ti_2O_3 layer. The useful results for ZnO growth on the buffer layer can be concluded that the phase-locking is performed not only by the lattice-matching but also by control of the surface chemistry on the mismatching substrate such as Ti_2O_3.

4. Conclusion

PAE-ZnO growth at low temperature of 340˚C was demonstrated on a-sapphire using a novel buffer layer consists of Ti_2O_3. (0001)Ti_2O_3 layer was epitaxially grown by LPCVD using TTIP and O_2 on a-sapphire in the in-plane relationship of [1-100]Ti_2O_3//[0001]Al_2O_3 at 320˚C. Surface roughness of the Ti_2O_3 layer was increased with the thickness, however, extremely smooth surface could be obtained on thin buffer layer below 1 nm-thickness. The ZnO layer could be epitaxally grown with single domain in the in-plane relationship of [1-100]ZnO//[1-100]Ti_2O_3 on the thin buffer layer. Hexagonal pyramid-shape ZnO grains by three-dimensional epitaxial growth were formed on the thin buffer layer in the in-plane epitaxial relationship with the significant high growth rate comparing to the two-dimensional growth. It was suggested from growth feature of the pyramid grains that epitaxial relationship in the three-dimensional growth was same as the two-dimensional growth and the high growth rate of the pyramid was caused by high sticking coefficient of adatoms on the three-dimensional surface. The number of the pyramid-grains could be obviously reduced by chemical wet etching of a-sapphire using hot $3H_2SO_4 + H_3PO_4$ solution. The total-epitaxial relationship was revealed as [1-100]ZnO//[1-100]Ti_2O_3//[0001]Al_2O_3 by RHEED observations. These crystallographic results can be concluded that Ti_2O_3 layer is a promised candidate for the buffer layer to achieve high-quality ZnO growth at low temperatures.

REFERENCES

[1] M. A. L. Johnson, *et al*., "MBE Growth and Properties of ZnO on Sapphire and SiC Substrates," *Journal of Electronic Materials*, Vol. 25, No. 5, 1996, pp. 855-862.

[2] A. B. M. A. Ashrafi, *et al*., "Nitrogen-Doped p-Type ZnO Layers Prepared with H_2O Vapor-Assisted Metalorganic Molecular-Beam Epitaxy," *Japanese Journal of Applied Physics*, Vol. 41, 2002, pp. L1281-L1284.

[3] R. D. Vispute, *et al*., "High Quality Crystalline ZnO Buffer Layers on Sapphire(001) by Pulsed Laser Deposition for III-V Nitrides," *Applied Physics Letters*, Vol. 70, No. 20, 1997, pp. 2735-2737.

[4] S. Yamauchi, *et al*., "Low Temperature Epitaxial Growth of ZnO Layer by Plasma-Assisted Epitaxy," *Thin Solid Films*, Vol. 345, No. 1, 1999, pp. 12-17.

[5] S. Yamauchi, *et al*., "Plasma-Assisted Epitaxial Growth of ZnO Layer on Sapphire," *Journal of Crystal Growth*, Vol. 214-215, 2000, pp. 63-67.

[6] S. Yamauchi, *et al*., "Surface Treatment of Si Using Hydrogen-Plasma to Improve Optoelectronic Property of ZnO on (111)Si," *Japanese Journal of Applied Physics*, Vol. 44, 2005, pp. 7801-7804.

[7] K. Minegishi, *et al*., "Growth of p-Type Zinc Oxide Films by Chemical Vapor Deposition," *Japanese Journal of Applied Physics*, Vol. 36, 1997, pp. L1453-L1455.

[8] M. Joseph, *et al*., "p-Type Electrical Conduction in ZnO Thin Films by Ga and N Codoping," *Japanese Journal of Applied Physics*, Vol. 38, 2001, pp. L1205-L1207.

[9] D. C. Look, *et al*., "Characterization of Homoepitaxial p-Type ZnO Grown by Molecular Beam Epitaxy," *Applied Physics Letters*, Vol. 81, No. 10, 2002, pp. 1830-

1832.

[10] S. Yamauchi, *et al.*, "Photoluminescence Studies of Undoped and Nitrogen-Doped ZnO Layers Grown by Plasma-Assisted Epitaxy," *Journal of Crystal Growth*, Vol. 260, No. 1-2, 2006, pp. 1-6.

[11] I. S. Hauksson, *et al.*, "Compensation Processes in Nitrogen Doped ZnSe," *Applied Physics Letters*, Vol. 61, No. 18, 1992, pp. 2208-2210.

[12] A. Ohtomo, *et al.*, "Lateral Grain Size and Electron Mobility in ZnO Epitaxial Films Grown on Sapphire Substrates," *Journal of Crystal Growth*, Vol. 214-215, 2000, pp. 284-288.

[13] Y. Chen, *et al.*, "Layer-by-Layer Growth of ZnO Epilayer on Al_2O_3(0001) by Using a MgO Buffer Layer," *Applied Physics Letters*, Vol. 76, No. 5, 2000, pp. 559-561.

[14] K. H. Ahna, *et al.*, "Kinetic and Mechanistic Study on the Chemical Vapor Deposition of Titanium Dioxide Thin Films by *in Situ* FT-IR Using TTIP," *Surface and Coatings Technology*, Vol. 171, No. 1-3, 2003, pp. 198-204.

[15] S. Tokita, *et al.*, "High-Rate Epitaxy of Anatase Films by Atmospheric Chemical Vapor Deposition," *Japanese Journal of Applied Physics*, Vol. 39, 2000, pp. L169-L171.

[16] S. Weisssmann, *et al.*, "Selected Powder Diffraction Data for Metals and Alloys," JCPDS, Card No. 10-63, 1978, p. 298.

[17] K. Nakahara, *et al.*, "Growth of Undoped ZnO Films with Improved Electrical Properties by Radical Source Molecular Beam Epitaxy," *Japanese Journal of Applied Physics*, Vol. 40, 2001, pp. 250-254.

[18] T. Fukuda and H. J. Scheel, "Crystal Growth Technology," Wiley, New York, 2003.

[19] I. Ohkubo, *et al.*, "In-Plane and Polar Orientations of ZnO Thin Films Grown on Atomically Flat Sapphire," *Surface Science*, Vol. 443, 1999, pp. L1043-L1048.

[20] H. Kato, *et al.*, "High-Quality ZnO Epilayers Grown on Zn-Face ZnO Substrates by Plasma-Assisted Molecular Beam Epitaxy," *Journal of Crystal Growth*, Vol. 265, No. 3-4, 2004, pp. 375-381.

Plasma-Assisted Chemical Vapor Deposition of TiO_2 Thin Films for Highly Hydrophilic Performance

Satoshi Yamauchi[1*], Yoh Imai[2]
[1]Department of Biomolecular Functional Engineering, Ibaraki University, Hitachi, Japan
[2]Department of Electric and Electronic Engineering, Ibaraki University, Hitachi, Japan

ABSTRACT

Titanium-oxide layer was grown on glass substrate by plasma-assisted chemical vapor deposition (PCVD) using oxygen gas plasma excited by radio-frequency power at 13.56 MHz in the pressure as low as 3mtorr at relatively low temperature below 400˚C, and studied on the crystallographic properties with the hydrophilic behavior comparing to the layer deposited by low-pressure chemical vapor deposition (LPCVD). Raman spectra indicated anatase-phase TiO_2 layer without amorphous-phase could be formed above 340˚C by simultaneous supply of plasma-cracked and non-cracked titanium-tetra-iso-propoxide (TTIP) used as preliminary precursor. Surface Scanning Electron Microscope images indicated the PCVD-layer consists of distinct nanometer-size plate-like columnar grains, in contrast to rugged micrometer-size grains in the LPCVD-layer. Extremely small water contact angle about 5˚ in dark and the quick conversion to super-hydrophilicity by UV-irradiation with a light-power density as low as 50 W/cm^2 were observed on the PCVD-layer grown at 380˚C, while the large initial contact angle was above 40˚ and the response for the UV-irradiation was gradual on the LPCVD-layer.

Keywords: PCVD; Titanium-Oxide Films; Anatase-TiO_2; Hydrophilicity

1. Introduction

Titanium dioxide (TiO_2) has been extensively investigated in view of photo-induced applications using the photo-catalytic reactions and the hydrophilicity on the surface [1,2], in addition to electronic and optoelectronic applications [3,4]. Commonly, anatase-phase TiO_2 is preferred for the photo-induced applications because of the efficient surface reaction by UV-irradiation compared to another crystal phases (brookite, rutile) [5]. In addition, the defects in the crystal should be also taken into account to control the reaction because it is recognized that the Ti^{3+} sites reduced from Ti^{4+} at the surface by photo-excited electrons accompanying oxygen vacancies generated by the photoexcited holes play an important role in the photo-induced surface reaction [6]. Therefore, a low of wet or dry process to fabricate anatase-TiO_2 layer has been advanced to control the crystallinity such as stoichiometric composition, impurity concentration, surface morphology, crystal orientation and so on. In such processes, wet process such as dip-coating, spray or sol-gel has been widely used for the photocatalytic and hydrophilic coating [7,8] with advantages of the low-cost and the wide-area coatings. In contrast, dry process such as reactive sputtering [9], electron beam evaporation [10] and metalorganic chemical vapor deposition [11] has been attractively studied to control the growth behavior. Plasma-assisted chemical vapor deposition (PCVD) has been candidate for the TiO_2 synthesis process for low temperature deposition, control of the grain structure and the surface morphology [12-14]. In the PCVD process, titanium tetra-iso-propoxide (TTIP) has been widely used as the preliminary precursor in order to reduce contamination in the layer [15], where O_2 gas is simultaneously used as the oxidant gas for the efficient reduction. The metal oxide dissociated in the plasma [13] brings about the low temperature deposition below 300˚C and the formation of highly dense columnar grains compared to the CVD-layers. However, the temperature to grow anatase-phase TiO_2 is increased above 450˚C, whereas the anatase-TiO_2 can be grown at the temperature above 300˚C by CVD [13]. The unfavorable increase of the growth temperature is probably come from hindering the crystal-phase formation by ion bombardment of excess energy particles [10] and poor surface migration of the metal-oxides on the growth surface. It is noted that the PCVD had been demonstrated in relatively high pressure above 0.1 torr by relatively high rf-power above 100 W [14] aiming at enough dissociation of the precursor in the

*Corresponding author.

plasma. On the other, the deposition enhanced chemical dissociation by reactive oxidants such as atomic-oxygen results in reduction of the TTIP-dissociation energy and the high growth rate as demonstrated by remote plasma-enhanced CVD [16]. The useful effects brought about the low-temperature growth including ana-tase-TiO_2, however, the layer grown at 350˚C was formed by relatively low dense grains with poor crystallinity compared to the CVD-layer. It is considered that the results indicated the thermal-dissociation of TTIP without enhancement by the reactive oxygen is more effective to growth anatase-phase TiO_2.

In this paper, PCVD of TiO_2 supporting the crystallization in anatase-phase by thermal-dissociation of TTIP is demonstrated at low-temperature below 400˚C and hydrophilic property on the anatase-phase layer is shown with the crystallinity and the surface morphology.

2. Experimental

Figure 1 shows a bell-jar type PCVD apparatus consists of a diffusion-pump (D.P.) and a rotary-pump (R.P.) for TiO_2 deposition. The back-pressure in the chamber was under 1×10^{-5} torr. An inductively coupled electrode to introduce radio-frequency (rf) power at 13.56 MHz was equipped in the chamber between a substrate holder and gas inlets. A Coil to apply DC-magnetic-field of 3000 gauss at the center of the chamber was also settled around the bell-jar to stabilize gas plasma excited by the rf-power, because PCVD in this study was performed in the pressure as low as 3mtorr and the low density plasma excited by 10 W rf-power. In the case of LPCVD, the rf-power was not induced during the growth. The temperatures of substrate holder and gas inlet for preliminary precursor were increased by resistive-heating and controlled by PID-systems.

Titanium tetra-iso-propoxide ($Ti(O\text{-}i\text{-}C_3H_7)_4$: 97%-purity) was used as preliminary precursor. The liquid-phase TTIP was charged in a quartz-cell and then purified in vacuum at 50˚C for 3hrs to remove volatile solvent in the liquid. The purified TTIP was vaporized at 70˚C and introduced into the chamber without any carrier gas through a stainless tube and a variable-valve at the temperatures about 90˚C. Pure oxygen gas (99.9999%-purity) was also introduced through a stainless tube and a variable-valve. The supply ratio of TTIP/O_2 was controlled by monitoring the chamber pressure when the TTIP and the O_2 were introduced into the chamber.

Quartz plates with mirror-surface used as substrates were rinsed in deionized water and dried by a spinner after removal of contaminations from the surface by organic solvents and a hot H_2SO_4 + H_2O_2, and then thermally cleaned at 400˚C in the chamber in low pressure below 1×10^{-5} torr for 30 min before the growth. Prior to the growth, inner-wall of the chamber was cleaned by oxygen plasma excited by 100 W rf-power to remove residual gas, and then titanium-oxide layers were grown at low temperatures ranging from 150˚C to 400˚C.

Thickness of the layer was checked by a contact-type surface profiler (DEKTAK150). The surface morphology was observed by SEM (HITACHI S-800) and the crystallinity was investigated using Raman spectrometer (JASCO NR-1100) using the 514.5 nm line of an Ar^+ laser (100 mW) as the excitation source. Hydrophilicity on the layer was evaluated by contact angle of water when deionized water (10 μl) was dropped on the layer in air at 20˚C with 50%-humidity. The photo-induced hydrophilicity was examined as a function of UV-irradiation time, in which a black-light peak at 365 nm with the low light power density of 50 $\mu W/cm^2$ was used as the light source.

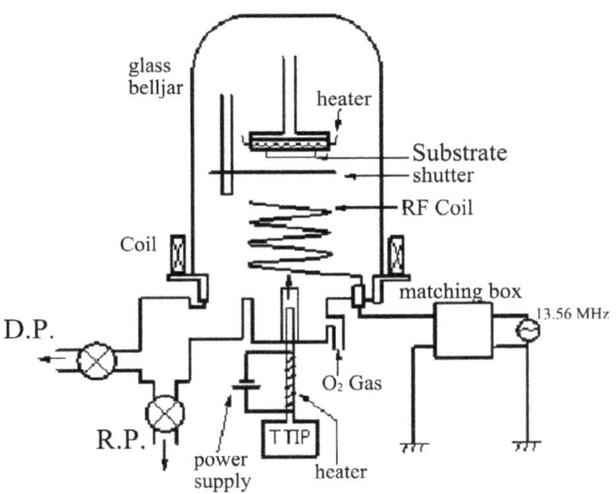

Figure 1. An apparatus of PCVD for TiO_2 deposition.

3. Results and Discussions

3.1. Growth Rate

Figure 2 shows growth rates of titanium-oxide layers by

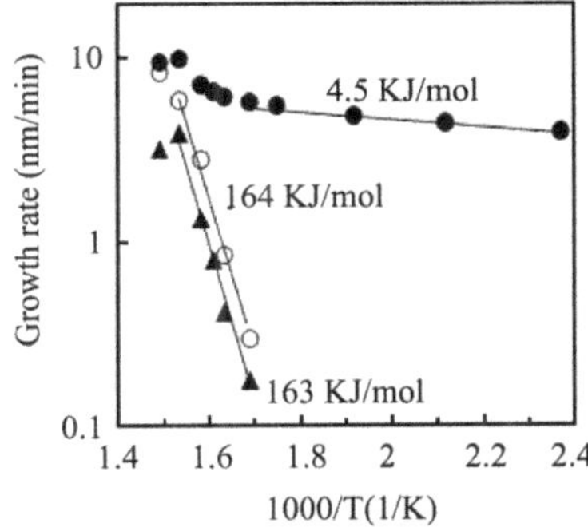

Figure 2. Growth rates of Titanium-oxide layer by PCVD (closed circles), LPCVD (open circles) and thermal-dissociation in PCVD (closed triangles).

PCVD (closed circles) and LPCVD (open circles) at various temperatures, where the O_2/TTIP supply ratio was controlled to 1. The rate by LPCVD was increased with the growth temperature above 320°C according to Arrhenius relationship and the activation energy was obtained as 164 kJ/mol. The energy value was seemed to be larger than the other reports [13,16] using TTIP and O_2 but similar to the value of 150 kJ/mol by CVD using TTIP in N_2-gas with the high flow rate [17]. It has been recognized that oxygen reduces the dissociation energy of TTIP, but the large activation energy indicates the dissociation of TTIP by LPCVD in such low pressure of 3mtorr with the small supply ratio of O_2/TTIP was owing to thermal-dissociation as follow [18],

$$Ti(OC_3H_7)_4 \rightarrow TiO_2 + 2C_3H_6 + 2HOC_3H_7$$

On the other hand, two activation energies were observed by PCVD. The activation energy in the low temperature below 300°C was 4.5 kJ/mol which was in good agreement with the previous report by PCVD [13] but much lower than that by thermal-dissociation enhanced by reactive-oxygen [16]. The significant low energy indicated that the preliminarily precursor was dissociated to titanium-oxides in the plasma as described elsewhere [13]. In contrast, the activation energy between 300°C and 380°C was determined as 163 kJ/mol (closed-triangles in **Figure 2**) after removal the growth rate extrapolated from the rate in the low temperature region performed by the plasma-cracked precursors. The activation energy above 300°C was coin-cident to that by LPCVD. It was reported that activation energy of TTIP-dissociation was reduced by plasma-excited oxygen [16], however, the enhanced dissociation was not observed in this work because the collision probability between TTIP and oxygen was relatively small in such low pressure with the small O_2/TTIP sup-ply ratio. The decreased growth rate above 400°C was believed to be caused by the depletion of the precursor at the growth surface due to the high desorption coefficient or the volumetric dissociation in the gas-phase. It should be concluded from these results that the PCVD above 300°C was performed by supply of the metal-oxides dissociated in the plasma, which was expected to form highly dense grain growth by the high-sticking coefficient, and thermal dissociation of TTIP, which was expected to enhance the crystallization into the anatasephase.

3.2. Raman Spectra

Figure 3 shows Raman spectra of 600 nm-thick titanium-oxide layers grown by PCVD at various temperatures. Typical peaks corresponding to anatase-TiO_2 were appeared at 144, 399, 514 and 639 cm^{-1} in the layers grown above 340°C, where the peaks were attributed to E_g, B_{1g}, B_{2g} and E_g vibration modes respectively [19]. However, any peaks except a weak broadband around 450 cm^{-1}, which was recognized due to amorphous phase of TiO_x [20], could not be observed for the layer grown at 250°C.

The broadband was decreased with the deposition temperature and disappeared in the layer grown above 340°C. The broadband was appeared for the layer grown at the temperature ranging from 300°C to 330°C but the weak E_g-band could be also observed. It is noted here that the low-temperature about 300°C to form anatase-TiO_2 was much lower than 450°C by PCVD reported previously [13], but consistent with the temperature for anatase-TiO_2 growth by LPCVD using TTIP as single precursor [21]. The significantly low temperature to form anatase-TiO_2 suggested that the crystallization could be supported by the non-cracked TTIP which was simultaneously supplied with the plasma-cracked precursors and thermally dissociated on the growth surface. **Figures 4(a)** and **(b)** show growth rate ratio by the CVD-mode ($GR_{thermal}$) and the PCVD-mode (GR_{plasma}) thermal-dissociation as a function of the growth temperature and integrated intensity of E_g-mode peak at 144 cm^{-1} as a function of the $GR_{thermal}/GR_{plasma}$ ratio, respectively. The ratio of $GR_{thermal}/GR_{plasma}$ except the open circle (grown at 400°C) was rapidly increased with the growth tempera

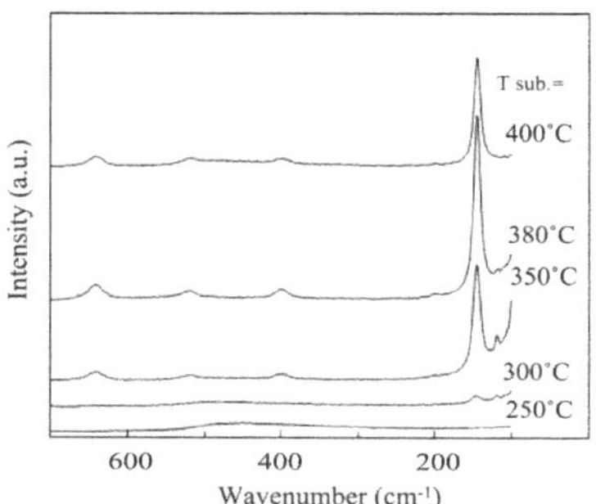

Figure 3. Raman spectra of PCVD-layers with 600 nm-thickness grown at various temperatures.

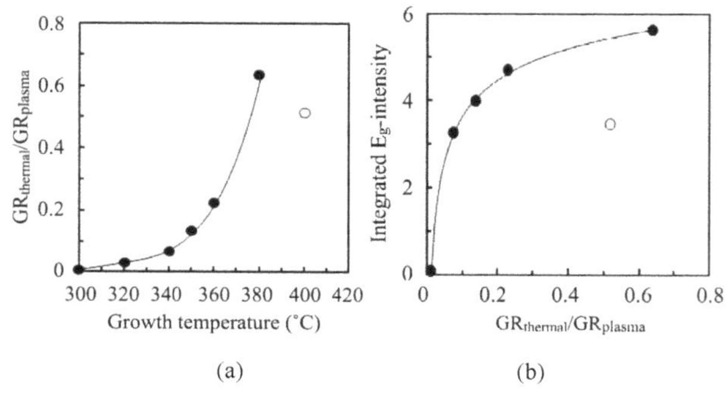

Figure 4. (a) Relative growth rate by thermal dissociation ($GR_{thermal}$) compared to growth by plasma cracked precursors (GR_{plasma}) as a function of the growth temperature; and (b) E_g-intensity as a function of $GR_{thermal}/GR_{plasma}$, where the open circles shows for the layer grown at 400°C.

ture above 300°C according to rapid increase of the thermal dissociation of TTIP. In contrast, the intensity of E_g-mode was rapidly increased and gradually saturated with the ratio of $GR_{thermal}/GR_{plasma}$, which indicated crystallization in anatase-phase was enhanced by relatively small amount of the thermally dissociated TTIP compared to the supply of the plasma-cracked precursors. In the case of the growth at 400°C, decrease of the E_g-band intensity (open circle in **Figure 4(b)**) and the weak broadband due to amorphous-phase (**Figure 3**) suggested the dissociated precursor in the gas-phase was simultaneously supplied on the growth surface.

Figure 5 shows the E_g-band spectra of PCVD- and LPCVD-TiO_2 layers grown at 380°C with 600 nm-thickness, where the background was numerically removed in the spectra. The spectrum with the FWHM of 10.7 cm^{-1} of the PCVD-layer was slightly sharper than that of the LPCVD-sample with the FWHM of 12.0 cm^{-1}. Additionally, Raman-spectrum shift of the PCVD-TiO_2 was small below 1 cm^{-1}, while the shift of the LPCVD-layer was about 2 cm^{-1}. It is recognized that the Raman-spectrum shift is originated from residual stress in the layer [22] on the substrate with different thermal-expansion coefficient. Previously, Alhomoudi *et al.* reported E_g-band of anatase-TiO_2 layer was shifted toward higher wavenumber above 3 cm^{-1} broadening about 23 cm^{-1} in the thick layer about 600 nm, where the layer was grown by reactive-sputtering around 300°C, and concluded the spectrum shift and broadening were caused by residual compressive stress and large distribution of the orientation of grains in the layer, respectively [23]. In contrast, both of the shift and the broadening were much smaller in the PCVD-TiO_2 layer as shown above, which clearly indicated the residual stress in the PCVD-layer was relaxed and speculated the layer consisted of uniform grains.

3.3. Surface Morphology

Figures 6(a) and **(b)** show surface SEM images of 600 nm-thick PCVD-layer and LPCVD-layer grown at 380°C, respectively. Relatively large and rugged grains with sub-micrometer size were found in the LPCVD-layer, which indicated the growth was performed by the low density nucleation at the initial stage and/or the secondary nucleation in the deep grain boundaries. In contrast, the PCVD-layer was formed by highly dense plate-like nano-grains with the width around 45 nm. It can be easily recognized that the significant difference of the grain feature compared to the LPCVD-layer was caused by highly dense nucleation at the initial stage and preferential growth along the thickness. The nucleation with high density could be performed by the plasma cracked metal-oxides with the high sticking coefficients, then the grains were grown by simultaneous supply of the plasma-cracked precursors and non-cracked TTIP, where the grain growth was governed by supply of the plasma-cracked precursors in PCVD-mode as shown in **Figure 4(b)**. In addition, it is considered that the distinct nano-grains in the PCVD-layer resulted in relaxation of the bi-axial stress observed in the Raman spectrum.

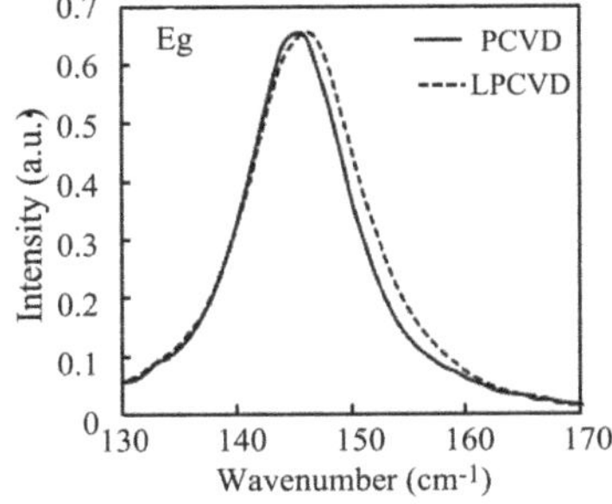

Figure 5. Eg-band spectra of PCVD-(solid-line) and LP-CVD-layer (dotted-line), where the both layers were grown at 380°C.

3.4. Hydrophilicity

Figure 7 shows contact angle of water on the PCVD-layers grown at various temperatures ranging from 250°C to 380°C as a function of irradiation time of UV-light

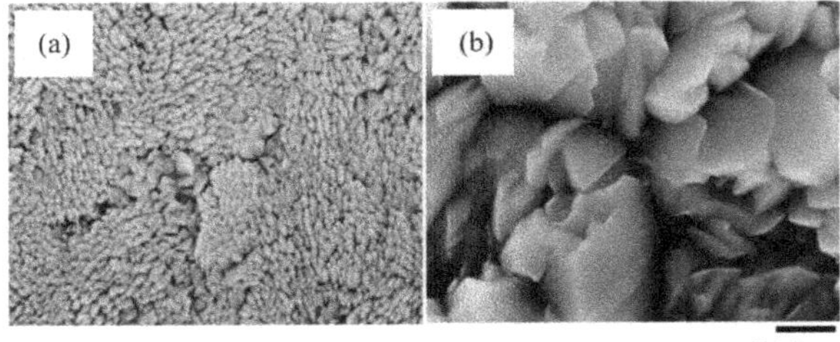

Figure 6. Surface SEM images of 600 nm-thick TiO_2 layers grown by (a) PCVD and (b) LPCVD at 380°C.

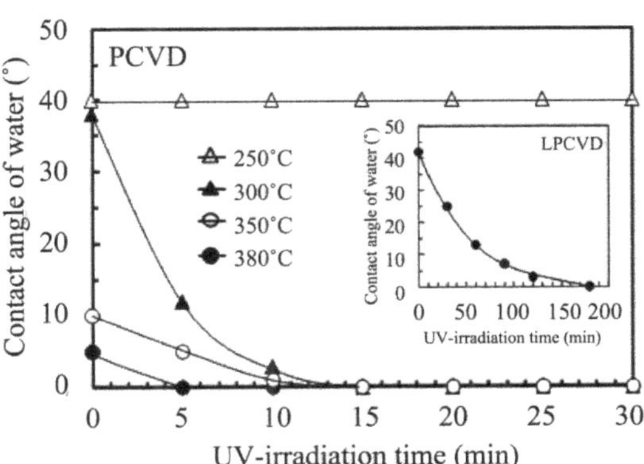

Figure 7. Contact angle of water on TiO_2 layers grown at various temperatures by PCVD as a function of UV-irradiation time. The inset shows variation of the angle on CVD-TiO_2 layer grown at 380°C for UV-irradiation time.

peak at 365 nm-wavelength with the low power density of 50 μW/cm^2. The inset shows the dependence of LPCVD-TiO_2 layer grown at 380˚C. Here, the grown layer was initially treated in atmosphere by the UV-irradiation for a few hours required to achieve super-hydrophilicity and exposed in air in dark for 1 month, and then the contact angle on the layers irradiated by the UV-light for each time was evaluated after the initial contact angle was checked before the UV-irradiation.

The hydrophilicity on the PCVD-layer including the initial angle and the hydrophilization feature by the UV-irradiation was significantly dependent on the growth temperature. The contact angle was around 40˚ before the UV-irradiation and not responsible to the irradiation on the layer grown at low temperature below 250˚C. Nakamura *et al.* previously showed the amorphous titanium-oxide deposited at low temperature below 200˚C by PCVD included hydroxyl which plays an important role of the hydrophilicity and the contact angle on the layer was responsible for UV-irradiation [24]. However, absorption due to -OH bond could not be observed in FTIR spectrum of the amorphous-phase PCVD-layer grown at 250˚C, and the contact angle was not responsible for the UV-irradiation. On the other, the contact angle on the layer consists of anatase- and amorphous-phase mixture grown at 300˚C was reduced by the UV-irradiation but the initial contact angle was similar to the amorphous-sample. In contrast, the initial contact angle on the layer grown above 340˚C, which consists of anatase-phase grains without amorphous-phase, was obviously decreased with the growth temperature and the prompt hydrophilization was observed by the UV-irradiation. Especially, the contact angle on the PCVD- layer grown at 380˚C was extremely small about 5˚ before the UV-irradiation and quickly converted to super-hydrophilicity within 5 min by the UV-irradiation. It is recognized that Ti^{3+} sites reduced from Ti^{4+} in TiO_2 crystal cause the hydrophilic conversion. The influence of surface morphology should be also considered in the hydrophilicity, however, the grain structure as shown in **Figure 3** was scarcely dependent on the growth temperature. Therefore, it can be concluded that the hydrophilic property dependent on the growth temperature was originated from the density of Ti^{3+} sites at the surface. On the other, the large initial contact angle about 40˚ and the gradual response for the UV-irradiation were observed on LPCVD-layer grown at 380˚C (inset of **Figure 7**), while the layer consists of anatase-TiO_2 without the amorphous-phase. Capillary effect in addition to chemical property on the surface should be taken into account to recognize the significantly different feature of hydrophilic properties. Previously, Katsumata *et al.* showed the hydrophilic conversion during UV-irradiation was enhanced on the sol-gel derived TiO_2 layer consists of nano-grains with the size around 50 nm by comparison of the property on the layer consists of relatively large 150 - 200 nm grains [25]. They found out that the difference of hydrophilicizing rate by UV-irradiation was originated from homogeneity of the surface and suggested the hydrphilicizing was enhanced by two-dimensional capillary effect on the homogeneous microstructure. Although the grain-structure in the PCVD layer as shown in **Figure 6(a)** was different from the sol-gel derived layer, the surface consists of nano-grains was homogeneous compared to remarkably heterogeneous surface of the LPCVD as shown in **Figure 6(b)**. It has been also recognized that the hydrophilicity after UV-irradiation is gradually degraded in dark since the life-time of -OH bond at the surface is not so long, however, long-time storage behavior for the super-hydrophilicity above 5 hrs was observed on the PCVD-TiO_2 layer. This useful behavior was speculated by percolation of water into the nano-scale slits between the grains as shown in **Figure 6(a)**, where the percolated water was probably kept in the pores for long-time avoiding the evaporation compared on the surface. Of course, the storage behavior could not be observed on the LPCVD-layer, in which the contact angle was gradually increased to 40˚ within 1 hr after exposure in dark.

4. Conclusion

Titanium-oxide layers were grown on quartz substrates by PCVD using TTIP and oxygen mixed gas plasma with O_2/TTIP supply ratio of 1 in the pressure as low as 3 mtorr at low temperatures below 400˚C, and then characterized by the growth behavior and the hydrophilicity comparing to the layer by LPCVD. The PCVD was performed by plasma-cracked precursors and thermally dissociated TTIP, while the LPCVD was owing to thermal dissociation of TTIP. Raman spectra investigated anatase-TiO_2 could be grown above 300˚C by the PCVD. Further, it was revealed that the crystallinity of anatase-TiO_2 was significantly dependent on the ratio of the PCVD-mode and the CVD-mode during the growth, where non-cracked TTIP played a role to enhance the crystallization of the plasma-cracked precursors on the growth surface. SEM observations showed that the PCVD anatase-phase TiO_2 layer was formed by homogeneous nano-size columnar grains, whereas the LPCVD layer consists of heterogeneous micro-size grains. Initial contact angle of water before UV-light irradiation was dependent on the growth temperature and drastically decreased on the layer grown above 340˚C. The initial contact angle about 5˚ on the PCVD layer grown at 380˚C showed the excellent hydrophilicity without UV-irradiation was performed by the PCVD. The contact angle on the PCVD-layer was quickly reduced and showed super-hydrophilicity within 5 min by UV-light irradiation with the low light power density of 50 $\mu W/cm^2$, while the long time about 3 hrs was required for the LPCVD layer. The PCVD-layer also showed interesting storage behavior of the super-hydrophilicity as long as 5 hrs in dark after the UV-irradiation, whereas the contact angle on the LPCVD-layer was increased up to 40° within 1 hr in dark.

REFERENCES

[1] R. Wang, K. Hashimoto and A. Fujishima, "Light-Induced Amphiphilic Surfaces," *Nature*, Vol. 388, No. 6641, 1997, pp. 431-432.

[2] A. Mills, A. Lepre, N. Elliott, A. Bhopal, I. P. Parkin and S. A. Neill, "Characterization of the Photocatalyst Pilkington Activ^{TM}: A Reference Film Photocatalyst?" *Journal of Photochemistry and Photobiology A: Chemistry*, Vol. 160, No. 3, 2003, pp. 213-224.

[3] S. A. Campbell, H. S. Kim, D. C. Gilmer, B. He, T. Ma and W. L. Gladfelter, "Titanium Dioxide (TiO_2)-Based Gate Insulators," *IBM Journal of Research and Development*, Vol. 43, No. 3, 1999, pp. 383-392.

[4] C. Martinet, V. Paillard, A. Gagnaire and J. Joseph, "Deposition of SiO_2 and TiO_2 Thin Films by Plasma Enhanced Chemical Vapor Deposition for Antireflection Coating," *Journal of Non-Crystalline Solids*, Vol. 216, No. 1, 1997, pp. 77-82.

[5] H. Gerischer and H. Heller, "The Role of Oxygen in Photooxidation of Organic Molecules on Semiconductor Particles," *The Journal of Physical Chemistry*, Vol. 95, No. 13, 1991, pp. 5261-5267.

[6] R. Wang, K. Hashimoto, A. Fujishima, M. Chikuni, E. Kojima, A. Kitamura, M. Shimohigoshi and T. Watanabe, "Photogeneration of Highly Amphiphilic TiO_2 Surfaces," *Advanced Materials*, Vol. 10, No. 3, 1998, pp. 135-138.

[7] N. A. Kotov, F. C. Meldrum and J. H. Fendler, "Monoparticulate Layers of Titanium Dioxide Nanocrystallites with Controllable Interparticle Distances," *The Journal of Physical Chemistry*, Vol. 98, No. 36, 1994, pp. 8827-8830.

[8] M. Okuya, N. A. Prokudina, K. Mushika and S. Kaneko, "TiO_2 Thin Films Synthesized by the Spray Pyrolysis Deposition (SPD) Technique," *Journal of the European Ceramic Society*, Vol. 19, No. 6-7, 1999, pp. 903-906.

[9] M. H. Suhail, G. Mohan Rao and S. Mohan, "Dc Reactive Magnetron Sputtering of Titanium—Structural and Optical Characterization of TiO_2 Films," *Journal of Applied Physics*, Vol. 71, No. 3, 1992, pp. 1421-1427.

[10] P. Lobl, M. Huppertz and D. Mergel, "Nucleation and Growth in TiO_2 Films Prepared by Sputtering and Evaporation," *Thin Solid Films*, Vol. 251, No. 1, 1994, pp. 72-79.

[11] V. Gauthier, S. Bourgeois, P. Sibillot, M. Maglione and M. Sacilotti, "Growth and Characterization of AP-MOCVD Iron Doped Titanium Dioxide Thin Films," *Thin Solid Films*, Vol. 340, No. 1, 1999, pp. 175-182.

[12] W. G. Lee, S. I. Woo, J. C. Kim, S. H. Choi and K. H. Oh, "Preparation and Properties of Amorphous TiO_2 Thin Films by Plasma Enhanced Chemical Vapor Deposition," *Thin Solid Films*, Vol. 237, No. 1-2, 1994, pp. 105-111.

[13] W. Yang and C. A. Wolden, "Plasma-Enhanced Chemical Vapor Deposition of TiO_2 Thin Films for Dielectric Applications," *Thin Solid Films*, Vol. 515, No. 4, 2006, pp. 1708-1713.

[14] S. Mathur and P. Kuhn, "CVD of Titanium Oxide Coatings: Comparative Evaluation of Thermal and Plasma Assisted Processes," *Surface and Coatings Technology*, Vol. 201, No. 3-4, 2006, pp. 807-814.

[15] E. Fredriksson and J. O. Carlsson, "Chemical Vapour Deposition of TiO and Ti_2O_3 from $TiCl_4/H_2/CO_2$ Gas Mixtures," *Surface and Coatings Technology*, Vol. 73, No. 3, 1995, pp. 160-169.

[16] K. H. Ahn, Y. B. Park, D. W. Park, "Kinetic and mechanistic study on the chemical vapor deposition of titanium dioxide thin films by in situ FT-IR using TTIP," Surf. Coat. Technol., Vol. 171, 2003, pp. 198-204.

[17] M. Yokozawa, H. Iwasa and I. Teramoto, "Vapor Deposition of TiO_2," *Japanese Journal of Applied Physics*, Vol. 7, 1968, pp. 96-97.

[18] C. P. Fictorie, J. F. Evans and W. L. Gladfelter, "Kinetic and Mechanistic Study of the Chemical Vapor Deposition of Titanium Dioxide Thin Films Using Tetrakis-(Isopropoxo)-Titanium(IV)," *Journal of Vacuum Science & Technology A*, Vol. 12, No. 4, 1994, pp. 1108-1113.

[19] T. Ohsaka, F. Izumi and Y. Fujiki, "Raman Spectrum of Anatase, TiO_2," *Journal of Raman Spectroscopy*, Vol. 7, No. 6, 1978, pp. 321-324.

[20] A. G. Gaynor, R. J. Gonzalez, R. M. Davis and R. Zallen, "Characterization of Nanophase Titania Particles Synthesized Using *in Situ* Steric Stabilization," *Journal of Materials Research*, Vol. 12, No. 7, 1997, pp. 1755-1765.

[21] B. C. Kang, S. B. Lee and J. H. Boo, "Growth of TiO_2 Thin Films on Si(100) Substrates Using Single Molecular Precursors by Metal Organic Chemical Vapor Deposition," *Surface and Coatings Technology*, Vol. 131, No. 1-3, 2000, pp. 88-92.

[22] S. J. Fonash, "Effects of Stress on Metal-Oxide-Semiconductor Structures," *Journal of Applied Physics*, Vol. 44, No. 10, 1973, pp. 4607-4615.

[23] I. A. Alhomoudi and G. Newaz, "Residual Stresses and Raman Shift Relation in Anatase TiO_2 Thin Film," *Thin Solid Films*, Vol. 517, No. 23, 2009, pp. 4372-4378.

[24] M. Nakamura, S. Kato, T. Aoki, L. Sirghi and Y. Hata-

naka, “Role of Terminal OH Groups on the Electrical and Hydrophilic Properties of Hydro-Oxygenated Amorphous TiO_x: OH Thin Films,” *Journal of Applied Physics*, Vol. 90, No. 7, 2001, pp. 3391-3395.

[25] K. Katsumata, A. Nakajima, H. Yoshikawa, T. Shiota, N. Yoshida, T. Watanabe, Y. Kameshima and K. Okada, “Effect of Microstructure on Photoinduced Hydrophilicity of Transparent Anatase Thin Films,” *Surface Science*, Vol. 579, No. 2-3, 2005, pp. 123-130.

9

Synthesis and Structural Study of Triphenylbismuth Bis (Salicylate)

Kheira Feham[1], Abdelkarim Benkadari[2], Abdelkader Chouaih[2*], Abdellah Miloudi[1,3], Gérard Boyer[4], Douniazed El Abed[1]

[1]Laboratoire de Chimie Fine (LCF), Université d'Oran, Es-Senia, Algérie
[2]Laboratoire SEA2M, Université de Mostaganem, Mostaganem, Algérie
[3]Département de Physique-Chimie, ENSET d'Oran, Oran, Algérie
[4]Laboratoire HIT, Université Paul Cezanne, Marseille, France

ABSTRACT

The crystal of triphenylbismuth bis (Salicylate) pentavalent was synthesized from the reaction of triphenylbismuth dichloride with salicylic acid dissolved in methylene chloride at room temperature. The molecular and crystal structures of triphenylbismuth bis (Salicylate) were determined by X-ray diffraction analysis. This compound crystallizes in the triclinic space group $P\overline{1}$ with crystallographic parameters: $a = 11.2937$ (3) Å, $b = 14.6516$ (3) Å, $c = 17.8253$ (4) Å, $\alpha = 78.2958$ (7)°, $\beta = 76.232$ (6)°, $\gamma = 85.351$ (6)°, $\mu = 6.332$ mm, $V = 2803.59$ (11) Å, $Z = 2$, $Dc = 1.693$ g/cm, $F(000) = 1392$, $T = 293(2)$K. The final residual factor is 0.0602 for 5806 reflexions with $I > 2\sigma(I)$. The bismuth atom of the compound has a distorted trigonal-bipyramidal configuration.

Keywords: Organobismuth; Structure; X-Ray Diffraction; Salicylate; Synthesis

1. Introduction

Bismuth is nontoxic and relatively cheap, and bismuth compounds have been widely used in catalysis and organic synthesis [1]. Inorganic bismuth compounds such as bismuth halides have been used as Lewis acid catalysts in a number of organic reactions [1-5]. However, the utilization of organobismuth compounds trivalent and pentavalent for organic synthesis is rarely reported partly due to the unstable nature of the Bi-C bonds [6,7]. Recent developments show that the incorporation of a bulky substituent in pentavalent organobismuth complex can result of organobismuth compounds that have stable Bi-C bonds [8-10]. In our research group, we have been working on the synthesis of stable pentavalent organobismuth compounds. In previous works we have synthesized organobismuth compounds such as pentavalent triphenylbismuth dichloride, triphenylbismuth diacetate and triphenylbismuth bis (thiophene carboxylate) [7,11]. In this work, we report synthesis and structure determination of the organobismuth compound triphenylbismuth bis (salicylate), **I**, where **I** is the pentavalent complex, bulky and more stable at room temperature.

*Corresponding author.

2. Materials and Methods

2.1. Synthesis

The compound triphenylbismuth bis (salicylate) **I** was prepared from the triphenyl dichloride and salicylic acid in solution of methylene chloride. The mixture is heated to react for one hour of stirring. At the end of the reaction, the mixture is removed and then recrystallized from dichloromethane/pentane (1.1) [12] (**Scheme 1**).

The obtained compound has the empirical formula $C_{32}H_{25}BiO_6$ containing an OH and carboxylate groups that were chosen for our investigation for the following considerations: 1) the hydroxyl group can act as both proton acceptor to promote the formation of intermolecular hydrogen bond; 2) the oxygen atoms of the car-

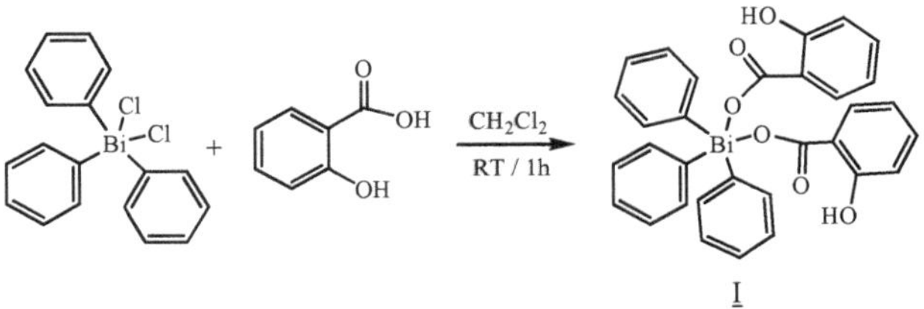

Scheme 1. The synthetic route of the triphenylbismuth bis (salicylate) compound, I.

bismuth (V) centre, hence it could act as a proton acbonyl of carboxylate ligand is weakly associated with theceptor and participate in the formation of hydrogen bonds; 3) The oxygen atom of the carboxylate links to the central atom of bismuth, forming a stable coordination bond. The resulting compound is studied using X-ray diffraction to obtain molecular crystal structure.

2.2. Single Crystal X-Ray Analysis

The crystal structure of **I** was determined by single crystal X-ray diffraction method. X-ray data were collected at room temperature (293 K) on a Kappa CCD diffractometer (Bruker Nonius, 1998) [13] using MoK$_\alpha$ radiation (k = 0.71073 Å), at a voltage of 50 kV and current of 20 mA. Cell parameters were obtained from refinement of 25 reflections collected from a random searching. Data reduction was performed with Denzo and Scalepack software [14]. Data obtained were processed with the WinGX integrated system software package for single crystal X-ray diffraction data solution, refinement and analysis [15]. The crystal structure was solved by direct methods and was refined by full-matrix least-squares refinement on F^2 using the software package SHELX-97 [16]. Molecular graphics were done with ORTEP-3 [17]. In the absence of significant anomalous scattering effects Friedel pairs have been merged. The crystal data and structure refinement details are listed in **Table 1**.

3. Results and Discussion

3.1. Bond Length and Angles

In **Figure 1**, the central bismuth-containing part exhibits a distorted pseudo-trigonal-bipyramidal structure.

The average values of bond distances and angles in different aryl rings are in agreement with the literature

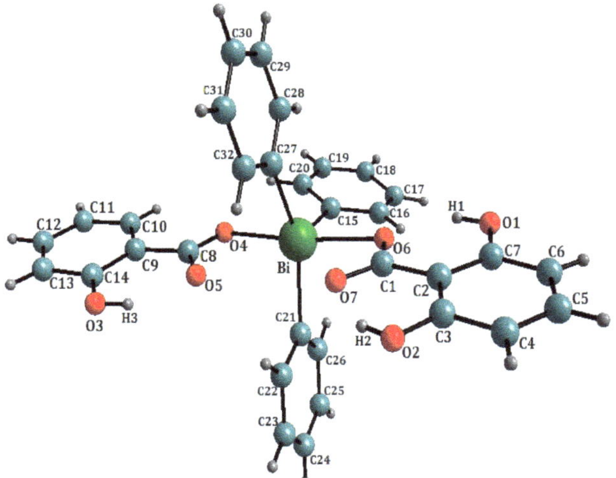

Figure 1. Perspective view of the molecule of triphenylbismuth bis (salicylate), I showing the labelling scheme.

Table 1. Crystal data and structure refinement details for $C_{32}H_{25}BiO_6$ compound.

Chemical formula	$C_{32}H_{25}BiO_6$
CCDC deposit	737,905
Colour	colourless
Crystal dimensions (mm)	0.15 × 0.10 × 0.10
Formula weight	1429
Crystal system	Triclinic
Space group	$P\bar{1}$
Unit cell dimensions	
a (Å)	11.2937 (3)
b (Å)	14.6516 (3)
c (Å)	17.8253 (4)
α (˚)	78.2958 (7)
β (˚)	76.232 (6)
γ (˚)	85.351 (6)
Cell volume (Å^3)	2803.59 (11)
Z	2
Temperature (K)	293 (2)
Density (g/cm^3)	1.693
Absorption coefficient (mm^{-1})	6.332
Diffractometer	Kappa CCD
Radiation, λ (Å)	MoK$_\alpha$, 0.71073
Theta min-max (˚)	1.68 - 27.89
Reflections collected/unique	32748/12722
Range of h, k, l	$-13 \le h \le 14$, $-18 \le k \le 19$, $0 \le l \le 23$
Absorption correction	Sortav (Blessing, 1997)
Data/restraints/parameters	5806/0/712
Goodness of fit on F^2	0.857
Final R indices $R\left[F^2 > 2\sigma\left(F^2\right)\right]$	0.0602
R indices (all data)	0.1564

values [18]. The bond distances and valence angles of molecule **I** are shown in **Table 2**.

According to X-ray diffraction data of compound **I** and as shown in **Figure 2**, Bi atom has distorted trigonal-bipyramidal coordination with oxygen atoms in apical positions. The O(1) and O(4) atoms are located at the apical positions and the C(21), C(27) and C(15) atoms are at the equatorial positions. The sum of angles in

Table 2. Bond distances and bond angles for non-hydrogen atoms (e.s.d.'s are given in parenthesis).

Bond distance (Å)		Bond angles (°)	
Bi-O1	2.303 (6)	O1-Bi-O4	172.6 (2)
Bi-O4	2.313 (6)	O1-Bi-C15	93.4 (3)
Bi-C15	2.215 (9)	O1-Bi-C21	87.8 (3)
Bi-C21	2.238 (8)	O1-Bi-C27	87.5 (3)
Bi-C27	2.200 (9)	O4-Bi-C15	91.9 (3)
O1-C1	1.303 (11)	O4-Bi-C21	86.0 (3)
O2-C1	1.246 (12)	O4-Bi-C27	91.1 (3)
O3-C3	1.380 (14)	C15-Bi-C21	104.1 (4)
O4-C8	1.289 (11)	C15-Bi-C27	144.4 (4)
O5-C8	1.239 (11)	C21-Bi-C27	111.5 (4)
O6-C10	1.358 (13)	Bi-O1-C1	105.4 (6)
C1-C2	1.501 (14)	B-O4-C8	107.9 (4)
C2-C3	1.415 (15)	O1-C1-O2	121.0 (9)
C2-C7	1.386 (16)	O1-C1-C2	114.5 (11)
C3-C4	1.395 (17)	O2-C1-C2	124.3 (11)
C4-C5	1.33 (2)	C1-C2-C3	117.0 (12)
C5-C6	1.36 (2)	C1-C2-C7	122.8 (11)
C6-C7	1.365 (16)	O3-C3-C2	121.7 (11)
C8-C9	1.472 (13)	O3-C3-C4	119.0 (14)
C9-C10	1.368 (14)	C2-C3-C4	119.3 (14)
C9-C14	1.403 (13)	C3-C4-C5	119.5 (17)
C10-C11	1.393 (17)	C4-C5-C6	120.5 (16)
		C-C6-C7	123.9 (16)
		C2-C7-C6	116.5 (14)
		O4-C8-O5	120.5 (9)
		O4-C8-C9	116.3 (10)
		O5-C8-C9	123.2 (10)
		C8-C9-C10	118.1 (10)
		C10-C9-C14	119.6 (10)

equatorial plan and the axial angle O-Bi-O for the title compound are 360° (144.4°, 104.1° and 111.5°) and 172.6°, respectively.

$$w = 1/\left[\sigma^2\left(Fo^2\right)+\left(0.0437P\right)^2+2.0512P\right]$$

where

$$P=\left(Fo^2+2Fc^2\right)/3$$

The Bi atom does not almost extend from the equatorial plan as we can view in **Figure 2**.

However, the (Bi-C(15), Bi-C(21), Bi-C(27)) distances are (2.215 (9), 2.238 (8) and 2.200 (9) Å) respectively. The Bi-O(1) and Bi-O(4) distances are (2.303 (6) Å) and (2.313 (6) Å) respectively. We note that, from **Figure 3** that the ligands of carboxylate groups are in cis-position compared to the phenyl group. In the other hand, intermolecular interaction between the bismuth atom and the two carbonyl groups forms the Van-Der-Waals bond.

The distances between valence-non-bonded Bi-O(2) and Bi-O(5) are respectively 2.814 (4) and 2.861 (3) Å (**Figure 3**) which indicate that oxygen atoms are weakly coordinated with the bismuth atom and form a cis conformation together.

This conformation apparently causes significant deviation of the C-Bi-C angles from 120° in the equatorial plan.

However, **Figure 4** shows, in the equatorial plane, that the two aromatic rings A and B are perpendicular to

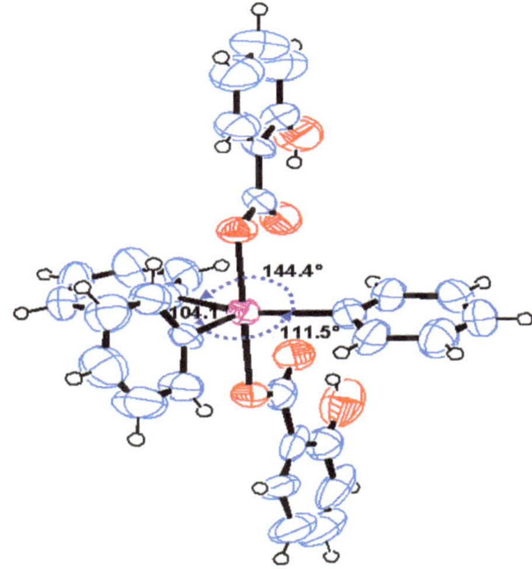

Figure 2. Three dimensional structure of compound I showing different angles in equatorial plan.

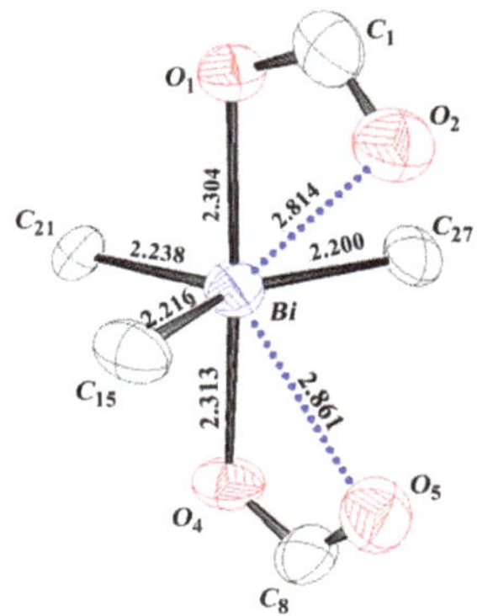

Figure 3. Bond lengths of atoms surrounding the bismuth atom.

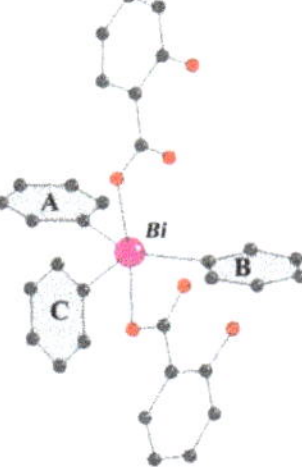

Figure 4. A, B and C rings compared to Bi-O bond apical position. A and B rings are in the same plan.

Bi-O bond while the third aromatic ring C is parallel to Bi-O. The axial positions are occupied by electronegative atoms and the equatorial ones by carbon atoms of the aromatic groups. The distortion of the coordination geometry at bismuth is mainly due to the restrictions imposed by the chelate rings.

The bond distances and valence angles of carbon (C15, C21 and C27) and oxygen (O2 and O4) atoms surrounding the bismuth central atom are shown in **Figure 5**.

3.2. Dihedral Angles

Torsion angles are reported in **Table 3**. We can see that, the carboxyl groups are coplanar (Bi-O(1)-C(1)-O(2) and Bi-O(4)-C(8)-O(5)). These two dihedral angles are about -8.01° and -1.83° respectively (**Table 3**).

3.3. Hydrogen Bonds

The bond distance between hydrogen and carbon, hydrogen and oxygen noted X-H (Å) and the possible intramolecular interactions by hydrogen bonds noted X-H (Å) of molecule **I** is shown in **Table 4**.
These different hydrogen bonds showed in **Figure 6** are responsible of the molecular packing in the unit cell.

3.4. Crystal Packing

Hydrogen bonds played crucial roles in the crystal packing. **Figure 7** shows the presence of two molecules in the unit cell $(Z = 2)$ which corresponds to the centrosymmetric triclinic space group $P1$ with two general positions (x, y, z) and $(-x, -y, -z)$.

4. Conclusions

The new pentavalent triphenylbismuth bis (salicylate) was synthesized with a good yield and its crystal structure was determined by X-ray diffraction analysis at room temperature. This work shows that coordination by the salicylate group is more stable, when compared to pentaphenylbismuth.

This is caused by rigidity of the ligand between the bismuth and oxygen atoms, and the presence of hydrogen bonding and the Van-Der-Waals bonding. The three-di-

Table 3. Dihedral angles in degrees with e.s.d.'s given in parenthesis.

Dihedral angles (°)	
O4-Bi-O1-C1	−148.12
C15-Bi-O1-C1	75.50
C21-Bi-O1-C1	179.52
C27-Bi-O1-C1	−68.82
O1- Bi-O4-C8	155.21
C15-Bi-O4-C8	−68.34
C21-Bi-O4-C8	−172.36
C27-Bi-O4-C8	76.14
O1-Bi-C15-C16	−101.51
O1-Bi-C15-C20	77.26
O4-Bi-C15-C16	83.58
O4-Bi-C15-C20	−97.65
C21-Bi-C15-C16	169.89
C21-Bi-C15-C20	−11.34
C27-Bi-C15-C16	−11.01
C27-Bi-C15-C20	167.76
O1-Bi-C21-C22	−165.63
O1-Bi-C21-C26	10.43
O4-Bi-C21-C22	18.32
O4-Bi-C21-C26	−165.61
C15-Bi-C21-C22	−72.63
C15-Bi-C21-C26	103.44
C27-Bi-C21-C22	107.94
C27-Bi-C21-C26	−76.00
O1-Bi-C27-C32	−76.02
O4-Bi-C27-C28	−85.34
O4-Bi-C27-C32	96.72
C21-Bi-C27-C28	−171.43
C21-Bi-C27-C32	10.62
Bi-O4-C8-O5	−1.83
Bi-O4-C8-C9	178.75
O1-C1-C2-C3	177.38
O1-C1-C2-C7	−4.16
O2-C1-C2-C3	1.62
O2-C1-C2-C7	−179.92
C1-C2-C3-O3	3.34
C1-C2-C3-C4	−179.12
C7-C2-C3-O3	−175.16
C7-C2-C3-C4	2.37
C1-C2-C7-C6	178.78
C3-C2-C7-C6	−2.81
O3-C3-C4-C5	175.31
C2-C3-C4-C5	−2.29

Continued

C3-C4-C5-C6	2.76
C4-C5-C6-C7	−3.45
C5-C6-C7-C2	3.39
O4-C8-C9-C10	−177.60
O4-C8-C9-C14	2.98
O5-C8-C9-C10	3.00
C8-C9-C10-O6	2.49
C8-C9-C10-C11	−179.81
C14-C9-C10-O6	−178.08
C14-C9-C10-C11	−0.37
C8-C9-C14-C13	−179.33
C10-C9-C14-C13	1.26
O6-C10-C11-C12	177.42
C9-C10-C11-C12	−0.40
C10-C11-C12-C13	0.29
C11-C12-C13-C14	0.60
C12-C13-C14-C9	−1.36
Bi-C15-C16-C17	−179.16
C20-C15-C16-C17	2.17
Bi-C15-C20-C19	178.95
C16-C15-C20-C19	−2.24
C15-C16-C17-C18	−0.17
C16-C17-C18-C19	−1.73
C17-C18-C19-C20	1.64
C18-C19-C20-C15	0.29
Bi-C21-C22-C23	175.91
C26-C21-C22-C23	−0.20
Bi-C21-C26-C25	−176.08
C22-C21-C26-C25	−0.05
C21-C22-C23-C24	0.53
C22-C23-C24-C25	−0.62
C23-C24-C25-C26	0.37
C24-C25-C26-C21	−0.04
Bi-C27-C28-C29	178.00
Bi-C27-C32-C31	−178.10

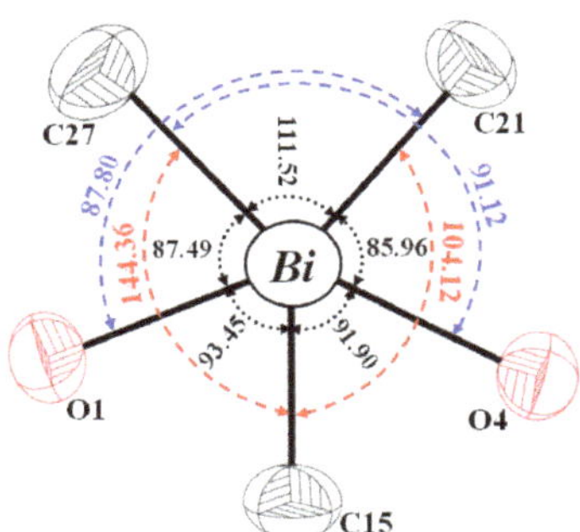

Figure 5. Bond distances and valence angles around the bismuth atom.

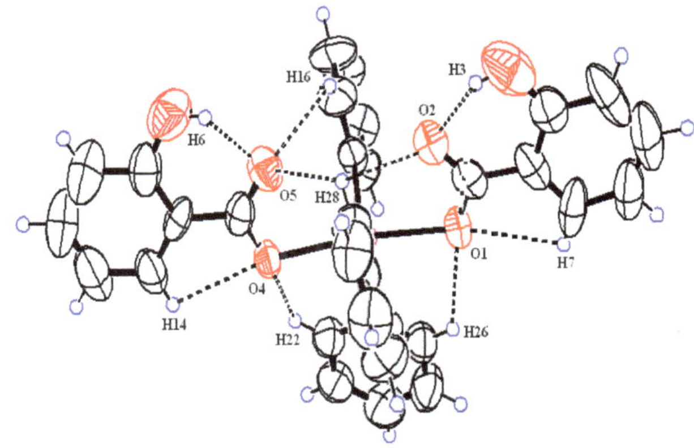

Figure 6. Hydrogen bonds representation ensuring the molecular packing in the unit cell of compound I.

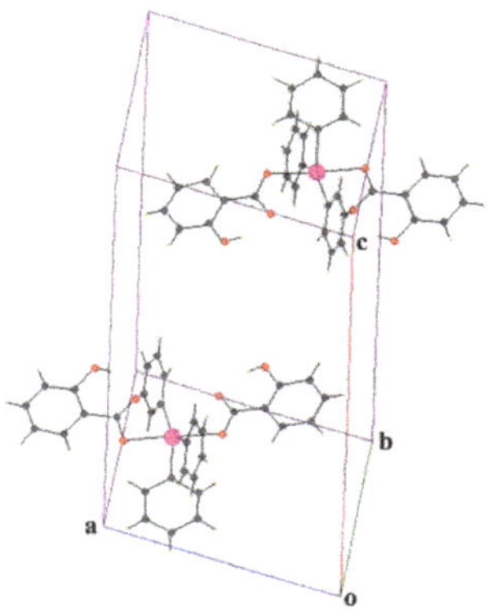

Figure 7. Molecular packing in the unit cell showing two molecules in symmetric positions (*x, y, z*) and (−*x*, −*y*, −*z*). Hydrogen atoms are shown as small spheres for clarity.

Table 4. Hydrogen bonds X-H and X···H of molecule I.

Distance of virtual bond X···H (Å)		Distance of actual bond X-H (Å)	
H3…O2	1.822	O3-H3	0.820
H6…O5	1.798	O6-H6	0.820
H7…O1	2.533	C7-H7	0.929
H14…O4	2.513	C14-H14	0.930
H16…O5	2.556	C16-H16	0.931
H22…O4	2.369	C22-H22	0.930
H26…O1	2.453	C26-H26	0.929
H28…O2	2.777	C28-H28	0.930
H28…O5	2.636		

bonding and the Van-Der-Waals bonding. The three-dimensional representation of the molecule shows the axial position of the two salicylate groups and the equatorial position of the three phenyls. This shows that the crystal structure of triphenylbismuth bis (Salicylate) pentavalent has a shape distorted trigonal-bipyramid.

REFERENCES

[1] In: H. Suzuki and Y. Matano, Eds., *Organobismuth Chemistry*, Elsevier, Amsterdam, 2001.

[2] In: H. Yamamoto and H. Ishihara, Eds., *Acid Catalysis in Modern Organic Synthesis*, Wiley-VCH, Weinheim, 2008.

[3] H. Gaspard-Iloughmane and C. Le Roux, "Bismuth(III) Triflate in Organic Synthesis," *European Journal of Organic Chemistry*, Vol. 2004, No. 12, 2004, pp. 2517-2532.

[4] R. M. Hua, "Recent Advances in Bismuth-Catalyzed Organic Synthesis," *Current Organic Synthesis*, Vol. 5, No. 1, 2008, pp. 1-27.

[5] R. M. A. Pinto, J. A. R. Salvador, C. Le Roux and J. A. Paixão, "Bismuth(III) Triflate-Catalyzed Direct Conversion of Corticosteroids into Highly Functionalized 17-Ketosteroids by Cleavage of the C17-Dihydroxyacetone Side Chain," *Journal of Organic Chemistry*, Vol. 74, No. 21, 2009, pp. 8488-8491.

[6] M. Bao, E. Hayashi and S. Shimada, "Cationic Organobismuth Complex with 5,6,7,12-Tetra Hydrodibenz [c,f][1,5] Azabismocine Framework and Its Coordination Complexes with Neutral Molecules," *Organometallics*, Vol. 26, No. 7, 2007, pp. 1816-1822.

[7] A. Miloudi, D. El-Abad, G. Boyer, J.-P., Galy and J.-P., Finet, "Synthesis, Structure and Reactivity of Triphenylbismuth Bis(2-Thiophenecarboxylate)," *Main Group Metal Chemistry*, Vol. 24, No. 11, 2001, pp. 767-774.

[8] M. Chovancová, R. Jambor, A. Růžička, R. Jirásko, I. Císařová and L. Dostál, "Synthesis, Structure, and Reactivity of Intramolecularly Coordinated Organoantimony and Organobismuth Sulfides," *Organometallics*, Vol. 28, No. 6, 2009, pp. 1934-1941.

[9] P. Simon, F. de Proft, R. Jambor, A. Ruzicka and L. Dostál, "Monomeric Organoantimony(I) and Organobismuth(I) Compounds Stabilized by an NCN Chelating Ligand: Syntheses and Structures," *Angewandte Chemie International Edition*, Vol. 49, No. 32, 2010, pp. 5468-5471.

[10] A. Soran, H. J. Breunig, V. Lippolis, M. Arca and C. Silvestru, "Syntheses, Solid-State Structures, Solution Behavior of Hypervalent Organobismuth(III) Compounds $[2\text{-}(Et_2NCH_2)C_6H_4]_nBiX_{3-n}$ and DFT Characterization of $[2\text{-}(Me_2NCH_2)C_6H_4]_nBiX_{3-n}$ [X = Cl, Br, I; n = 1–3]," *Journal of Organometallic Chemistry*, Vol. 695, No. 6, 2010, pp. 850-862.

[11] A. Miloudi, D. ElAbad, G. Boyer, J.-P., Galy, J.-P., Finet and S. Didier, "Reactivity of 2-Aminothiazole and 2- or 6-Aminobenzothiazole Derivatives towards the Triphenylbismuth Diacetate/Catalytic Copper Diacetate Phenylation System," *European Journal of Organic Chemistry*, Vol. 2004, No. 7, 2004, pp. 1509-1516.

[12] T. Arnauld, D. H. R. Barton and E. Doris, "The Chemistry of Pentavalent Organobismuth Reagents. Part 14. Recent Advances in the Copper-Catalyzed Phenylation of Amines," *Tetrahedron*, Vol. 53, No. 12, 1997, pp. 4137-4144.

[13] E. Nonius, "CAD-4 Express Software," Delft, 1996.

[14] Z. Otwinowski and W. Minor, "Processing of X-Ray Diffraction Data Collected in Oscillation Mode," *Methods in Enzymology*, Vol. 276, 1997, pp. 307-326.

[15] L. J. Farrugia, "WinGX Suite for Small-Molecule Single-Crystal Crystallography," *Journal of Applied Crystallography*, Vol. 32, 1999, pp. 837-838.

[16] G. M. Sheldrick, "A Short History of SHELX," *Acta Crystallographica*, Vol. 64, No. 1, 2008, pp. 112-122.

[17] L. J. Farrugia, "ORTEP-3 for Windows—A Version of ORTEP-III with a Graphical User Interface (GUI)," *Journal of Applied Crystallography*, Vol. 30, 1997, pp. 565-567.

[18] E. Prince and A. J. C. Wilson, "International Tables for X-ray Crystallography," 2nd Edition, Kluwer Academic Press, Boston, 1992.

Kinetics and Mechanism of Spontaneous Crystallization of Potassium Nitrate from Its Supersaturated Aqueous Solutions

Oleg D. Linnikov*, Irina V. Rodina, Igor G. Grigorov, Evgeniy V. Polyakov
Institute of Solid State Chemistry, Ural Branch of the Russian Academy of Sciences, Ekaterinburg, Russia

ABSTRACT

Kinetics of spontaneous crystallization of potassium nitrate from its supersaturated aqueous solutions has been studied simultaneously by electrical conductance and optical transmittance methods. It was found that spontaneous crystallization of potassium nitrate was accompanied by aggregation of crystals. Growth of salt crystals was in the kinetic mode of the growth process, and was described by the equation of the first order regarding supersaturation of solution. The mechanism of aggregation and intergrowth of crystals during bulk crystallization via formation of nucleus-bridges between crystals found earlier for several salts was confirmed. Specific surface energy of potassium nitrate was evaluated on the basis of the above mechanism of aggregation and intergrowth of crystals. The established value of the specific surface energy was reasonable and agreed satisfactorily with the available literature data. Examination of crystal deposit after completion of crystallization allowed detecting crystal agglomerates of freakish and irregular forms, which may be considered as the direct confirmation of the above mechanism of intergrowth of crystals. Kinetics of crystallization, aggregation and size distribution of salt crystals after completion of crystallization have been satisfactory described by the earlier proposed model of the crystallization process. An excellent agreement was established between the experimental data on potassium nitrate solubility in aqueous solutions found in the present work and those available in the literature.

Keywords: Potassium Nitrate; Aggregation; Crystallization; Solubility; Activation Energy; Crystal Growth; Intergrowth; Agglomerates

1. Introduction

Very scarce information about crystallization kinetics of potassium nitrate in aqueous solutions is available in the literature. There are only a few works where crystallization of potassium nitrate from aqueous solutions was studied [1-3]. So, Helt and Larson [1] investigated the crystallization of KNO_3 using MSMPR crystallizer. The operation temperature was from 10°C to 25°C, and a degree of supersaturation was varied within 0.005 - 0.01. The order of the crystal growth process was found to be 1 and the activation energy of the growth process was 31 $kJ{\cdot}mol^{-1}$ [1]. The crystal growth rate ranged in the interval $(3.5 - 10.7)\times10^{-8}\,m\cdot s^{-1}$. Later, Graber *et al.* [2] studied the growth of seed crystals of sodium and potassium nitrates from aqueous solutions in a perfectly stirred batch crystallizer. It was also found that the order of the crystal growth process was 1 for both salts. The maximum crystal growth rate was directly proportional to the seed crystal size. The activation energy of the growth process for potassium nitrate was 24.512 $kJ{\cdot}mol^{-1}$, and the process was under diffusion control [2].

The growth rate of different faces of single potassium nitrate crystals from aqueous solution was studied by Lacmann *et al.* with the aid of optical microscopical method [3]. For some crystals, the surface morphology was determined by means of light microscopy, scanning electron microscopy, Michelson interferomtry or scanning force microscopy [3]. Crystals were grown under constant supersaturation at 15°C - 20°C. The relative supersaturation was in the range from 0.3% to 1.8%. Intense dispersion of the crystal growth rates was observed. For all crystal faces, the order of the crystal growth process was between 1.15 and 1.60 [3]. Examination of the surface morphology showed that crystal faces had a flat structure; no significant objects were found.

The present work deals with spontaneous crystallization of potassium nitrate from its supersaturated aqueous solutions. The objective of this work was to study (*in situ*) kinetics of spontaneous crystallization of potassium nitrate from its supersaturated aqueous solutions using si-

*Corresponding author.

multaneously two methods of measurement: 1) electrical conductivity; and 2) optical transmittance of a supersaturated solution in salt crystallization.

2. Experimental

2.1. Experimental Setup

The experiments were carried out on a laboratory setup with a glass crystallization vessel of 250 cm^3 capacity. The volume of a supersaturated solution of KNO_3 was 165 cm^3. The laboratory setup was analogous to that used in [4], and it allowed to conduct simultaneously and automatically computer records of electrical conductivity W and optical transmittance I of the supersaturated solution during salt crystallization. The concentration of the salt during crystallization was determined from electrical conductivity of the solution using calibration curves. The principal diagram of the laboratory setup is shown in **Figure 1**.

2.2. Experimental Procedure

Supersaturation in investigated salt solution was created by thermic method. For this, supersaturated solution of KNO_3 in the crystallization vessel was heated on 5˚C - 10˚C above the saturation temperature in order to dissolve small salt crystals, and then it was stayed at this temperature about 0.5 h. After that the solution was cooled to the temperature below the saturation temperature. When the required temperature was achieved, a magnet stirrer was turned on, and spontaneous crystallization of KNO_3 started after a certain time (induction period). Simultaneously, the electrical conductance and optical transmittance of the solution during salt crystallization were recorded automatically.

After the crystallization was completed, the stirrer was switched off and a variation in optical transmittance of the solution was registered during sedimentation of formed potassium nitrate crystals. Analogous curves of variation of optical transmittance of solutions during sedimentation are used in photo-sedimentation method for determination of particle-size distributions of powders [5]. Therefore, the obtained curve also was used to calculate the particle-size distribution of the formed salt crystals, and for determination of the optical absorption constant K_c of the salt [5].

After sedimentation, the most part of the solution was decanted, and salt crystals were rapidly (during 3 - 5 s) filtered under vacuum-suction using a ceramic filter, and then on the filter immediately were washed with ethanol saturated by KNO_3. Then salt crystals on the filter were air-dried under vacuum-suction. These operations made it possible to remove remainders of the solution from crystals and prevent their intergrowth during air-drying. The crystalline product formed in each run was imaged with optical and SE microscopy.

Two liquid samples (5 ml) were collected from the solution. The first probe was taken from the supersaturated solution at the beginning of the crystallization process, and the second one-after the end of the crystallization. The samples were evaporated and dried at 80˚C, and then weighed. The salt concentrations in the solution were subsequently calculated from the mass of anhydrous solid and the volume (5 ml) of samples.

In the process of crystallization, the solution was kept (thermostated) in the crystallization vessel at constant temperature (± 0.1˚C) and vigorously agitated with a magnetic stirrer. The rotation frequency of the stirrer was 450 rpm. The error in optical transmittance measurements was ± 1.5%; the root-mean-square deviation of the measured electrical conductivity was 0.2%.

The solutions were prepared using analytical grade salt and distilled water. For additional purification, the solutions were recrystallized by thermal evaporation. Experiments were carried out at temperature 16.3˚C - 39.6˚C in the range of $C/C_0 = 1.01 - 1.06$.

The experimental data were mathematically treated using the following equations [4]:

$$-\frac{dC}{d\tau} = \frac{S\rho_T}{v} K\left(C - C_0\right)^z \quad (1)$$

$$W = a + bC \quad (2)$$

$$D_0 = \log\frac{I_0}{I} = K_c S \quad (3)$$

$$N = \frac{1}{\pi}\left(\frac{D_0 \Psi}{K_c}\right)^3 \left[\frac{\rho_T}{6\left(C_i - C\right)v}\right]^2 \quad (4)$$

$$\bar{d} = \left[\frac{6\left(C_i - C\right)v}{\pi \cdot N \cdot \rho_T}\right]^{1/3} \quad (5)$$

where $\bar{d}$ is the average diameter of salt crystals formed in the solution (m); τ is the time (s); ρ_T is the density of salt crystals $\left(\mathrm{kg \cdot m^{-3}}\right)$; S is the total surface area of growing salt crystals (m^2); v is the volume of the solution $\left(\mathrm{m^3}\right)$; C_0 is the concentration of the saturated salt solution $\left(\mathrm{kg \cdot m^{-3}}\right)$; C is the concentration of the supersaturated salt solution at the moment of time τ $\left(\mathrm{kg \cdot m^{-3}}\right)$; C_i is the initial concentration (before crystallization, at $\tau = 0$) of the supersaturated salt solution $\left(\mathrm{kg \cdot m^{-3}}\right)$; K is the average coefficient of the growth rate of salt crystals; z is the order of the salt crystal growth process (usually $z = 1$ - 2); W is the electrical conductivity of the salt solution in a relative scale of measurement (mV); a, b are the regression coefficients; D_0 is the optical density of the salt solution in the course of crystallization; I_0, I are the optical transmittance of the salt solution at the initial moment of time and at moment of time τ, respectively (%); K_c is the optical absorption constant of the salt (m^{-2}); N is the total number of growing salt crystals in the solution; Ψ is the spherical coefficient of salt crystals [6].

Formula (1) is the well-known empirical equation, which describes the rate of salt crystallization from supersaturated solutions after the end of the induction period when nucleation of salt crystals in the solution is usually completed [2,4,7-9]. Equation (2) describes electrical conductivity of salt solutions in the examined concentration range. Formula (3) represents the Wagner's law [5], according to which the optical density of a suspension with a small content of the solid phase is directly proportional to the total surface area S of solid particles (here, the total surface area of growing salt crystals) contained in the solution. This law is valid when the size of solid particles exceeds 2 μm. In every run of this study, the Wagner's law was valid. This was verified in special runs. The optical absorption constant K_c in Equation (3) was determined at sedimentation of formed salt crystals after the crystallization was completed.

Equations (4) and (5) make it possible to calculate the total number and the average diameter of the crystals formed during crystallization [4]. So, if m is the mass of salt precipitated from solution (kg) and N is the total number of salt crystals in the solution, then the average mass m_c of one salt crystal is $m_c = m/N$ and the average volume of one salt crystal v_c is $v_c = m_c/\rho_T$. Therefore, under assumption that the shape of the crystal is nearly spherical, its average diameter is determined by the equation

$$\bar{d} = (6 \cdot v_c/\pi)^{1/3} \quad (6)$$

From this, the average surface area of one salt crystal $\bar{S}$ is

$$\bar{S} = \pi \bar{d}^2 = \pi (6 \cdot v_c/\pi)^{2/3} \quad (7)$$

Expressing v_c in terms of a change in the solution concentration during crystallization and the total number N of formed salt crystals, we obtain formula (5).

Often in the description of processes occurring in solutions (crystallization, sedimentation, etc.) it is convenient to consider crystals as ball-shaped particles. However, the shape of most crystals differs sharply from a sphere. In this case a correction factor (a spherical coefficient) should be introduced [6]:

$$\Psi = \frac{S_s}{S_c} \quad (8)$$

where S_c is the surface area of a crystal (m^2); S_s is the surface area of a sphere calculated under condition that the volume of the sphere equals to the volume of the crystal (m^2).

The spherical coefficient Ψ makes it possible to consider crystals of different shapes as ball-shaped particles. With account of the fact that $S = \bar{S} \cdot N$, Equation (3) may be transformed by using formulas (5)-(8) into

$$D_0 = \log\frac{I_0}{I} = K_c S = K_c \frac{N\pi\bar{d}^2}{\Psi} = K_c \frac{N\pi}{\Psi}\left[\frac{6(C_i - C)v}{\pi \cdot \rho_T \cdot N}\right]^{2/3} \quad (9)$$

Then we derive Equation (4) from expression (9).

Integration of Equation (1) for the simplest case of $z = 1$ and the subsequent joint solution with Equations (2) and (3) gives the average coefficient of the growth rate of salt crystals

$$K = -\frac{K_c v}{D_0 \rho_T (\tau - \tau_0)} \ln\frac{C - C_0}{C_i^* - C_0} \quad (10)$$

where τ_0 is the time, after which only crystal growth process occurs in the solution (nucleation of crystals is completed or negligible); C_i^* is the concentration of the supersaturated salt solution at the moment of time τ_0 (kg·m^{-3}).

As was mentioned above, the crystal-size distribution was found when the crystallization was completed using curves of variation of optical transmittance of solutions during sedimentation of formed salt crystals. The diameters of the formed crystals during their sedimentation were calculated using the equation [5,6]:

$$d_i = \sqrt{\frac{18\mu H}{0.843 g(\rho_T - \rho)\tau_i \log(\Psi/0.065)}} \quad (11)$$

where μ is the dynamic viscosity of the saturated salt solution (Pa·s); ρ is the density of the saturated salt solution (kg·m^{-3}); H is the sedimentation height (see **Figure 1**) of crystals (m); τ_i is the time required for sedimentation of crystals having the diameter $d_i\,(\text{s})$. The viscosity and density of the saturated salt solutions were measured in special runs using an efflux viscometer and a set of areometers.

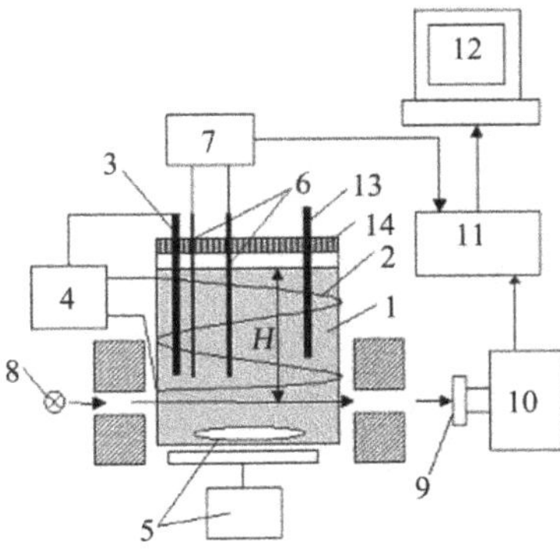

Figure 1. Principal diagram of a laboratory setup: 1—crystallization vessel containing supersaturated salt solution; 2—electric heater; 3—thermometer probe; 4—thermo regulator; 5—magnetic stirrer; 6—platinum conductometric electrodes; 7—electrical conductance meter; 8—light; 9—light-sensitive device; 10—amplifier; 11—analog-digital converter; 12—computer; 13—thermometer; 14—cover; H-sedimentation height.

3. Results and Discussion

3.1. Spherical Coefficient

To determine the spherical coefficient, 514 images of potassium nitrate crystals formed in the present experiments were measured. We could not make microphotographs of crystals in all dimensions and, therefore, we had no possibility to measure the size (height) of formed potassium nitrate crystals perpendicularly to a plane of photos. So for its evaluation we considered the parameters of the unit cell for a modification of KNO_3 with orthorhombic structure. There is a following basis for this. First, although in reality the crystallographic cell does not define the crystal shape of crystal, but relations between dimensions of a crystal often are approximately like in the cell (consider, for example, such salts like NaCl or KCl). Second, at aggregation of crystals at bulk crystallization there is always a morphological selection of formed agglomerates of crystals. Most of aggregates have a correct form with low hydrodynamic resistance, and what is more, the most agglomerates look like monocrystals although in actual fact they are pseudo-monocrystals. Hence, for these agglomerates the relation between their dimensions is the same like for individual crystal.

At measurements the crystals were imagined as parallelepipeds, and a ratio between length and width of them was calculated. The ratio found was 1.73 and was close to its theoretical value 1.43 for the parameters of the unit cell for modification of KNO_3 with orthorhombic structure (a = 0.54119 nm, b = 0.91567 nm, c = 0.64213 nm [3], $b/c = 1.43$). As was mentioned above, it was not possible to measure the height of formed potassium nitrate crystals, so for calculations of the spherical coefficient, the ratio $a:b:c = 0.8:1.73:1$ was used, it was close to the above structural parameters. The calculated spherical coefficient was 0.77. This value was used in all further calculations. It should be noted, however, that, taking into account accepted above assumption about relations between dimensions of salt crystal, this calculated value of the spherical coefficient was approximate.

3.2. Initial Period of Spontaneous Crystallization

As was mentioned above, the initial supersaturation of KNO_3 solutions was very small. Therefore, spontaneous crystallization of potassium nitrate was accompanied by a long induction period. Typical experimental curves describing the variation of the optical transmittance I and electrical conductivity W of the salt solution in the course of the crystallization are shown in **Figure 2**.

As is seen from **Figure 2**, in the initial period of crystallization, the optical transmittance of the solution slightly changes, whereas the electrical conductivity remains unchanged during about 70 s. Then some dramatic change in the optical transmittance and electrical conductivity of the solution occurs after about 80 and 70 s, respectively. This corresponds to an induction period of the salt crystallization. Different duration of the induction period for I (about 80 s) and W (about 70 s) reflects different sensitivities of the employed measurement techniques.

3.3. Variation of the Average Diameter and the Total Number of Salt Crystals during Crystallization

Typical curves of the variation of the average diameter and the total number of salt crystals in solution during crystallization are shown in **Figure 3**. The curves were calculated from experimental data using Equations (4) and (5).

It is seen from **Figure 3** that the calculated dependences of the average diameter and the total number of salt crystals in the solution change non-monotonously with time during crystallization. According to these calculations in the initial period of the crystallization, the salt crystals are large, and their average diameter is about 140 μm. And there are only a few salt crystals in the solution. Then a drastic reduction in the average diameter of salt crystals takes place (in 100 s), and the total number of salt crystals in the solution sharply increases. But after about 120 s the average diameter of salt crystals

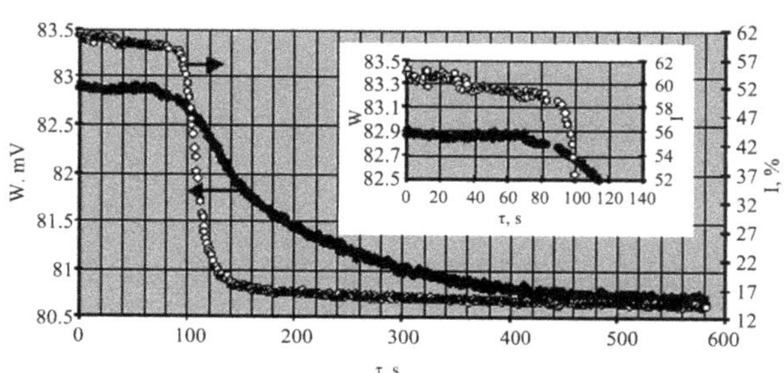

Figure 2. Electrical conductivity W (in a relative scale of measurement) and optical transmittance I of the salt solution during spontaneous crystallization (run 1) vs. time.

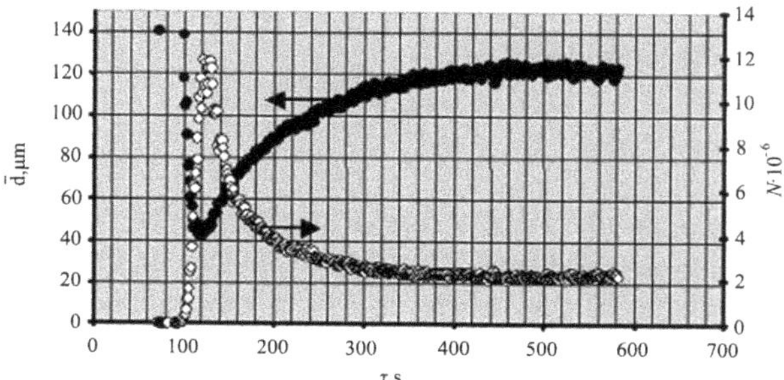

Figure 3. Calculated dependences of the average diameter $\bar{d}$ and the total number of salt crystals N during crystallization (run 1) vs. time.

begins to increase and the total number of salt crystals in the solution monotonously decreases.

Obviously, the initial parts of curves in **Figure 3** (to 120 s) are in contradiction with the modern experimental and theoretical data on crystallization, and this is a result of incorrect calculations with formulas (4) and (5). In fact, salt crystals cannot be so large at the beginning of the crystallization process and then to decrease during crystallization. The obtained dependences may be explained by understated values of optical transmittance of solution in the initial stage of salt crystallization (to 120 s). As was mentioned above, Equation (3) is valid when the size of solid particles in solution exceeds 2 μm. In this study, the size of salt crystals formed in the initial period of crystallization was very small and, as a result of diffraction of light, the optical transmittance of solutions was also very small (understated). Therefore, the calculations by means of expressions (4) and (5) were incorrect. A similar effect during spontaneous crystallization of salts was observed also in [4,10].

From **Figure 3** it is seen that after 120 s calculations with formulas (4) and (5) give "true" values of d and N. From this moment, the average diameter of salt crystals increases monotonously, and their total number in the solution decreases. This indicates that after 120 s only crystal growth process takes place in the solution (nucleation of crystals is completed or negligible). Therefore, these experimental points may be used to determine the growth parameters of salt crystals. Moreover, the reduction in the total number of salt crystals formed in the solution demonstrates that aggregation of crystals takes place during crystallization. N is largest at the beginning of the crystallization (after 120 s, **Figure 3**). Earlier, the aggregation of crystals was observed during spontaneous crystallization of other salts [4,10,11], but their induction periods were short and the largest total number of salt crystals was in solutions immediately after the completion of the induction period. Aggregation of potassium nitrate crystals during bulk crystallization was also found in [12].

3.4. Solubility of Potassium Nitrate

The obtained experimental results allow us to ascertain the data on the solubility of potassium nitrate in aqueous solutions. All obtained experimental results on the solubility and the data of other authors are visualized in **Figure 4**.

From **Figure 4** it is seen that the experimental results found in the present work agree excellently with the data of [13]. In the range 0˚C - 40˚C the solubility of KNO_3 in water and the density of the saturated salt solution may be described by the empirical equations (correlation coefficients 0.999):

$$C_0 = 132.17 + 2.52 \cdot t \cdot \ln t \qquad (12)$$

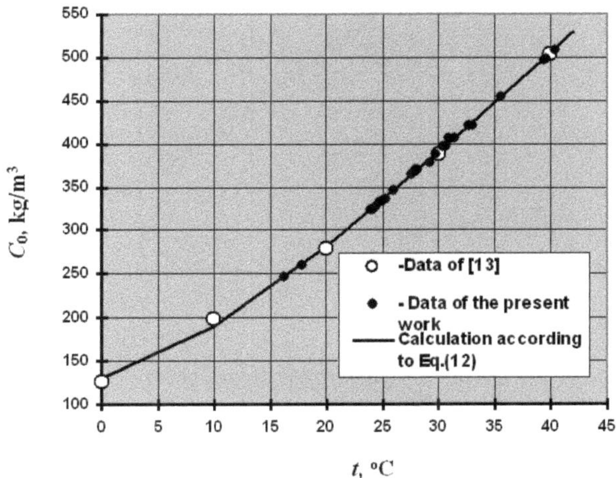

Figure 4. Effect of temperature on solubility of potassium nitrate in aqueous solution.

$$\rho = 1045.7 + 5.8297 \cdot t \qquad (13)$$

where t is the temperature (˚C).

3.5. Growth of Crystals

Integration of Equation (1) at $z = 1$ gives

$$\ln \frac{C - C_0}{C_i^* - C_0} = -\frac{S\rho_T}{v} K(\tau - \tau_0) \qquad (14)$$

and at $z = 2$

$$\frac{1}{C - C_0} = \frac{1}{C_i^* - C_0} + \frac{S\rho_T}{v} K(\tau - \tau_0) \qquad (15)$$

These two equations are usually used at determination of the order of crystal growth process at bulk crystallization. Examination of the obtained experimental data on crystallization of KNO_3 showed that in all runs the crystallization process gives best possible fit with Equation (14). The typical kinetic dependences of one and the same run in the coordinates of Equations (14) and (15) are displayed in **Figure 5**.

It is seen that the experimental points are well plotted on a straight line only in the coordinates of Equation (14) (see **Figure 5(a)**). In the coordinates of Equation (15) a curvilinear relation takes place (see **Figure 5(b)**). Calculation of the average coefficient of the growth rate of salt crystals with Equation (10) showed that K was constant during the crystallization process. This also confirms the first order $(z = 1)$ of Equation (1). The linear dependence of natural-logarithmic fitting for the obtained average coefficients K of the growth rate of KNO_3 crystals versus the inverse temperature $1/T$ according to the well-known Arrhenius equation is graphically presented in **Figure 6**.

It is seen a great scattering of experimental points is observed in **Figure 6**. Probably this is a result of some

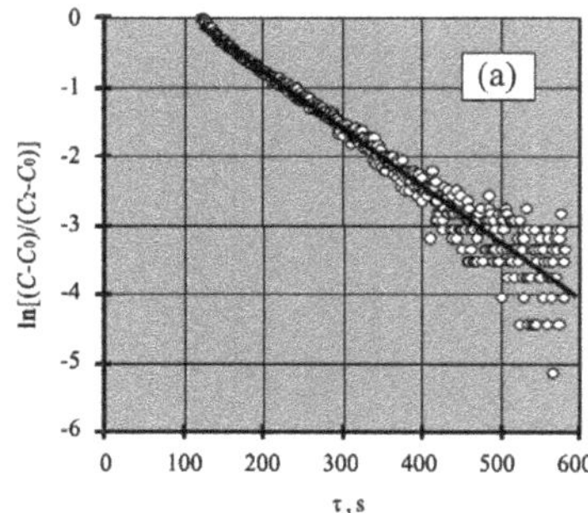

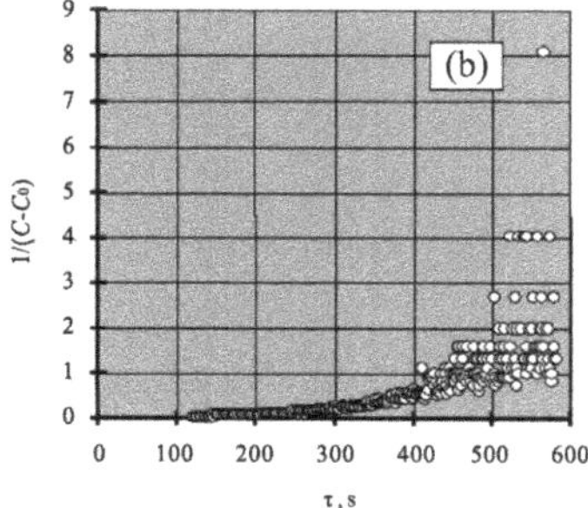

Figure 5. Experimental data on crystallization of one and the same run (run 1) in the coordinates of Equation (1) after integration at z = 1 - 2; a – z = 1 [Equation (14)]; b – z = 2 [Equation (15)].

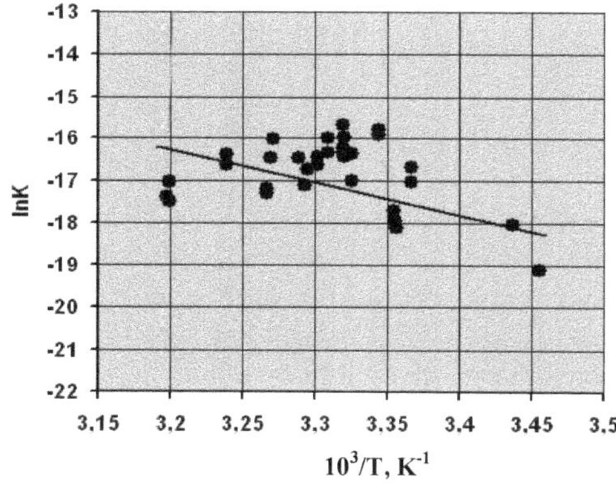

Figure 6. Natural-logarithmic fitting of the obtained values of the average coefficients *K* of the growth rate of KNO_3 crystals against the inverse temperature $1/T$ according to the Arrhenius equation.

experimental errors. It should be noted also that only the average coefficients K of the growth rate of KNO_3 crystals for every run are presented in **Figure 6**. But the total number of experimental points in all experiments was 4134. From **Figure 6** one may find that variations with temperature of the average coefficient of the growth rate of KNO_3 crystals is described by equation:

$$K = 3547.2 \cdot \exp\left[-\frac{(63.5 \pm 3.2) \times 10^3}{RT}\right] \quad (16)$$

where R is the universal gas constant ($J \cdot K^{-1} \cdot mol^{-1}$); T is the temperature (K).

The correlation factor of Equation (16) is 0.52. The Fisher test showed that Equation (16) adequately described the experimental data. From Equation (16) one can see that the activation energy of the growth process of KNO_3 crystals is significantly higher than those derived in [1,2], it is quite high and has reasonable value $63.5\ kJ \cdot mol^{-1}$ (the standard deviation of the activation energy is $1.61\ kJ \cdot mol^{-1}$). Such high value of the activation energy is characteristic of the kinetic mode of the growth process of salt crystals [8]. In work [2] it was found that crystal growth process was under diffusion control. Therefore, it may be concluded that growth of KNO_3 crystals in the experimental conditions of the present work was in the kinetic mode.

3.6. Kinetics of Aggregation and Mechanism of Intergrowth of Salt Crystals

The kinetic of aggregation of crystals during spontaneous crystallization of KNO_3 may be formally described by the Smoluchowski equation for coagulation of colloidal particles [14]. The differential form of the equation for aggregation of crystals looks as

$$-\frac{dN}{N^2} = K_a d\tau \quad (17)$$

where K_a is the aggregation constant.

A typical curve of variation of the total number of crystals in solution during the spontaneous crystallization of potassium nitrate in one of the runs is shown in **Figure 7**. The data are presented in the coordinates of transformed Equation (17) after integration (the integration was performed under the condition that K_a was constant). As can be seen, at the initial period of the crystallization the dependence is almost linear. This confirms that Equation (17) is valid during this stage of the crystallization. But further a curvilinear dependence takes place. Apparently, the reason of this is consecutive decreasing of the aggregation constant K_a due to falling of the supersaturation in solution. The analogous curves were also found in other runs and at crystallization of other salts [4,10,11].

The presented results in **Figure 7** allow to find the values of the aggregation constant K_a at the different time moments of the crystallization process. So, the experimental points in **Figure 7** with high accuracy may be approximated by the following empirical regression formula:

$$\frac{1}{N} = 10^{-9}\left(594.09434 - \frac{64497.476}{\tau}\right) \quad (18)$$

Differentiation of expression (18) allows to find the aggregation constant K_a at any moment of time. It should be noted, however, Equation (18) is an empirical fitting function $Y = f(X)$, where Y is $1/N$, and X is τ. So, parameters of expression (18) have no physical sense, and they should not be compared with Equation (17).

In works [4,15] was found that aggregation and intergrowth of salt crystals into agglomerates during bulk crystallization occur owing to the formation of nu-cleus-bridges between crystals according to the theory put forward by Polak [16-18]. These nucleus-bridges connect crystals in aggregates and form chemical bonds between them (**Figure 8**).

From Polak's theory follows that nucleation of nucleus-bridges is more advantageous in comparison with nucleation in the volume of solution and even nucleation on its own substrate. So, formation of nucleus-bridges in the narrow gap between two crystals takes place even in absence of nucleation in the volume of solution and on the surface of a crystal. For example, according to classic nucleation theory the work of formation of a nucleus on its own substrate is [8]

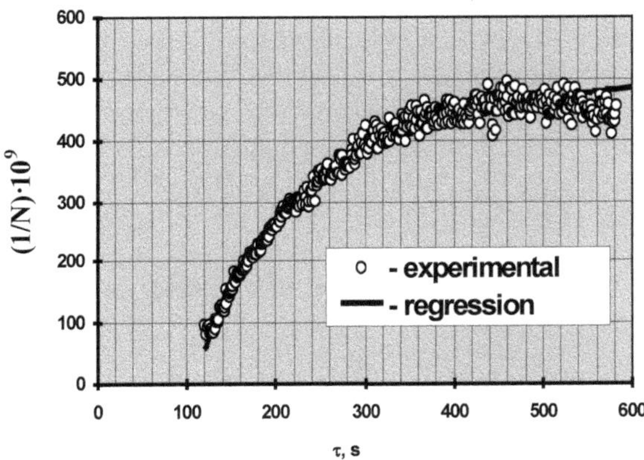

Figure 7. Aggregation kinetics of potassium nitrate crystals during spontaneous crystallization (run 1) in the coordinates of Equation (17) after integration.

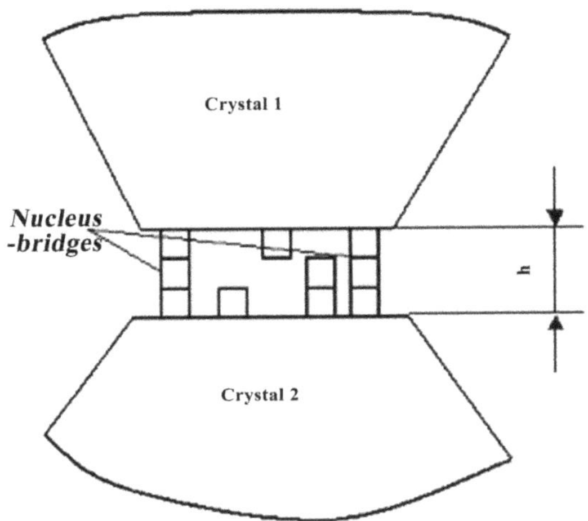

Figure 8. Scheme of intergrowth of two crystals proposed by Polak [16-18], where h is the height of a nucleus-bridge connecting two connivent crystals (m).

$$A_s = \frac{\pi M \sigma^2 a}{\rho_T RT \ln\left(\frac{C}{C_0}\right)} \tag{19}$$

where A_s is the work of formation of a nucleus in the form of a disc on its own substrate (J); M is the molar weight of a salt (kg·mol^{-1}); σ is the specific surface energy (J·m^{-2}); a is the parameter of crystal lattice (m).

Polak found that the work A_b of formation of a nucleus-bridge between the nearest faces of connivent crystals is [16-18]

$$A_b = \frac{4h^2\sigma^2}{h\rho_T \frac{RT}{M}\ln\left(\frac{C}{C_0}\right) + 2\sigma} \tag{20}$$

From this one can evaluate relations between the work of formation of a nucleus on its own substrate A_s and the work A_b of formation of a nucleus-bridge between the nearest faces of connivent crystals:

$$\frac{A_s}{A_b} = \frac{\pi a}{2h}\left[\frac{1}{2} + \frac{\sigma}{h\rho RT \ln\left(\frac{C}{C_0}\right)}\right] \tag{21}$$

From Equation (21) it follows that relation A_s/A_b is increasing when supersaturation of solution is decreasing. Let's make some evaluations. For example, for $CaSO_4 \cdot 2H_2O$ in [15] was found $h \approx 1\times10^{-9}\,\text{m}$, hence at $\sigma \approx 17\times10^{-3}\,\text{J/m}^2$, $\rho_T = 2330\,\text{kg/m}^3$, $a = 4.75\times10^{-10}\,\text{m}$, $T = 298\text{K}, C/C_0 = 3.05$ one can calculate $A_s/A_b \approx 2.34$.

Thus, formation of nucleus-bridges in the narrow gap between two crystals takes place even in absence of nucleation on surface of crystal.

According to results of [4,15] variation of the aggregation constant K_a during crystallization may be described by the expressions:

$$K_a = K_1\left\{1 - \exp\left[-D_3 \cdot \exp\left(-\frac{D_1}{\ln\left(\frac{C}{C_0}\right) + D_2}\right)\right]\right\} \approx K_1 D_3 \cdot \exp\left[-\frac{D_1}{\ln\left(\frac{C}{C_0}\right) + D_2}\right] \tag{22}$$

$$D_1 = \frac{4h\sigma^2 N_A M}{\rho_T R^2 T^2} \tag{23}$$

$$D_2 = \frac{2\sigma M}{h\rho_T RT} \tag{24}$$

$$D_3 = I_0 \tau_k \quad (25)$$

where K_1, D_1, D_2, D_3, I_0 are constants; h is the height of a nucleus-bridge connecting two connivent crystals (m); N_A is the Avogadro constant $\left(\text{mol}^{-1}\right)$; τ_k is the time during which crystals are in contact at their collision in solution (s). It should be noted that the aggregation constant K_a is a function of the aggregate size. But earlier it was shown that the dependence of K_a on the sizes of crystals should not be taken into consideration if a stirrer rate is constant and does not change during the crystallization process [4].

Values of the aggregation constant K_a, established from the experimental data, allow to verify Equation (22). So, in Equation (22) there are three unknown coefficients: D_1, D_2 and $K_1 \cdot D_3$. But at crystallization in isothermal conditions D_1 and $D_2 \approx \text{const}$. If agitation rate is constant and does not change during the crystallization process, hence, one can take $\tau_k \approx \text{const}$. Therefore, D_3 will be also approximately constant value, and $K_1 \cdot D_3 \approx$ const. After taking the log of expression (22) one can get the equation useful for following calculations:

$$\ln K_a = \ln\left(K_1 \cdot D_3\right) - \frac{D_1}{\ln \dfrac{C}{C_0} + D_2} \quad (26)$$

The known computer program Table Curve 2D v5.00 allows to find these unknown coefficients D_1, D_2 and $K_1 \cdot D_3$ for the linear dependence

$$\ln K_a - 1/\left[\ln\left(C/C_0\right) + D_2\right]$$

[please see Equation (26)]. Linearity of above equation is possible only for one found value of D_2. Therefore, this linearity may be considered like some evidence of validity of Equation (22). Then if from found values of D_1 and D_2 will be calculated reasonable values of h and σ, and σ will be near to the published data for this salt, then it will be additional evidence of validity of Equation (22).

Results of calculations according to Equation (26) are shown in **Figure 9**, where two typical obtained dependences are presented.

As can be seen, displayed in **Figure 9** linear dependences (correlation coefficients are in the range 0.75-0.99) confirm validity of Equation (22). Calculations with the help of Equations (23) and (24) give values of 2.19 $\text{mJ} \cdot \text{m}^{-2}$ for the specific surface energy (standard deviation -2.11 $\text{mJ} \cdot \text{m}^{-2}$, number of calculated points-4202) and 2.46×10^{-9} m for the height of nucleus-bridges (standard deviation, -0.67×10^{-9} number of calculated points-4202). Obviously, if one will take into account all procedures of these calculations, it is difficult to expect more precise values. However, the values obtained are reasonable and agree well with those found in [4,15,19]. For example, in two runs of [15] for spontaneous crystallization of KNO_3

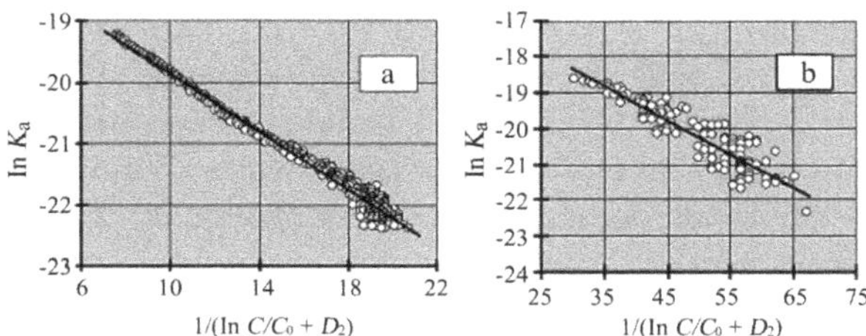

Figure 9. Aggregation constant versus degree of supersaturation of solution C/C_0 in the coordinates of Equation (26): a—run 1 and b—run 3.

from supersaturated aqueous-ethanol solutions σ was found to be 2.03 and $0.96 \text{mJ} \cdot \text{m}^{-2}$ and the height of nucleus-bridges was 1.83×10^{-9} and 2.34×10^{-9} m, respectively. It is interesting to note that respective substitutions in Equation (21) give $As/A_b \approx 7$.

The validity of Polak's mechanism of crystal intergrowth was also confirmed in spontaneous crystallization of KCl from supersaturated aqueous and aqueous-ethanol solutions [4], in bulk crystallization of $CaC_2O_4H_2O$ from supersaturated aqueous solutions (recalculation in [15] data of [20,21]), and in spontaneous crystallization of different salts from supersaturated aqueous-ethanol solutions: NaCl [19], $NaNO_3$, $CaSO_4 \cdot 2H_2O$, KNO_3, $NaNO_2$, K_2SO_4, $Ba(NO_3)_2$ [15].

3.7. Forms of Intergrowth of Crystals in Agglomerates

To obtain additional information about mechanism of intergrowth of crystals during bulk crystallization, the agglomerates of potassium nitrate crystals formed in the present runs were analyzed by means of optical and SE microscopy. So, the mechanism of intergrowth of crystals proposed by Polak [16-18] in case of its realization at bulk crystallization assumes that formation of nucleus-bridges connecting crystals occurs at the moment of collision of the crystals in solution (formation of nucleus-bridges is nucleation process, therefore, there is always some probability that formation of nucleus-bridges between crystals may take place immediately at their collision or during very short time interval when the crystals are near each other). Only this, in our opinion, may explain intergrowth of crystals at bulk crystallization. Really, according to results of [22] the contact time of crystals at their collision in a solution must be near to 0.05 s. It is clear that it is not enough for intergrowth of crystals by means of crystal growth, especially, for sparingly soluble salts having low growth rates of crystals. Besides, a thin layer of a solution always exists between approaching crystals. Therefore, when two nearest faces of the crystals grow towards each other, wedging pressure between these faces does not allow them to come into contact for intergrowth.

If intergrowth of crystals occurs almost instantly, then some agglomerates of crystals of very freakish form should be found in a crystal product after finish of the crystallization process. Of course, the number of them will be small, as hydrodynamic conditions of bulk crystallization process lead to the morphological selection of formed agglomerates of crystals and prevalence among them agglomerates of the correct form with low hydrodynamic resistance, right up to pseudo-monocrystals [23, 24]. Hence, detection of agglomerates of crystals of the irregular form will serve as additional evidence that the mechanism of intergrowth of crystals proposed by Polak takes place in the conditions of bulk crystallization from solutions.

In **Figure 10(a)** a few images of agglomerates of potassium nitrate crystals of the irregular shape, which had been found in a deposit after the end of crystallization, are shown. More detailed optical and SE microscopy images of formed large agglomerates of two crystals are shown in **Figures 10(b)-(d)**.

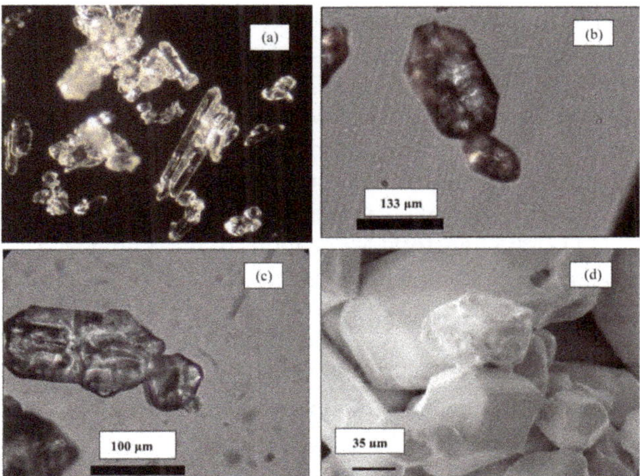

Figure 10. Optical and SE microscopy images (a)-(d) of potassium nitrate crystal agglomerates of irregular shape formed in runs of the present work.

It is clear that such agglomerates of irregular shape might be formed only, if the intergrowth of two crystals took place immediately at their collision in agitated solution. These images have been obtained after slight etching of a surface of crystals in 94% ethanol using ultrasound. This procedure allowed to destroy crystal pseudo-agglomerates, in which were only Van der Waals forces between crystals and there was no intergrowth due to chemical bonds.

Examination of published data showed that there are also many other works, where were found crystal agglomerates of irregular and freakish forms. For example, the intergrowth of potassium nitrate crystals during bulk crystallization was also found in [12].

It is interesting that in the photos presented in [12] one may see almost the same crystal agglomerates of potassium nitrate of irregular shape like in **Figure 10(a)** of the present work. In the photos of crystals of another salt $\left(K_4\left[Fe(CN)_6\right]\cdot 3H_2O\right)$ studied in [12] is also discovered agglomerates of irregular shapes. The agglomerates of freakish forms formed from relatively large crystals of potassium chloride may be found in **Figures 3** and **4** of work [25].

3.8. Model of Spontaneous Crystallization Process

The great interest for theory and practic of bulk crystallization present question about a possibility to predict variation of salt concentration and concentration of salt crystals, and also particle-size distribution of crystal product during and at the end of a crystallization process. Earlier in [26,27] a model of spontaneous crystallization was proposed. The model describes kinetics of the crystallization process after the end of the induction period of crystallization, when aggregation of salt crystals started and nucleation of new crystals was finished or was negligible. It is supposed that intergrowth of salt crystals during crystallization proceeds via the formation of nucleus-bridges between these crystals according to the above mechanism of crystal intergrowth. The model is described by a system of differential equations [26,27]:

$$-\frac{dC}{d\tau}=3.3\Psi^{-1}\left(\pi\cdot\rho_T\right)^{1/3}\cdot n^{1/3}\cdot K\left(C_i-C\right)^{2/3}\left(C-C_0\right)^z \quad (27)$$

$$\frac{dn}{d\tau}=-n^2\cdot v\cdot K_a \quad (28)$$

$$\frac{dn_1}{d\tau}=-2\cdot v\cdot K_a\cdot n_1\cdot n \quad (29)$$

$$\frac{dn_2}{d\tau}=2\cdot v\cdot K_a\left(0.5\cdot n_1^2-n_2\cdot n\right) \quad (30)$$

$$\frac{dn_3}{d\tau}=2\cdot v\cdot K_a\left(n_1\cdot n_2-n_3\cdot n\right) \quad (31)$$

$$\frac{dn_k}{d\tau}=v\cdot K_a\left(n_1\cdot n_{k-1}+n_2\cdot n_{k-2}+n_3\cdot n_{k-3}+\cdots+n_{k-1}\cdot n_1\right)-2\cdot v\cdot K_a\cdot n_k\cdot n \quad (32)$$

where n is the total concentration of growing salt crystals in the solution $\left(m^{-3}\right)$, $n=N\cdot v^{-1}$; n_1 is the concentration of initial salt crystals in the solution (m^{-3}), *i. e.* those crystals which were in the solution immediately after the beginning of the aggregation process, in particular, at this moment $n_1=n$; k is the number of crystals in a crystal aggregate; $n_2,n_3,\cdots,n_k$ are the concentrations of salt crystals (or rather crystal aggregates) consisting of two (n_2), three (n_3) and $k(n_k)$ initial salt crystals, respectively $\left(m^{-3}\right)$.

The first equation of systems (27)-(32) describes variation of a salt concentration in solution during crystallization. The equation takes into account the order of the growth process and the total surface area of growing salt crystals. The second equation of the system is transformed Equation (17). It describes the kinetics of aggregation of salt crystals during bulk crystallization. The aggregation constant K_a in this formula may be calculated with Equation (22). The others equations of systems (27)-(32) are transformed equations of the Smoluchowski theory [14]. They determine concentrations of crystal aggregates consisting of two (n_2), three (n_3) and more (n_k) initial salt crystals. The detailed derivation of systems (27)-(32) was given in [26]. It should be noted that here we introduce into the model some correction factor taking into account that most of salt crystals are not ball-shaped particles. This correction factor is the sphereсal coefficient already used above.

To find a crystal size distribution during or at the end of crystallization, Equation (22) and all equations of systems (27)-(32) must be solved simultaneously. Then, in the first approximation, diameters and weight fractions of formed crystal agglomerates may be found using equations [26,27]:

$$d_k = \left(\frac{6k^{2/3}(C_i - C)}{\pi \rho_T \sum_k n_k k^{2/3}} \right)^{1/3} \quad (33)$$

$$Q_k = \frac{100 n_k k^{2/3}}{\sum_k n_k k^{2/3}} \quad (34)$$

$$Q = \sum_k Q_k \quad (35)$$

where d_k is the average diameter of crystal agglomerates consisting of k initial salt crystals (m); Q_k is the weight fraction of crystal agglomerates consisting of k initial salt crystals in a crystal deposit of salt (%); Q is the total weight fraction of all crystal agglomerates consisting from 1 to k initial salt crystals in a deposit of salt (%). To estimate the crystal size distribution at the end of crystallization, C_0 instead of C must be used in Equation (33).

The typical results of calculations according to the above model for spontaneous crystallization of potassium nitrate studied in the present work are given in **Figures 11** and **12**. To establish the variations in the concentration C of the salt and the total number N of salt crystals in the solution during crystallization, the first two equations of systems (27)-(32) together with Equation (22) were solved simultaneously by a numerical method. For this the Runge-Kutta method and the found average crystal growth rate coefficients K_r and crystal aggregation parameters D_1, D_2, $K_1 \cdot D_3$ were used.

As is seen, **Figures 11** and **12** exhibit a satisfactory agreement between the corresponding experimental and theoretical data.

In order to estimate a crystal-size distribution at the end of the crystallization process, Equation (22) and Equations (33)-(35) together with full system of Equations (27)-(32) were also solved using the special computer program. The theoretical calculations were carried out for $k = 150$, *i.e.* for crystal agglomerates consisting of 150 initial salt crystals. Therefore, 2 + 150 equations of systems (27)-(32) and Equation (22) were solved simultaneously. Typical results of these calculations are displayed in **Figure 13**, where arrows point out the beginning of the theoretical curves.

From **Figure 13** it is seen that there are some differences between the experimental points and the theoretical curves. For large d this discrepancy is within experimenttal errors, whereas for small d the deviations are significant (see experimental points for small d, and arrows in **Figure 13**). In particular, the proposed model of spontaneous crystallization gives larger diameters of crystals for small d in comparison with those found experimentally (see arrows in **Figure 13**). This may be explained by secondary nucleation and breaking of salt crystals during crystallization. Nevertheless, the theoretical curves for the size distribution of potassium nitrate crystals, on the whole, are in satisfactory agreement with the experimental data.

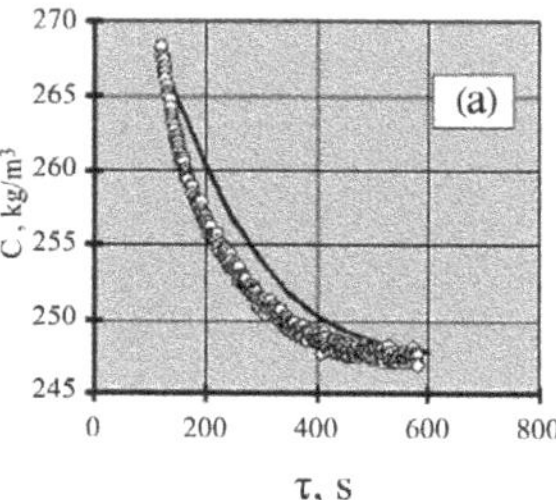

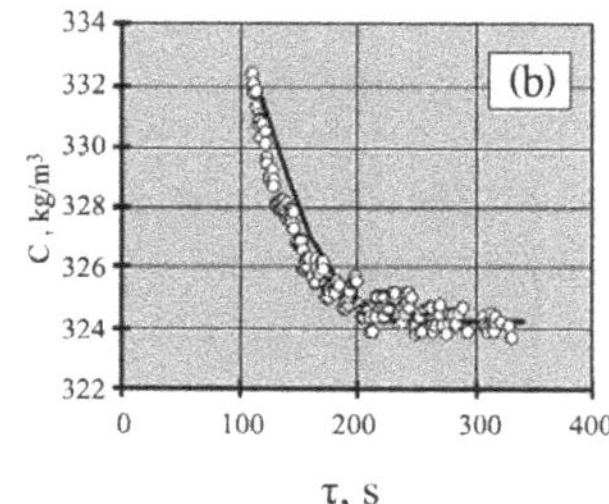

Figure 11. Variation of concentration C of potassium nitrate in solution during spontaneous crystallization: a—run 1; b—run 4; ○—experimental points; —theoretical calculation according to systems (27)-(32) and Equation (22).

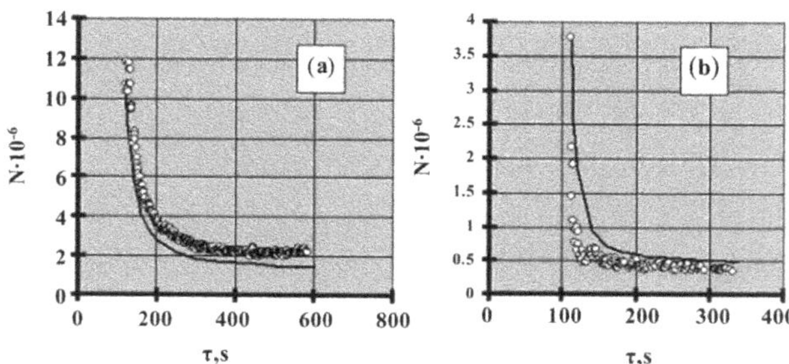

Figure 12. Variation of the total number N of salt crystals in solution during spontaneous crystallization of potassium nitrate: a—run 1; b—run 4; ○—experimental points; — theoretical calculation according to systems (27)-(32) and Equation (22).

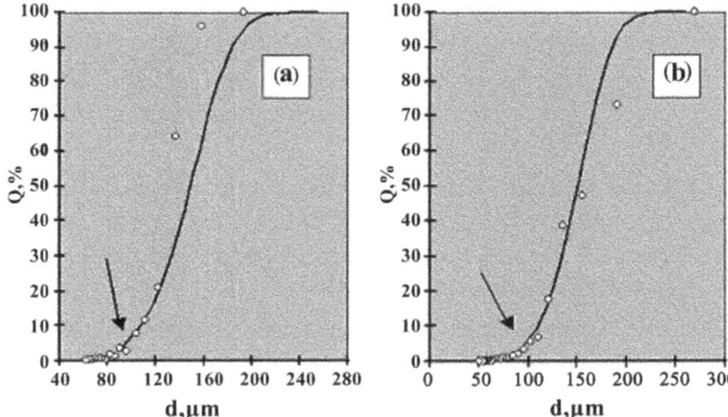

Figure 13. Size distribution of potassium nitrate crystals after the completion of spontaneous crystallization: a—run 1; b—run 4; ○—experimental points; —theoretical calculations according to systems (27)-(32), Equation (22) and Equations (33)-(35), where arrows point out the beginning of the theoretical curves; d is the diameter of potassium nitrate crystals (agglomerates).

4. Conclusion

Thus, the experiments performed showed that spontaneous crystallization of potassium nitrate from its supersaturated aqueous solutions is accompanied by aggregation of salt crystals. The growth of crystals during crystallization was described by the equation of the first order. The high value 63.5 $kJ \cdot mol^{-1}$ of the activation energy points out on the kinetic mode of the growth process.

Analysis of the experimental data on the aggregation kinetics showed that the intergrowth of crystals proceeded via formation of nucleus-bridges between crystals. This was confirmed by examination of equations of above mechanism on the obtained experimental data and by detection of crystal agglomerates of freakish and irregular shapes, which we consider like the direct confirmation of the mentioned mechanism of intergrowth of crystals. Besides, the specific surface energy of potassium nitrate evaluated on the basis of the above mechanism of aggregation and intergrowth of crystals had reasonable value, and agreed satisfactorily with the available literature data. This is additional evidence of validity of mechanism of intergrowth of crystals during bulk crystallization via formation of nucleus-bridges between crystals.

The earlier proposed model of spontaneous crystallization was also tested. It was found a satisfactory coincidence between the experimental data and theoretical curves on concentration of the salt and the total number of salt crystals in solution during spontaneous crystallization, and also between experimental data and theoretical curves of particle-size distribution of formed crystals after completion of the crystallization.

An excellent agreement was also established between the experimental data on solubility of potassium nitrate in aqueous solutions found in the present work and those available in the literature.

5. Acknowledgements

This work was financially supported by the Russian Foundation for Basic Research, Project No 05-08-01289-a.

REFERENCES

[1] J. E. Helt and M. A. Larson, "Effects of Temperature on the Crystallization of Potassium Nitrate by Direct Measurement of Supersaturation," *AIChE Journal*, Vol. 23, No. 6, 1977, pp. 826-828.

[2] T. A. Graber, M. E. Taboada, M. N. Alvarez and E. H. Schmidt, "Determination of Mass Transfer Coefficients for Crystal Growth of Nitrate Salts," *Crystal Research and Technology*, Vol. 34, No. 10, 1999, pp. 1269-1277.

[3] J. Rolfs, R. Lacmann and S. Kipp, "Crystallization of Potassium Nitrate (KNO_3) in Aqueous Solution I," *Journal of Crystal Growth*, Vol. 171, No. 1-2, 1997, pp. 174-182.

[4] O. D. Linnikov, "Spontaneous Crystallization of Potassium Chloride from Aqueous and Aqueous-Ethanol Solutions. Part 2: Mechanism of Aggregation and Coalescence of Crystals," *Crystal Research and Technology*, Vol. 39, No. 6, 2004, pp. 529-539.

[5] P. A. Kouzov, "Fundamentals of Analysis of the Dispersion Composition of Industrial Dusts and Powdered Materials," Chemistry, Leningrad, 1974.

[6] A. N. Verigin, I. A. Shchupliak and M. F. Mihalev, "Crystallization in Dispersible Systems," Chemistry, Leningrad, 1986, pp. 60-61.

[7] B. R. Smith and F. Sweet, "The Crystallization of Calcium Sulfate Dihydrate," *Journal of Colloid and Interface Science*, Vol. 37, No. 3, 1971, pp. 612-618.

[8] A. A. Chernov, "Modern Crystallography," Nauka, Moscow, 1980.

[9] J. W. Mullin, "Crystallization," 4th Edition, Butterworth-Heinemann, Oxford, 2001.

[10] O. D. Linnikov, "Kinetics of Bulk Crystallization of Salts from Aqueous Solutions of Ethanol," *Russian Journal of Applied Chemistry*, Vol. 74, No. 7, 2001, pp. 1106-1111.

[11] O. D. Linnikov, "Spontaneous Crystallization of Sodium Chloride from Aqueous-Ethanol Solutions. Part 1. Kinetics and Mechanism of the Crystallization Process," *Crystal Research and Technology*, Vol. 41, No. 1, 2006, pp. 10-17.

[12] L. N. Matusevich and K. N. Shabalin, "The Intensity of Stirring of a Solution and Size of Received Crystals," *Zhurnal Prikladnoi Khimii*, Vol. 25, No. 11, 1952, pp. 1157-1164.

[13] M. Tovbin and S. Krasnova, "The Stability of Supersaturated Salt Solutions," *The Journal of Physical Chemistry*,

Vol. 23, No. 7, 1949, pp. 863-870.

[14] M. Smoluchowski, "Versuch Einer Mathematischen Theorier der Koagulationskinetik Kolloider Lösungen," *Zeitschrift Fur Physikalische Chemie*, Vol. 92, 1917, pp. 129-168.

[15] O. D. Linnikov, "Mechanism of Aggregation and Intergrowth of Crystals during Bulk Crystallization from Solutions," *Crystal Research and Technology*, Vol. 43, No 12, 2008, pp. 1268-1277.

[16] A. F. Polak, "Hardening of Mono-Mineral Astringent Substances," Publishers of literatures on construction, Moscow, 1966.

[17] E. D. Shchukin, E. A. Amelina, R. K. Yusupov, V. P. Vaganov and P. A. Rebinder, "The Experimental Study of Influence of Solution Supersaturation and the Contact Time of Crystals on Intergrowth of Separate Crystals," *Soviet Physics—Doklady*, Vol. 213, No. 1, 1973, pp. 155-158.

[18] V. P. Vaganov, E. A. Amelina, R. K. Yusupov, E. D. Shchukin and P. A. Rebinder, "The Experimental Study of Formation of Crystal Contacts at Intergrowth of Separate Crystals," *Colloid Journal of the USSR*, Vol. 36, No. 3, 1974, pp. 436-442.

[19] O. D. Linnikov, "Spontaneous Crystallization of Sodium Chloride from Aqueous-Ethanol solutions. Part 2. Mechanism of Aggregation and Coalescence of Crystals," *Crystal Research and Technology*, Vol. 41, No. 2, 2006, pp. 138-144.

[20] A. S. Bramley, M. J. Hounslow and R. L. Ryall, "Aggregation during Precipitation from Solution. Kinetics for Calcium Oxalate Monohydrate," *Chemical Engineering Science*, Vol. 52, No. 5, 1997, pp. 747-757.

[21] M. J. Hounslow, A. S. Bramley and W. R. Paterson, "Aggregation during Precipitation from Solution. A Pore Diffusion-Reaction Model for Calcium Oxalate Monohydrate," *Journal of Colloid and Interface Science*, Vol. 203, No. 2, 1998, pp. 383-391.

[22] D. M. Levins and J. R. Glastonbury, "Particle-Liquid Hydrodynamics and Mass Transfer in a Stirred Vessel. Part 1," *Transactions of the Institution of Chemical Engineers*, Vol. 50, 1972, pp. 32-41.

[23] I. V. Melikhov and A. S. Kelebeev, "Coagulation Growth of Crystals of Barium Sulphate from Strong Supersaturated Solution," *Crystallography*, Vol. 24, No. 2, 1979, pp. 410-412.

[24] I. V. Melikhov, I. V. Miheeva and V. N. Rudin, "The Mechanism of Crystallization of Semihydrate of Calcium Sulphate in the Conditions Modelling Production of Phosphoric Acid by Hemihydrate Method," *Theoretical Foundation of Chemical engineering*, Vol. 19, No. 6, 1985, pp. 742-748.

[25] S. Sarig, C. L. Leci and N. Eidelman, "Agglomeration of Potassium Chloride Crystals from Supersaturated Aqueous Solutions," *Journal of Crystal Growth*, Vol. 47, No. 3, 1979, pp. 365-372.

[26] O. D. Linnikov, "Spontaneous Crystallization of Potassium Chloride from Aqueous and Aqueous-Ethanol Solutions. Part 3: Model of the Crystallization Process," *Crystal Research and Technology*, Vol. 41, No. 10, 2006, pp. 988-996.

[27] O. D. Linnikov, "Spontaneous Crystallization of Sodium Chloride from Aqueous-Ethanol Solutions. Part 3: Examination of a Model of the Crystallization Process," *Crystal Research and Technology*, Vol. 42, No. 8, 2007, pp. 758-765.

11

Lab Source Anomalous Scattering Using Cr $K\alpha$ Radiation

Sibi Narayanan, Devadasan Velmurugan*
Centre of Advanced Study in Crystallography and Biophysics, University of Madras,
Maraimalai (Guindy) Campus, Chennai, India

ABSTRACT

High-throughput crystallography requires a method by which the structures of proteins can be determined quickly and easily. Experimental phasing is an essential technique in determining the three-dimensional protein structures using single-crystal X-ray diffraction. In macromolecular crystallography, the phases are derived either by Molecular Replacement (MR) method using the atomic coordinates of a structurally similar protein or by locating the positions of heavy atoms that are intrinsic to the protein or that have been added (MIR, MIRAS, SIR, SIRAS, MAD and SAD). Availability of in-house lab data collection sources (Cu $K\alpha$ and Cr $K\alpha$ radiation), cryo-crystallography and improved software for heavy atom location and density modification have increased the ability to solve protein structures using SAD. SAD phasing using intrinsic anomalous scatterers like sulfur, chlorine, calcium, manganese and zinc, which are already present in the protein becomes increasingly attractive owing to the advanced phasing methods. An analysis of successful SAD phasing on three proteins, lysozyme, glucose isomerase and thermolysin based on the signal of weak anomalous scatterers such as sulfur atom and chloride ion have been carried out. This analysis also proves that even the anomalous signal provided or present naturally in a macromolecule is good enough to solve crystal structures successfully using lab source chromium-generated X-ray radiation.

Keywords: Anomalous Scattering; SAD Phasing; Lysozyme; Glucose Isomerase; Thermolysin

1. Introduction

The sequencing of complete genomes has spurred a growing need for structural information on the encoded gene products that has driven the development of high-throughput crystallography [1-7]. Anomalous scattering methods are currently widely used for phasing macromolecular structures [8]. Experimental phasing is an essential technique in determining the three-dimensional protein structures using single-crystal X-ray diffraction. The power of Single-wavelength Anomalous Diffraction (SAD) was illustrated even 25 years ago and the anomalous signal from six sulfur atoms was used to solve the structure of crambin, a 4.8 kDa protein [9,10]. In recent years, SAD has become the choice for phasing [11-13] after gaining advantage over the Multi-wavelength Anomalous Diffraction, MAD [14]. The SAD approach shares advantage with MAD that all measurements can be made on the same single crystal so that lack of isomorphism does not occur and a single data set at single wavelength is enough [15]. Availability of synchrotron radiation, cryo-crystallography and improved software for heavy atom location and density modification have increased the ability to solve protein structures using SAD. In the absence of synchrotron radiation, SAD method using rotating anode generator (Cu $K\alpha$ and Cr $K\alpha$) lab source data is mostly used to solve the phase problem in X-ray crystallography. Using high redundant data, sufficiently accurate anomalous data can now be collected in the laboratory using Cu $K\alpha$ and Cr $K\alpha$ radiations [16-20].

Longer wavelengths than Cu $K\alpha$ (1.54 Å) would produce a larger signal, but at the same time experimental difficulties may increase as does the noise level in the data [19]. It has been recently reported that data collection wavelengths in the range of $\lambda = 1.5 - 3.0$ Å are fairly easy to handle in a diffraction experiment and even at home sources using instant Cr $K\alpha$ radiation [20]. The use of chromium-anode X-ray radiation is very useful for SAD experiments. The anomalous scattering signal is more than doubled for various metals when compared to conventional copper characteristic wavelength. Further-

*Corresponding author.

more, naturally bound metals and atoms from crystallization solutions tend to show a significant increase in anomalous scattering with chromium radiation [21,22]. Improved data quality helps in exploiting the weak anomalous signal derived only from the sulfurs or in particular from halide ions incorporated by soaking [23].

Most of the sulfur SAD data have been collected on tunable synchrotron beamlines in order to make use of the appreciably larger f'' of sulfur at longer wavelengths in the range between 1.7 - 2.5 Å [24-30] although Cu $K\alpha$ is also feasible for sulfur SAD phasing [31-33]. Both copper and chromium anode wavelengths (1.54 and 2.29 Å) have been increasingly employed for the same purpose in lab X-ray sources with success [34-36]. Using in-house copper wavelengths and synchrotron beamlines, phasing of protein like lysozyme, glucose isomerase and thermolysin have been carried out [27,28,31].

NaCl and KCl are part of almost every protein buffer and crystallization cocktail. Sulfur atoms and chloride ions are present in almost all known proteins and already been successfully utilized for the SAD phasing of many proteins. On an average, a protein contains 2.4 methionine and 1.7 cysteine residues per 100 amino acids [36], corresponding to an estimated *Bijvoet* ratio of 1.4% at 2.0 Å wavelength (assuming 7.46 non-hydrogen atoms per amino acid). Many other heavy atoms have also been successfully derivatized for SAD phasing [37-47]. Only rarely are intrinsic S atoms the sole source of the anomalous signal [16,18,48]. However, heavy atom derivatives suffer from nonspecific binding, which results in low occupancy of the heavy-atom sites leading to weak anomalous signal and disruption of the crystal lattice and failure in derivatization.

Exploiting the anomalous signal already present in the native protein or in the solvent would eliminate the extra experimental work *via* derivatization and would also eliminate the risk of lack of isomorphism. The anomalous signal from different anomalous scatterers present within the SAD data (native lysozyme, native glucose isomerase and native thermolysin) collected at chromium wavelength is used for successful structure solution.

2. Methods

2.1. Crystallization

Hen Egg White Lysozyme (HEWL), Glucose Isomerase (GI) and Thermolysin (TL) were purchased from Hampton Research Ltd., USA and used without further purification. Crystallization was performed based on the earlier established protocols [49-51]. Crystals were obtained in two days by hanging-drop vapour diffusion method. HEWL crystallized in tetragonal space group, GI crystallized in orthorhombic space group and TL crystallized in hexagonal space group. Before data collection, the crystals were quickly soaked in mother liquor containing cryoprotectant and then mounted into a loop of the required size using a cold nitrogen stream at 100 K. Crystallization, cryoprotectant and soaking conditions are shown in **Table 1**.

2.2. Data Collection

Three datasets (native HEWL; native GI; native TL) were collected separately using Rigaku R-Axis IV++ image plate detector [52] equipped with Cr $K\alpha$ (2.29 Å) anode X-ray generator, at CCMB, Hyderabad, India. The crystals were mounted in a random orientation. Crystals diffracted upto 2.53 Å and 360 frames were collected with crystal to detector distance being 110 mm at 0.5° oscillation steps and 120 seconds exposure time per frame in each case. The intensity data were collected and analyzed for anomalous signal, substructure solution and binding sites. The intensities were integrated using *HKL*2000 [53], refining all parameters including crystal mosaicity.

2.3. Substructure Solution, Phase Determination and Model Building

Anomalous scatterers were located using the dual-space recycling algorithm enabled in *SHELXD* [54-56]. Patterson seeding was used in all cases, resulting in an appreciable improvement in the success rate. 100 cycles of dual-space recycling were performed for the peak search and structure factor calculation. For the location of substructures of anomalous scatterers with *SHELXD*, only the internal loop which relies on the strongest E magnitudes was used. The best solution obtained with the highest correlation coefficient was directly fed as an input to *SHELXE* [57] for phase calculation and improvement without further refinement of the atomic positions and their occupancies. The high resolution limit for each data was used for phasing. The poly ala model building

Table 1. Crystallization, soaking concentration and cryoprotectant details.

	Native HEWL	Native GI	Native TL
Crystallization Condition	50 mM Sodium Acetate and 1 M Sodium Chloride pH 4.7	200 mM Magnesium chloride and 100 mM Tris, pH 4.7	1.4 mM Calcium Acetate, 10 mM Zinc Acetate, 1 mM Sodium Nitrate and 50 mM Tris; pH 7.3
Cryoprotectant	30% Glycerol	25% MPD	25% Ethylene Glycol

was carried out using *SHELXE beta* version. The final quality *SHELXE* phases were fed as input to *ARP/wARP* [58] for main-chain tracing and side-chain docking. Clearly interpretable maps were obtained after the automated model building. Manual rebuilding was carried out using *COOT* [59] and the final model was refined using *REFMAC*5.0 [60] available in *CCP4i suite* [61]. Figures were prepared using *PyMOL* software [62].

3. Results and Discussion

3.1. Data Analysis

The data was processed using *HKL*2000, by separately merging the anomalous pairs as I^+ and I^-. The overall redundancy of the data collected to 2.53 Å for the complete datasets varied between 10.3 and 43.49 at high resolution bin. The intensity data were analyzed for the presence of anomalous signal using *HKL2MAP* [63]. The full resolution range of the data was found to have anomalous signal measurability greater than 14%. The mean signal to noise ratio for all the datasets were found between 19.9 and 57.6.

The completeness of the datasets varied between 98% and 100%. Both the redundancy and completeness play a major role in model building after substructure determination. The substructure solution, phasing and refinement statistics are shown in **Table 2**. The overall *figure of merit* ranged from 0.53 to 0.58 for all the datasets. The map correlation coefficient was also found to be higher for all the datasets. The model obtained was manually corrected using *COOT* and refined using *REFMAC*5.0. The B-factor (thermal factor) in all the cases or datasets ranged from 10.3 to 39.7. The final model of all the datasets were refined to reasonable R_{work} and R_{free} values.

All the three datasets were processed and scaled using *HKL*2000. The scala output obtained was given as input for *SHELXC/D/E* phasing using *CCP4i.* At the chromium wavelength of 2.29 Å, a sulfur atom has an f'' value of 1.14 $\bar{e}$ and for a chlorine atom, $f'' = 2.01$ $\bar{e}$ as estimated from program *CROSSEC* [64]. During the substructure solution of HEWL dataset using *SHELXD*, fifteen peaks were identified out of which eight peaks correspond to sulfur as there are four disulphide bridges in HEWL. Similarly, seven chloride ions were identified by comparing the chloride ion sites in the earlier reported HEWL structure [31]. The density modified phases were used for model building using the *ARP/wARP* program web server [53] and refined using *REFMAC*5.0 of *CCP4i* software. For the dataset with redundancy 10.6, the automated solution and the model building by *ARP/wARP* resulted in building 126 amino acid residues, in which 110 were with side chains indicating that the dataset with reflections collected upto 2.53 Å was sufficient for structure solution and 10 fold redundancy is enough for substructure solution. Final refined model has R_{work} = 19.4% and R_{free} = 23.6%. The disulphide bonds

Table 2. Crystal Data Statistics, phasing and model building details.

	Native HEWL	Native GI	Native TL
Cell Parameters	a = b = 77.67 Å, c = 37.43 Å	a = 92.78 Å, b = 97.31 Å, c = 102.64 Å	a = b = 92.76 Å, c = 128.36 Å, γ = 120°
Space Group	$P4_32_12$	I222	$P6_122$
Mosaicity	0.76	0.49	0.6
Resolution range	27.48 - 2.53 Å	27.48 - 2.53 Å	38.33 - 2.53 Å
Redundancy	10.6 (9.98)	10.3 (9.73)	43.49 (42.84)
Completeness (%)	99.5 (97.8)	98.1 (94.8)	100.0 (100.0)
$I/\sigma(I)$	57.6 (13.3)	17.4 (4.8)	19.9 (7.1)
Anomalous signal (%)	17.1	18.3	14.6
FOM	0.56	0.58	0.53
Map Correlation	0.76	0.81	0.79
B_{Wilson}	10.53	12.79	39.69
Total Residues Built	122	388	316
Side Chains	110	383	314
R_{work} (%)	19.4	18.2	18.6
R_{free} (%)	23.6	22.8	21.9
Number of solvent molecules	102	140	145

showed clear elliptical density and Met sulfurs showed spherical density. Anomalous map of monomeric HEWL with eight sulfur peaks and seven chloride ions at 5σ with water molecules is shown in **Figure 1**.

In the native GI data, the anomalous difference map showed significant density for nine sulfur atoms within the protein and an additional peak corresponding to one manganese ion. At the chromium wavelength of 2.29 Å, a manganese ion has a f″ value of 0.97 $\overline{e}$ as estimated from program *CROSSEC*. Out of nine sulfur atom peaks, eight were of MET sulfurs and one peak was of the CYS. The position of the manganese ion peak was identified by comparing the manganese position or occu--pied site in the earlier reported GI structures [27]. They were given as input into *SHELXE* for obtaining the electron density maps. The final maps obtained after density modification were subjected to analysis by *ARP/wARP* web server for automatic chain tracing and model building. The electron density map allowed *ARP/wARP* program to build 388 residues, in which 383 were built with side chains. The dataset has a redundancy of 10.3. The dataset with reflections collected upto 2.53 Å was sufficient for structure solution and 10 fold redundancy is enough for substructure solution. Final refined model has R_{work} = 18.2% and R_{free} = 22.8%. Anomalous map of monomeric GI with nine sulfur atoms and one manganese ion at 5σ with water molecules is shown in **Figure 2**.

For native TL data, anomalous signals for three calcium ions, one zinc ion and two sulfur atoms were located using *SHELXD*. At the chromium wavelength of 2.29 Å, a calcium ion has a f″ value of 3.62 $\overline{e}$ and for a zinc ion, f″ = 2.05 $\overline{e}$ as estimated from program *CROSSEC*. The two sulfur atom peaks belonged to the MET sulfurs present in the TL. The positions of the four calcium ion peaks and one zinc ion peak were identified by comparing their respective positions in the earlier reported TL structures [28]. The density modified phases were used for model building using the *ARP/wARP* program web server and refined using *REFMAC*5.0 of *CCP4i* software. TL dataset had a redundancy of 43.49. Final refined model has R_{work} = 18.6% and R_{free} = 21.9%. 316 residues were built totally, in which 314 were built with side chains, by the iterative free-atom density modification and model building procedure. Anomalous map of monomeric TL with four calcium ions, one zinc ion and two sulfur atoms at 5σ with water molecules is shown in **Figure 3**. Crystal data statistics, phasing and model building details are given in **Table 2**.

SAD phasing has been carried out earlier on proteins like HEWL, GI and TL using synchrotron beamlines and in-house copper wavelengths [27,28,31]. Using in-house chromium X-ray source, HEWL structure was attempted using different mounting techniques [29]. Using in-house chromium wavelength, glucose isomerase structure was also attempted using different solving techniques like *ARP/wARP* and *OASIS*2004 [30], but the complete structure could not be solved. Our attempt to collect en-

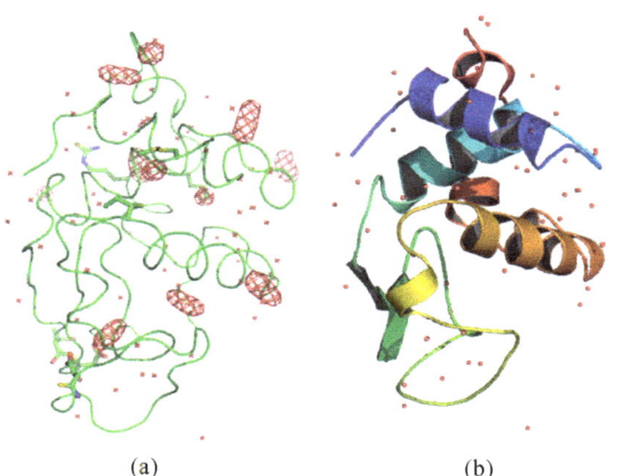

(a) (b)

Figure 1. Anomalous map (a) and Cartoon representation (b) of monomeric HEWL with eight sulfur peaks and seven chloride ions at 5σ with water molecules.

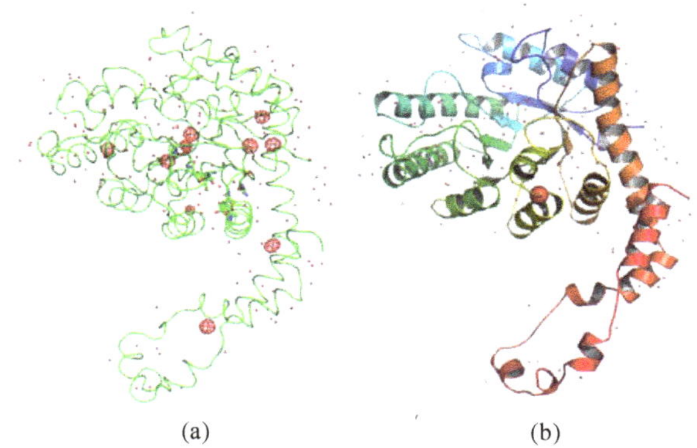

(a) (b)

Figure 2. (a) Anomalous map and (b) Cartoon representation of monomeric GI with nine sulfur atoms and one manganese ion at 5σ with water molecules.

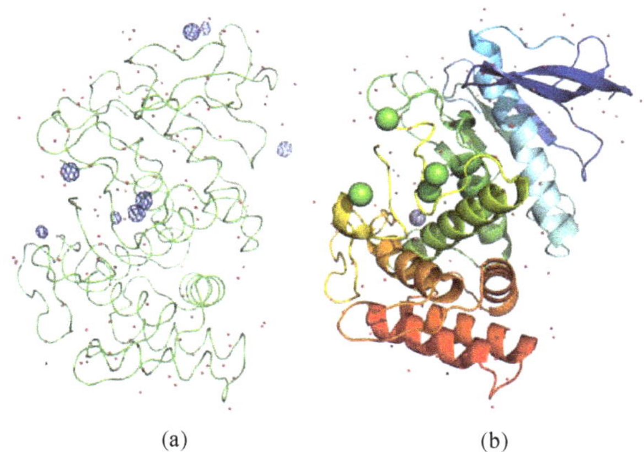

(a) (b)

Figure 3. (a) Anomalous map and (b) Cartoon representation of monomeric TL with four calcium ions, one zinc ion and two sulfur atoms at 5σ with water molecules.

hanced anomalous scattering data of TL using Cr *Kα* radiation has not been attempted before.

The obstacle so far has been the quality of the data required to take advantage of the vanishingly small signal of the sulfur atom and chloride ions. The location of the S atoms and other metal ions with the integrated Patterson and direct methods implemented in *SHELXD* is relatively straightforward. The results reveal that anomalous signal enhancers like chlorine and sulfur, which are naturally bound metals can very well contribute and allow easier substructure solution for sulfur SAD phasing. Access to in-house chromium generated X-ray radiation makes it easier to solve protein structures using bound lighter atoms and metals.

4. Conclusions

High-throughput crystallography requires a method by which the structures of proteins can be determined quickly and easily. SAD phasing using intrinsic anomalous scatterers like sulfur, chlorine, calcium, manganese and zinc becomes increasingly attractive owing to the advanced phasing methods. Due to the large anomalous scattering coefficient of the heavy metal ions used with respect to copper wavelength, they prove to be excellent heavy atom markers for automated phasing. Metal ions present in the soaking solution in molar concentrations are well enough to rapidly diffuse or enter into protein crystals. The anomalous scattering signal obtained from such crystals through soaking or co-crystallization within the diffraction data can be utilized for structure solution. The approach is quick as well as efficient. In all the cases discussed above, it was possible not only to locate the anomalous scatterers, but also subsequently to solve the protein model by SAD phasing.

All the collected datasets are of good quality and they have close to 100% completeness. Longer wave- lengths not only provide an increased anomalous signal for phase determination, but also allow a much clearer definition of substructures their positions and occupan- cies, which may turn out to be very important for eluci- dating the function of a molecule. Data quality is deci- sive for successful location of the anomalous substruc- ture. The example of successful SAD phasing based on the signal of weak anomalous scatterers such as sulfur atom and chloride ion prove that even the anomalous signal provided or present naturally in a macromolecule is good enough to solve crystal structures successfully using lab source chromium-generated X-ray radiation.

5. Acknowledgements

SN and DV thank UGC (SAP), Government of India for the financial support for this research. DV thanks DST-FIST and UGC-SAP for funding facilities to the Centre for Advanced Study in Crystallography and Biophysics. Chromium datasets were collected at X-ray facility, CCMB, Hyderabad funded by CSIR Facility Creation Project (FAC0004) as part of Eleventh Five Year Plan. SN and DV thank Dr. R. Shankaranarayanan for extending his lab facilities to collect anomalous scattering datasets using Cr *Kα* radiation.

REFERENCES

[1] U. Heinmann, G. Illing and H. Oschkinat, "High-Througput Three-Dimensional Protein Structure Determination," *Current Opinion in Biotechnology*, Vol. 12, No. 4, 2001, pp. 348-354.

[2] P. Kuhn, K. Wilson, M. G. Patch and R. C. Stevens, "The Genesis of High-Throughput Structure-Based Drug Discovery Using Protein Crystallography," *Current Opinion in Chemical Biology*, Vol. 6, No. 5, 2002, pp. 704-710

[3] T. L. Blundell, H. Jhoti and C. Abell, "High-Throughput Crystallography for Lead Discovery in Drug Design," *Nature Reviews Drug Discovery*, Vol. 1, No. 1, 2002, pp. 45-54.

[4] J. S. Brunzelle, P. Shafaee, X. Yang, S. Weigand, Z. Ren and W. F. Anderson, "Automated Crystallographic System for High-Throughput Protein Structure Determination," *Acta Crystallographica*, Vol. D59, No. 7, 2003, pp. 1138-1144.

[5] B. Rupp, "High-Throughput Crystallography at an Affordable Cost: The TB Structural Genomics Consortium Crystallization Facility," *Accounts of Chemical Research*, Vol. 36, No. 3, 2003, pp. 173-181.

[6] T. L. Blundell and S. Patel, "High-Throughput X-Ray Crystallography for Drug Discovery," *Current Opinion in Pharmacology*, Vol. 4, No. 5, 2004, pp. 490-496.

[7] N. S. Ganapathy and P. K. Andrew, "In-House Sulfur SAD Phasing: A Case Study of the Effects of Data Quality and Resolution Cutoffs," *Acta Crystallographica*, Vol. D62, No. 7, 2006, pp. 707-716.

[8] B.-C. Wang, "Resolution of Phase Ambiguity in Macromolecular Crystallography," *Methods in Enzymology*, Vol. 115, 1985, pp. 90-111.

[9] E. J. Dodson, G. G. Dodson, A. Lewitoya and M. Sabesan, "Zinc-Free Cubic Pig Insulin: Crystallization and Structure Determination," *Journal of Molecular Biology*, Vol. 125, No. 3, 1978, pp. 387-396.

[10] W. A. Hendrickson and M. M. Teeter, "Structure of the Hydrophobic Protein Crambin Determined Directly from the Anomalous Scattering of Sulphur," *Nature*, Vol. 290, No. 5802, 1981, pp. 107-113.

[11] L. M. Rice, T. N. Earnest and A. T. Brunger, "Single-

Wavelength Anomalous Diffraction Phasing Revisited," *Acta Crystallographica*, Vol. D56, No. 11, 2000, pp. 1413-1420.

[12] A. Goulet, G. Vestergaard, F. C. Rodrigues, V. Campanacci, R. A. Garrett, C. Cambillau and L. M. Ortiz, "Getting the Best out of Long-Wavelength X-Rays: *De Novo* Chlorine/Sulfur SAD Phasing of a Structural Protein from ATV," *Acta Crystallographica*, Vol. D66, No. 3, 2010, pp. 304-308.

[13] J. Doutch, M. A. Hough, S. S. Hasnain and R. W. Strange, "Challenges of Sulfur SAD Phasing as a Routine Method in Macromolecular Crystallography," *Journal of Synchrotron Radiation*, Vol. 19, No. 1, 2012, pp. 19-29.

[14] R. Fourme, W. Shepard and R. Kahn, "Application of the Anomalous Dispersion of X-Rays to Macromolecular Crystallography," *Progress in Biophysics and Molecular Biology*, Vol. 64, No. 2, 1995, pp. 167-199.

[15] E. D. Judit, B. Gabor, Ma. Qingjun, B. Heiko and G. M. Sheldrick, "In-House Measurement of the Sulfur Anomalous Signal and Its Use for Phasing," *Acta Crystallographica*, Vol. D59, No. 4, 2003, pp. 688-696.

[16] C. Yang and J. W. Pflugrath, "Applications of Anomalous Scattering from S Atoms for Improved Phasing of Protein Diffraction Data Collected at Cu *Kα* Wavelength," *Acta Crystallographica*, Vol. D57, No. 10, 2001, pp. 1480-1490.

[17] C. T. Lemke, G. D. Smith and P. L. Howell, "S-SAD, Se-SAD and S/Se-SIRAS Using Cu *Kα* Radiation: Why Wait for Synchrotron Time?" *Acta Crystallographica*, Vol. D58, Vol. 12, 2002, pp. 2096-2101.

[18] J. E. Debreczeni, G. Bunkóczi, B. Girmann and G. M. Sheldrick, "In-House Phase Determination of the Lima-Bean Trypsin Inhibitor: A Low-Resolution Sulfur-SAD Case," *Acta Crystallographica*, Vol. D59, No. 2, 2003, pp. 393-395.

[19] J. G. Olsen, C. Flensberg, O. Olsen, G. Bricogne and A. Henriksen, "Solving the Structure of the Bubble Protein Using the Anomalous Sulfur Signal from Single-Crystal In-House Cu *Kα* Diffraction Data Only," *Acta Crystallographica*, Vol. D60, No. 2, 2004, pp. 250-255.

[20] M. S. Weiss, P. A. Tucker, S. Panjikar, K. D. Carugo, M. Polentarutti and C. Muller-Dieckmann, "On the Routine Use of Soft X-Rays in Macromolecular Crystallography. Part II. Data-Collection Wavelength and Scaling Models," *Acta Crystallographica*, Vol. D60, 2004, pp. 28-38.

[21] O. Carugo and K. Djinovic Carugo, "When X-Rays Modify the Protein Structure: Radiation Damage at Work," *Trends Biochemical Sciences*, Vol. 30, No. 4, 2005, pp. 213-219.

[22] C. Yang, J. W. Pflugarth, D. A. Courville, C. N. Stence and J. D. Ferrara, "Away from the Edge: SAD Phasing from the Sulfur Anomalous Signal Measured In-House with Chromium Radiation," *Acta Crystallographica*, Vol. D59, No. 11, 2003, pp. 1943-1957.

[23] J. H. Naismith, *et al.*, "Refined Structure of Cadmium-Substituted Concanavalin A at 2.0 Å Resolution," *Acta Crystallographica*, Vol. D49, No. 6, 1993, pp. 561-571.

[24] J. Brown, R. M. Esnouf, M. A. Jones, J. Linnell, K. Harlos, A. B. Hassan and E. Y. Jones, "Structure of a Functional IGF2R Fragment Determined from the Anomalous Scattering of Sulfur," *The EMBO Journal*, Vol. 21, No. 5, 2002, pp. 1054-1062.

[25] E. J. Gordon, G. A. Leonard, S. McSweeney and P. F. Zagalsky, "The C1 Subunit of Alpha-Crustacyanin: The *de Novo* Phasing of the Crystal Structure of a 40 kDa Homodimeric Protein Using the Anomalous Scattering from S Atoms Combined with Direct Methods," *Acta Crystallographica*, Vol. D57, No. 9, 2001, pp. 1230-1237.

[26] Z. J. Liu, E. S. Vysotski, C. J. Chen, J. P. Rose, J. Lee and B. C. Wang, "Structure of the Ca^{2+} Regulated Photoprotein Obelin at 1.7 Å Resolution Determined Directly from Its Sulfur Substructure," *Protein Science*, Vol. 9, No. 11, 2000, pp. 2085-2093.

[27] U. A. Ramgopal, M. Dauter and Z. Dauter, "SAD Manganese in Two Crystal Forms of Glucose Isomerase," *Acta Crystallographica*, Vol. D59, No. 5, 2003, pp. 868-875.

[28] M. S. Weiss, T. Sicker and R. Hilgenfield, "Soft X-Rays, High Redundancy, and Proper Scaling: A New Procedure for Automated Protein Structure Determination via SAS," *Structure*, Vol. 9, No. 9, 2001, pp. 771-777.

[29] Y. Kitago, N. Watanabe and I. Tanaka, "Structure Determination of a Novel Protein by Sulfur SAD Using Chromium Radiation in Combination with a New Crystal-Mounting Method," *Acta Crystallographica*, Vol. D61, No. 8, 2005, pp. 1013-1021.

[30] N. Watanabe, Y. Kitago, I. Tanaka, J-W. Wang, Y.-X. Gu, C.-D. Zheng and H.-F. Fan, "Comparison of Phasing Methods for Sulfur-SAD Using In-House Chromium Radiation: Case Studies for Standard Proteins and a 69 kDa Protein," *Acta Crystallographica*, D61, No. 11, 2005, pp. 1533-1540.

[31] Z. Dauter, M. Dauter, E. de La Fortelle, G. Bricogne and G. M. Sheldrick, "Can Anomalous Signal of Sulfur Become a Tool for Solving Protein Crystal Structures?" *Journal of Molecular Biology*, Vol. 289, No. 1, 1999, pp. 83-92.

[32] K. Sekar, V. Rajakannan, D. Velmurugan, T. Tamane, R. Thirumurugan, M. Dauter and Z. Dauter, "A Redetermination of the Structure of the Triple Mutant (K53, 56, 120M) of Phospholipase A2 at 1.6 Å Resolution Using Sulfur-SAS at 1.54 Å Wavelength," *Acta Crystallographica*, Vol. D60, No. 9, 2004, pp. 1586-1590.

[33] D. Roeser, A. Dickmanns, K. Gasow and M. G. Rudolph, "*De Novo* Calcium/Sulfur SAD Phasing of the Human Formylglycine-Generating Enzyme Using In-House

Data," *Acta Crystallographica*, Vol. D61, No. 8, 2005, pp. 1057-1066.

[34] C. T. Lemke, G. D. Smith and P. L. Howell, "S-SAD, Se-SAD and S/Se-SIRAR Using Cu Kalpha Radiation: Why Wait for Synchrotron Time?" *Acta Crystallographica*, Vol. D58, No. 12, 2002, pp. 2096-2101.

[35] G. N. Sarma and P. A. Karplus, "In-House Sulfur SAD Phasing: A Case Study of the Effects of Data Quality and Resolution Cutoffs," *Acta Crystallographica*, Vol. D62, No. 7, 2006, pp. 707-716.

[36] J. Nan, Y. Zhou, C. Yang, E. Brostromer, O. Kristensen and X.-D. Su, "Structure of a Fatty-Acid-Binding Protein from Bacillus Subtilis Determined by Sulfur-SAD Phasing Using In-House Chromium Radiation," *Acta Crystallograhica*, Vol. D65, No. 5, 2009, pp. 440-448.

[37] M. Yogavel, *et al*., "Structural Analysis of Actinidin and a Comparison of Cadmium and Sulfur Anomalous Signals from Actinidin Crystals Measured Using In-House Copper- and Chromium-Anode X-Ray Sources," *Acta Crystallographica*, Vol. D66, No. 12, 2010, pp. 1323-1333.

[38] I. Uson, *et al*., "Locating the Anomalous Scatterer Substructures in Halide and Sulfur Phasing," *Acta Crystallographica*, Vol. D59, No. 1, 2003, pp. 57-66.

[39] P. McCaldon and P. Argos, "Oligopeptide Biases in Protein Sequences and Their Use in Predicting Protein Coding Regions in Nucleotide Sequences," *Proteins*, Vol. 4, No. 2, 1988, pp. 99-122.

[40] Z. Dauter and M. Dauter, "Anomalous Signal of Solvent Bromides Used for Phasing of Lysozyme," *Journal of Molecular Biology*, Vol. 289, No. 1, 1999, pp. 93-101.

[41] Z. Dauter, M. Dauter and K. R. Rajashankar, "Novel Approach to Phasing Proteins: Derivatization by Short Cryo-Soaking with Halides,"*Acta Crystallographica*, Vol. D56, No. 2, 2000, pp. 232-237.

[42] Z. Dauter and M. Dauter, "Entering a New Phase: Using Solvent Halide Ions in Protein Structure Determination," *Structure*, Vol. 9, No. 2, 2001, pp. R21-R26.

[43] S. Korolev, *et al*., "Using Surface-Bound Rubidium Ions for Protein Phasing," *Acta Crystallographica*, Vol. D57, No. 7, 2001, pp. 1008-1012.

[44] R. A. P. Nagem, Z. Dauter and I. Polikarpov, "Protein Crystal Structure Solution by Fast Incorporation of Negatively and Positively Charged Anomalous Scatterers," *Acta Crystallographica*, Vol. D57, No. 7, 2001, pp. 996-1002.

[45] T. C. Appleby, *et al.*,"Structure of Human Uridine-Cytidine Kinase 2 Determined by SIRAS Using a Rotating-Anode X-Ray Generator and a Single Samarium Derivative," *Acta Crystallographica*, Vol. D61, No. 3, 2005, pp. 278-284.

[46] J. Jakoncic, *et al.*,"Anomalous Diffraction at Ultra-High Energy for Protein Crystallography," *Journal of Applied Crystallography*, Vol. 39, No. 6, 2006, pp. 831-841.

[47] G. Guncar, *et al*., "The Use of Co^{2+} for Crystallization and Structure Determination, Using a Conventional Monochromatic X-Ray Source, of Flax Rust Avirulence Protein," *Acta Crystallographica*, Vol. F63, No. 3, 2007, pp. 209-213.

[48] M. Yogavel,,, *et al*., "Structural Analysis of Actinidin and a Comparison of Cadmium and Sulfur Anomalous Signals from Actinidin Crystals Measured Using In-House Copper- and Chromium-Anode X-Ray Sources," *Acta Crystallographica*, Vol. D66, No. 12, 2010, pp. 1323-1333.

[49] J. E. Debreczeni, G. Bunkoczi, Q. Ma, H. Blaser and G. M. Sheldrick, "In-House Measurement of the Sulfur Anomalous Signal and Its Use for Phasing," *Acta Crystallographica*, Vol. D59, No. 4, 2003, pp. 688-696.

[50] K. N. Vennila and D. Velmurugan, "In-House SAD Phasing with Surface-Bound Cerium Ions," *Acta Crystallographica*, Vol. F67, No. 12, 2011, pp. 1662-1665.

[51] V. Dhanasekaran and D. Velmurugan, "SAD Phasing with In-House Cu *Kα* Radiation Using Barium as Anomalous Scatterer," *Indian Journal of Biochemistry and Biophysics*, Vol. 48, No. 6, 2011, pp. 399-405.

[52] Z. Dauter, M. Dauter, E. de La Fortelle, G. Bricogne and G. M. Sheldrick, "Can Anomalous Signal of Sulfur Become a Tool for Solving Protein Structures?" *Journal of Molecular Biology*, Vol. 28, No. 1, 1999, pp. 83-92.

[53] H. L. Carrell, H. Hoier and J. P. Glusker, "Modes of Binding Substrates and Their Analogues to the Enzyme D-Xylose Isomerase," *Acta Crystallographica*, Vol. D50, No. 113, 1994, pp. 113-123.

[54] C. Mueller-Dieckmann, *et al*., "On the Routine Use of Soft X-Rays in Macromolecular Crystallography. Part IV. Efficient Determination of Anomalous Substructures in Biomacromolecules Using Longer X-Ray Wavelengths," *Acta Crystallographica*, Vol. D63, 2007, pp. 366-380.

[55] Rigaku Americas, "New Trails Drive," The Woodlands, Texas, pp. 77381-5209.

[56] Z. Otwinowski and W. Minor, "Processing of X-Ray Diffraction Data Collected in Oscillation Mode," In: C. W. Carter Jr. and R. M. Sweet, Eds., *Methods in Enzymology*, Vol. 276, Macromolecular Crystallography, Part A, Academic Press, New York, 1997, pp. 307-326.

[57] G. M. Sheldrick, H. A. Hauptman, C. M. Weeks, M. Miller and I. Uson, "*Ab Initio* Phasing," In: E. Arnold and M. Rossman, Eds., *International Tables for Crystallography*, Vol. F, Chapter 16.1, IUCr and Kluwer Academic Publishers, Dordrecht, 2001, pp. 333-345.

[58] I. Uson and G. M. Sheldrick, "Advances in Direct Meth-

ods for Protein Crystallography," *Current Opinion in Structure Biology*, Vol. 9, No. 5, 1999, pp. 643-648.

[59] G. G. Langer, S. X. Cohen, A. Perrakis and V. S. Lamzin, "Automated Macromolecular Model Building for X-Ray Crystallography Using ARP/wARP Version 7," *Nature Protocols*, Vol. 3, No. 7, 2008, pp. 1171-1179.

[60] P. Emsley and K. Cowtan, "Coot: Model-Building Tools for Molecular Graphics," *Acta Crystallographica*, Vol. D60, No. 12, 2004, pp. 2126-2132.

[61] A. A. Vagin, *et al*., "REFMAC5 Dictionary: Organization of Prior Chemical Knowledge and Guidelines for Its Use," *Acta Crystallographica*, Vol. D60, No. 12, 2004, pp. 2284-2295.

[62] CCP4 (Collaborative Computational Project, Number 4), "The CCP4 Suite: Programs for Protein Crystallography," *Acta Crystallographica*, Vol. D50, No. 5, 1994, pp. 760-763.

[63] W. L. DeLano, "The PyMOL Molecular Graphics System," DeLano Scientific, San Carlos, 1998.

[64] T. Pape and T. R. Schneider, "HKL2MAP: A Graphical User Interface for Macromolecular Phasing with SHELX Programs," *Journal of Applied Crystallography*, Vol. 37, No. 5, 2004, pp. 843-844.

[65] T. C. Don, "Calculation of Anomalous Scattering Factors at Arbitrary Wavelengths," *Journal of Applied Crystallography*, Vol. 16, No. 4, 1983, pp. 437.

12

Method to Study the Primary Nucleation for Solid Solution: Application to Uranium-Neodymium Oxalate Coprecipitation

Delphine Parmentier[1], Murielle Bertrand[1], Edouard Plasari[2], Pascal Baron[1]
[1]Commissariat à l'Energie Atomique et aux Energies Alternatives,
Radiochemistry and Process Department, Bagnols-sur-Cèze, France
[2]Reactions and Chemical Engineering Laboratory-CNRS, Nancy, France

ABSTRACT

Actinides co-precipitation is currently investigated in order to synthesize solid solutions of actinides mixed oxalates. This paper deals with the thermodynamic and kinetic study of the precipitation of uranium-neodymium oxalate system. Based on an analysis of the theories developed in the literature, a new expression for the determination of the supersaturation ratio for the solid solutions is presented. An experimental study of the nucleation kinetics was performed on the mixed uranium-neodymium oxalates. Homogeneous and heterogeneous primary nucleation laws are obtained using a specific stopped flow apparatus. The experimental results are consistent with the classical behaviour of nucleation phenomena. The values of the kinetic parameters for the solid solution point out that the formation of the uranium-neodymium mixed oxalates is kinetically favoured compared with the simple uranium and neodymium oxalates.

Keywords: Coprecipitation; Solid Solution; Thermodynamics; Kinetics; Supersaturation; Nucleation; Uranium

1. Introduction

Current concepts for future nuclear systems aim at improving the fuel cycle with the co-management of actinides in order to enhance the fuel performance and to reduce the proliferation risk. The actinide co-conversion processes, such as COEX^{TM} process [1], play an important role by producing mixed actinide compounds used as starting materials for fuel re-fabrication. The co-conversion of actinides into oxide can be achieved by different processes such as oxalic co-precipitation, thermal co-denitration or co-gelation. The Oxalic co-precipitation is one investigated way to synthesize solid solutions of actinide mixed oxalates which have to meet strict requirements [2]. Thus to support the process development, the determination of mechanisms and kinetic laws associated with the co-precipitation appears to be a key issue. This article focuses on the kinetic study in order to determine the primary nucleation rate for the uranium-neodymium oxalate system, which simulates the uranium-actinide system behaviour.

Compared with simple salts, mechanisms involved in the formation of mixed crystals are much more complex and remain poorly understood. The phenomenological theories are not still clarified in the literature. Equilibrium conditions for a solid solution-aqueous solution (SS-AS) system can be described using the approach developed by Lippmann who proposes an extension of the solubility product concept [3]. Several both experimental and theoretical studies have recently focused on the thermodynamics of mixed crystals [4-13] contrary to the kinetic studies which are reported in very few papers only [14,15]. The key point in the precipitation of solid solutions is the definition of the driving force. All authors agree to describe the difficulty in determining a rigorous expression of the supersaturation which is a fundamental crystallization parameter. The difficulty arises from the need to express supersaturation using the chemical potentials of components in both solid and aqueous solutions which are mutually independent. The way that the driving force should be defined is still controversial [16,17]. Shtukenberg *et al.* [11] propose to evaluate the supersaturation using Lippmann's approach, whereas Astilleros *et al.* [8] developed two formulations, the first one based on the stoichiometric saturation concept [18] and the second one based on the actual activity concept

[8].

This study focuses on the uranium(IV)-neodymium(III) oxalate solid solutions described by the following precipitation reaction, where can vary from 0 to 0.5 [2]:

$$\begin{aligned} &0.6N_2H_5^+ + (0.4+\alpha)H_3O^+ \\ &\quad + (1-\alpha)U^{4+} + \alpha Nd^{3+} + 2.5C_2O_4^{2-} \\ &\rightarrow (N_2H_5^+)_{0.6}(H_3O^+)_{0.4+\alpha}U_{1-\alpha}Nd_\alpha(C_2O_4^{2-})_{2.5}\cdot 7H_2O \end{aligned} \quad (1)$$

The solid solutions composed of U(IV)-Nd(III) oxalates are complex as the stoichiometric coefficients are different for each component. Solid solutions usually described in the literature have simple formula such as (Ba, Sr)SO_4 or (Ba, Sr)CO_3 [19-21] where cations and anions have the same charge, while in our case (see Equation (1)) the formation of the solid solution is much more complex and needs a more profound investigation not previously encountered in the literature. Thus, the aim of this work is to study experimentally the formation of this kind of complex solid solutions and to find adequate nucleation rate expressions in order to use them in modelling of precipitation processes in nuclear industry. The first part of this paper is concerned with the discussion of the Lippmann model in the case of complex solid solutions. A new suitable expression for the relative supersaturation ratio is then presented. Section 3 describes the experimental study of the nucleation kinetics performed on the mixed uranium-neodymium oxalate. The experimental results obtained are presented in section 4 and are confronted with the new model of the supersaturation.

2. Supersaturation Concept in SS-AS Systems

In precipitation, the supersaturation concept describes in a quantitative way the deviation of a given aqueous solution from the equilibrium state. For a pure solid of the form BA, the solubility product is defined as the product of activities of species in solution at equilibrium conditions:

$$B^{z+} + A^{z-} \leftrightarrow BA$$
$$K_s = \frac{a_{B,eq}\cdot a_{A,eq}}{a_{BA,eq}} \quad (2)$$

where K_s is the solubility product of BA, a_A and a_B the activity of the ions A^{z-} and B^{z+} in the liquid phase, a_{AB} the activity of BA in the solid phase and z the positive and negative charge of ions, eq means at equilibrium state.

The driving force of the nucleation process, the relative supersaturation ratio, can be calculated by the well-known general expression:

$$S = \left(\frac{\prod a_i^{\upsilon_i}}{K_s}\right)^{\frac{1}{\sum \upsilon_i}} \quad [19] \quad (3)$$

where S is the supersaturation ratio, $\prod a_i^{\upsilon_i}$ the product of the ionic activities and υ_i the stoichiometric number of ion i.

Regarding the solid solutions, the application of the supersaturation concept (for example in the case of the coprecipitation of BA and CA compounds, see Equations (4) and (5)) requires first to define the equilibrium conditions.

2.1. Lippmann's Approach

Lippmann [3] proposed to extent the solubility product concept to the solid solution based on an analogy to liquid-vapour systems. We consider a solid solution-aqueous solution system where the cations B^{z+} and C^{z+} and the anions A^{z-} are presented in the aqueous phase and react to lead to the solid solutions $(BA)_{X_1}(CA)_{X_2} = B_{X_1}C_{X_2}A$ formed by the pure end-members BA and CA. The equilibrium conditions between the solid solution and the aqueous solution are given by two expressions provided by the following reaction schemes:

$$B^{z+} + A^{z-} \leftrightarrow BA, K_1 = \frac{a_{B,eq}\cdot a_{A,eq}}{a_{BA,eq}} = \frac{\gamma_{\pm,1}^2\cdot C_{B,eq}\cdot C_{A,eq}}{\gamma_1 X_1} \quad [3] \quad (4)$$

$$C^{z+} + A^{z-} \leftrightarrow CA, K_2 = \frac{a_{C,eq}\cdot a_{A,eq}}{a_{CA,eq}} = \frac{\gamma_{\pm,2}^2\cdot C_{C,eq}\cdot C_{A,eq}}{\gamma_2 X_2} \quad [3] \quad (5)$$

where K_1 and K_2 are the solubility product of the end-members, X_1 and X_2 the molar fractions of endmembers in the solid solution, a_A, a_B and a_C the activities of the ions A^{z-}, B^{z+} and C^{z+} in the aqueous solution, a_{BA} and a_{CA} the activities of BA and CA in the solid solution, C_A, C_B and C_C the concentrations of the ions A^{z-}, B^{z+} and C^{z+}, γ_1 and γ_2 the activity coefficients in the solid phase and $\gamma_{\pm 1}$ and $\gamma_{\pm 2}$ the mean activity coefficients in the liquid phase.

These reactions are coupled only through the mass balance that implies:

$$X_1 + X_2 = 1 \quad [3] \quad (6)$$

We note BA the endmember 1 and CA the endmember 2 with conventionally $K_1 < K_2$ [3].

According to Lippmann, in the same way as the total vapour pressure of a liquid mixture is the sum of the partial vapour pressure of each component, the total solubility product can be defined as the sum of the partial solubility products of each endmember of the solid solution:

$$\begin{aligned} &\sum \Pi_{eq} \\ &= a_{B,eq}\cdot a_{A,eq} + a_{C,eq}\cdot a_{A,eq} \\ &= K_1\gamma_1 X_1 + K_2\gamma_2 X_2 = K_1\gamma_1 X_1 + K_2\gamma_2(1-X_1) \end{aligned} \quad [3]\ (7)$$

where $\sum \Pi_{eq}$ is the total solubility product.

The products $a_{B,eq} \cdot a_{A,eq}$ and $a_{C,eq} \cdot a_{A,eq}$ are analogous to partial pressures of a vapour pressure diagram of a binary system and can be assimilated to partial solubility products.

The total solubility product as a function of the solid solution composition $X\left(X = X_1\right)$ leads to the Lippmann solidus equation. Similarly, the total solubility product as a function of the aqueous solution composition $x\left(x = x_1\right)$ leads to the Lippmann solutus equation. Therefore a phase diagram is obtained by analogy to the vapour pressure diagram.

Using the Lippmann approach, the driving force can be expressed as follows [11]:

$$S(X) = \frac{\sum \Pi}{\sum \Pi_{eq}} = \frac{a_B \cdot a_A + a_C \cdot a_A}{a_{B,eq} \cdot a_{A,eq} + a_{C,eq} \cdot a_{A,eq}} = \frac{a_B \cdot a_A + a_C \cdot a_A}{K_1 \cdot a_{BA,eq} + K_2 \cdot a_{CA,eq}} = \frac{a_B \cdot a_A + a_C \cdot a_A}{K_1 \cdot \gamma_1 X + K_2 \cdot \gamma_2 (1 - X)} \quad [11] \quad (8)$$

where S is the supersaturation ratio depending on X.

In the literature, this approach described by Lippmann was applied by several authors for bivalent cations and anions and the considered systems can thus be described by Equations (4) and (5) where $\left|z^+\right| = \left|z^-\right| = 2$. The classical systems studied were (Ba, Sr)SO_4, (Ba, Sr)CO_3, (Ca, Mg)CO_3, etc. There are however more complex solid solutions in which the endmembers present different stoichiometric coefficients: this specially applies to the nuclear industry where different actinide solid solutions are formed by coprecipitation [22-24].

For solid systems with endmembers of different stoichiometry, the Lippmann formalism needs to be adjusted. To illustrate these cases, we consider now a simple SS-AS system where the cations B^+ and C^{2+} and the anion A^{2-} are presented in the aqueous phase and react to lead to solid solutions $B_{X_1}C_{X_2}A$ formed by the pure endmembers B_2A and CA. The equilibrium conditions of this binary system are as follows:

$$2B^+ + A^{2-} \leftrightarrow B_2A$$

$$K_1 = \frac{a_{B,eq}^2 \cdot a_{A,eq}}{a_{BA,eq}} = \frac{\gamma_{\pm,1}^3 \cdot C_{B,eq}^2 \cdot C_{A,eq}}{\gamma_1 X_1} \quad (9)$$

$$C^{2+} + A^{2-} \leftrightarrow CA$$

$$K_2 = \frac{a_{C,eq} \cdot a_{A,eq}}{a_{CA,eq}} = \frac{\gamma_{\pm,2}^2 \cdot C_{C,eq} \cdot C_{A,eq}}{\gamma_2 X_2} \quad (10)$$

According to the Lippmann proposition, the total solubility product would be obtained by adding Equations (9) and (10):

$$\sum \Pi_{eq} = a_{B,eq}^2 \cdot a_{A,eq} + a_{C,eq} \cdot a_{A,eq} = K_1 \gamma_1 X_1 + K_2 \gamma_2 X_2 \quad (11)$$

This equation is not adequate as it is composed of terms with different dimensionalities: the first endmember solubility product K_1 is expressed in $(\text{mol}\cdot\text{m}^{-3})^3$, whereas the second endmember solubility product K_2 is in $(\text{mol}\cdot\text{m}^{-3})^2$. This example points out that the Lippmann model for a complex solid solution cannot be directly applicable since it fails to satisfy the fundamental dimensionality principle. To our opinion, the Lippmann theoretical approach is right, but his mathematical approach needs to be generalized. We accept the Lippmann similarity between liquid phase-vapour phase equilibrium and solid solution-aqueous solution equilibrium, but a new mathematical expression consistent with equilibrium relations can be proposed. For a liquid-vapour equilibrium, assuming a thermodynamically ideal liquid solution, the total vapour pressure is given as follows:

$$P_{eq} = P_{1,eq} \cdot X_1 + P_{2,eq} \cdot X_2 \quad (12)$$

while Equation (7) proposed by Lippmann is equivalent to the following expression:

$$P_{eq}^2 = P_{1,eq}^2 \cdot X_1 + P_{2,eq}^2 \cdot X_2 \quad (13)$$

which is not a correct equation.

In these equations, $P_{1,eq}$ and $P_{2,eq}$ are the partial vapour pressure, X_1 and X_2 the molar fractions of the components and P_{eq} the total vapour pressure of the liquid mixture. Whatever the system considered, the terms added in Equation (12) have always the same dimensionality as a pressure. The concentration dimensionality is directly proportional to a pressure. Therefore, to be consistent with the vapour-liquid system, the total equilibrium term should have always the same dimensionality as a concentration and should be defined as the sum of terms with a concentration dimensionality (in thermodynamics, in the domain of multi-component phase equilibrium, only terms having pressure or concentration dimensionalities in power 1 can be additional terms).

In addition, if we consider the expression of the supersaturation ratio defined by Shtukenberg *et al.* [11] (see Equation (8)), we can notice that for the pure endmembers this expression doesn't converge to the general expression of the supersaturation ratio (see Equation (3)).

2.2. New Expression

Based on these considerations, we propose a new expression to determine the total solubility term at equilibrium. For a system defined by Equations (4) and (5), the new formulation is:

$$\sum\Pi_{eq}^{1/2}=\sqrt{a_{B,eq}\cdot a_{A,eq}}+\sqrt{a_{C,eq}\cdot a_{A,eq}} \quad (14)$$

Based on equilibrium Equations (4) and (5), it follows:

$$\sum\Pi_{eq}^{1/2} =\gamma_{\pm,1}\sqrt{C_{B,eq}\cdot C_{A,eq}}+\gamma_{\pm,2}\sqrt{C_{C,eq}\cdot C_{A,eq}} =\sqrt{K_1\gamma_1X_1}+\sqrt{K_2\gamma_2X_2} \quad (15)$$

Therefore, the absolute supersaturation will be $(X=X_1)$:

$$s(X) =\sum\Pi^{1/2}-\sum\Pi_{eq}^{1/2} =\sqrt{a_B\cdot a_A}+\sqrt{a_C\cdot a_A}-\sum\Pi_{eq}^{1/2} =\gamma_{\pm,1}\sqrt{C_B\cdot C_A}+\gamma_{\pm,2}\sqrt{C_C\cdot C_A}-\sqrt{K_1\gamma_1X}-\sqrt{K_2\gamma_2(1-X)} \quad (16)$$

and the supersaturation ratio:

$$S(X) =\frac{\sum\Pi^{1/2}}{\sum\Pi_{eq}^{1/2}}=\frac{\sqrt{a_B\cdot a_A}+\sqrt{a_C\cdot a_A}}{\sqrt{a_{B,eq}\cdot a_{A,eq}}+\sqrt{a_{C,eq}\cdot a_{A,eq}}} =\frac{\gamma_{\pm,1}\sqrt{C_B\cdot C_A}+\gamma_{\pm,2}\sqrt{C_C\cdot C_A}}{\sqrt{K_1\gamma_1X}+\sqrt{K_2\gamma_2(1-X)}} \quad (17)$$

We can also note that the Expressions (16) and (17) converge to the right expressions for the single component systems when $X\to1$ or $X\to0$, contrary to Lippmann's approach (see Equation (3)).

More generally, for any kind of solid solution of the form $B_{m_1\cdot X_1}C_{m_2\cdot X_2}A_{n_1+n_2}$ obtained from the endmembers $B_{m_1}A_{n_1}$ and $C_{m_2}A_{n_2}$, equilibrium is defined by the following mass action equations:

$$m_1B^{n_1+}+n_1A^{m_1-}\Leftrightarrow B_{m_1}A_{n_1}$$

$$K_1=\frac{a_{B,eq}^{m_1}\cdot a_{A,eq}^{n_1}}{a_{B_{m_1}A_{n_1},eq}}=\frac{\gamma_{\pm,1}^{m_1+n_1}\cdot C_{B,eq}^{m_1}\cdot C_{A,eq}^{n_1}}{\gamma_1X_1} \quad (18)$$

$$m_2C^{n_2+}+n_2A^{m_2-}\Leftrightarrow C_{m_2}A_{n_2}$$

$$K_2=\frac{a_{C,eq}^{m_2}\cdot a_{A,eq}^{n_2}}{a_{C_{m_2}A_{n_2},eq}}=\frac{\gamma_{\pm,2}^{m_2+n_2}\cdot C_{C,eq}^{m_2}\cdot C_{A,eq}^{n_2}}{\gamma_2X_2} \quad (19)$$

Using the new expression, the total solubility term and the supersaturation ratio are defined as follows:

$$\sum\Pi_{eq}^{1/\sum\nu_i} =\left(a_{B,eq}^{m_1}\cdot a_{A,eq}^{n_1}\right)^{1/(m_1+n_1)}+\left(a_{C,eq}^{m_2}\cdot a_{A,eq}^{n_2}\right)^{1/(m_2+n_2)} =\left(K_1\gamma_1X_1\right)^{1/(m_1+n_1)}+\left(K_2\gamma_2X_2\right)^{1/(m_2+n_2)} \quad (20)$$

and

$$S(X) =\frac{\sum\Pi^{1/\sum\nu_i}}{\sum\Pi_{eq}^{1/\sum\nu_i}}=\frac{\left(a_B^{m_1}\cdot a_A^{n_1}\right)^{1/(m_1+n_1)}+\left(a_C^{m_2}\cdot a_A^{n_2}\right)^{1/(m_2+n_2)}}{\left(a_{B,eq}^{m_1}\cdot a_{A,eq}^{n_1}\right)^{1/(m_1+n_1)}+\left(a_{C,eq}^{m_2}\cdot a_{A,eq}^{n_2}\right)^{1/(m_2+n_2)}} =\frac{\left(a_B^{m_1}\cdot a_A^{n_1}\right)^{1/(m_1+n_1)}+\left(a_C^{m_2}\cdot a_A^{n_2}\right)^{1/(m_2+n_2)}}{\left(K_1\gamma_1X\right)^{1/(m_1+n_1)}+\left[K_2\gamma_2(1-X)\right]^{1/(m_2+n_2)}} \quad (21)$$

2.3. Other Expression of the Supersaturation

Applying the concept of the stoichiometric saturation to a solid solution [18] and the Lippmann approach [3], Pina *et al.* [7,9,19] derived the supersaturation ratio expression based simultaneously on equilibrium relations of Reactions (3) and (4) and on the formation of the solid solution according to the single reaction:

$$X_1B^{z+}+X_2C^{z+}+A^{z-}\Leftrightarrow B_{X_1}C_{X_2}A \quad \text{with } K_s=a_{B,eq}^{X_1}\cdot a_{C,eq}^{X_2}\cdot a_{A,eq} \quad (22)$$

Using equilibrium Equations (3), (4) and (22), they propose the following expression $(X_1=X)$:

$$S(X) =\sqrt{\frac{a_B^X\cdot a_C^{1-X}\cdot a_A}{\left(K_1\cdot a_{BA}\right)^X\cdot\left(K_2\cdot a_{CA}\right)^{1-X}}} =\sqrt{\frac{a_B^X\cdot a_C^{1-X}\cdot a_A}{\left(K_1\gamma_1X\right)^X\cdot\left[K_2\gamma_2(1-X)\right]^{1-X}}} \quad [19] \quad (23)$$

The weak point of the Putnis *et al.*'s formalism [20-25] is the dualism used to obtain the equilibrium expressions. Indeed the authors accept simultaneously the solid phase as composed of two endmembers *BA* and *CA* (Reactions (3) and (4)) and as composed of a single compound $B_{X1}C_{X2}A$ (Reaction (22)). We have compared our formalism proposed in this study with the one proposed by Putnis *et al.* in the case of the coprecipitation of (Ba, Sr)SO_4 where $K_{SrSO_4}=K_2$ is three orders of magnitude higher than $K_{BaSO_4}=K_1$. The calculations of the supersaturation ratio for X_1 values ranging from 0 to 1 (not shown here) give a maximum difference of 2.4 times higher for our expression than for the Putnis *et al.* expression. This difference is very sensible to the ratio K_2/K_1. When K_1 and K_2 are of the same order of magnitude (which is the case in the nuclear industry), the values of supersaturation ratio obtained by the two approaches are almost the same [26]. Here, we can also note that the mathematical calculations based on the Putnis *et al.* formalism are much more complex than those corresponding to the formalism we propose. For all these reasons, in this article, only the new expression based on the analogy between liquid-vapour equilibrium and SS-AS equilibrium is used to calculate the supersaturation.

2.4. Determination of the Supersaturation Ratio Based Only on Liquid Phase Composition

As shown in Equations (16) and (17), the supersaturation is calculated as a function of the composition of the solid phase X. All authors in their studies have used the molar composition X to express the supersaturation ratio. However, during the precipitation reaction, the variation of the liquid phase composition is easily measured or known, while the composition of the solid phase can be known only at the end of precipitation process. For this reason, it is interesting to express the supersaturation based only on the composition of the liquid phase. Let us illustrate it using first the simple case expressed by Equations (4) and (5) for which the supersaturation ratio is given by Equation (17).

Considering a solid solution defined by Equations (4) and (5), the mole fractions of endmembers BA and CA in the solid solution are given by:

$$X_1 = \frac{\gamma_{\pm,1}^2 \cdot C_{B,eq} \cdot C_{A,eq}}{\gamma_1 \cdot K_1} \tag{24}$$

$$X_2 = \frac{\gamma_{\pm,2}^2 \cdot C_{C,eq} \cdot C_{A,eq}}{\gamma_2 \cdot K_2} \tag{25}$$

Knowing that $X_1 + X_2 = 1$, the concentration of the reactant A can be deduced from Equations (24) and (25):

$$C_{A,eq} = \frac{1}{\frac{\gamma_{\pm 1}^2 \cdot C_{B,eq}}{\gamma_1 \cdot K_1} + \frac{\gamma_{\pm 2}^2 \cdot C_{C,eq}}{\gamma_2 \cdot K_2}} \tag{26}$$

The total solubility term expressed as a function of the liquid composition is given by Equation (15).

Substituting Equation (26) in Equation (15) and dividing the numerator and the denominator by $\left(C_{B,eq} + C_{C,eq}\right)^{1/2}$ lead to the relationship between the mole fraction of the solid phase X_1 and the mole fraction in the liquid phase x_1:

$$\sqrt{K_1\gamma_1 X_1} + \sqrt{K_2\gamma_2\left(1-X_1\right)} = \frac{\gamma_{\pm 1}\sqrt{x_1} + \gamma_{\pm 2}\sqrt{1-x_1}}{\sqrt{\frac{\gamma_{\pm 1}^2 \cdot x_1}{\gamma_1 \cdot K_1} + \frac{\gamma_{\pm 2}^2 \cdot \left(1-x_1\right)}{\gamma_2 \cdot K_2}}} \tag{27}$$

where x_1 and x_2 are the mole fractions of the endmembers in the liquid phase at equilibrium.

According to the Lippmann phase diagram, the aqueous solution at equilibrium has the same aqueous activity fraction as the supersaturated solution [3,8,11]. Consequently, x_1 and x_2 are defined as:

$$x_1 = \frac{C_B}{C_B + C_C} = x \tag{28}$$

$$x_2 = \frac{C_C}{C_B + C_C} = 1 - x \tag{29}$$

Equation (27) is very important since it links the relation liquid and solid phase compositions at any time of precipitation reaction. Finally, from Equation (27), the analytical solution of the relative supersaturation is obtained as a function of the composition of the aqueous solution only:

$$S(x) = \frac{\sum \Pi^{1/2}}{\sum \Pi_{eq}^{1/2}} = \frac{\gamma_{\pm,1}\sqrt{C_B \cdot C_A} + \gamma_{\pm,2}\sqrt{C_C \cdot C_A}}{\frac{\gamma_{\pm,1}\sqrt{x} + \gamma_{\pm,2}\sqrt{1-x}}{\sqrt{\frac{\gamma_{\pm,1}^2 x}{K_1\gamma_1} + \frac{\gamma_{\pm,2}^2\left(1-x\right)}{K_2\gamma_2}}}} \tag{30}$$

Let us use the same methodology to find the supersaturation ratio expression in our more complex case. A series of solid solutions can be synthesized by mixing an acid solution of uranium(IV) and neodymium(III) nitrate with an oxalic acid solution [22-27]. The solid solutions are formed by the following isostructural endmembers:

Endmember 1:

$$\left(N_2H_5^+\right)_{0.6}\left(H_3O^+\right)_{0.4} U\left(C_2O_4^{2-}\right)_{2.5} \cdot 7H_2O \tag{31}$$

Endmember 2:

$$\left(N_2H_5^+\right)_{0.6}\left(H_3O^+\right)_{0.9} U_{0.5}Nd_{0.5}\left(C_2O_4^{2-}\right)_{2.5} \cdot 7H_2O \tag{32}$$

The mixed oxalates can be obtained only under specific experimental conditions used in the nuclear industry [2,23]. Three series of mixed lanthanide(III) and actinide(IV) were studied by Chapelet-Arab *et al.*: the hexagonal, the triclinic and the quadratic series. The originality of these structures is the existence of a mixed crystallographic site for tetravalent actinide and for trivalent lanthanide or actinide. The charge balance depending on the $\mathrm{Ln(III)}/\left(\mathrm{Ln(III)} + \mathrm{U(IV)}\right)$ mole ratio is ensured by the adjustment of single charged ions within the structure $\left(H_3O^+, N_2H_5^+\right)$.

The solubility products of the endmembers 1 and 2 were experimentally determined by Costenoble under experimental conditions encountered in nuclear industry given in **Table 1** [23-28].

The precipitation reactions of the endmembers are:

Table 1. Experimental conditions for obtaining the solid solutions of endmembers 1 and 2.

HNO_3 (mol·L^{-1})	1
Hydrazine (mol·L^{-1})	0.10
Oxalic acid excess (mol·L^{-1})	0.15

$$U^{4+} + 0.6N_2H_5^+ + 0.4H_3O^+ + 2.5C_2O_4^{2-} \Leftrightarrow (N_2H_5)_{0.6}(H_3O)_{0.4}U(C_2O_4)_{2.5}\cdot 7H_2O \quad (33)$$

$$0.5U^{4+} + 0.5Nd^{3+} + 0.6N_2H_5^+ + 0.9H_3O^+ + 2.5C_2O_4^{2-} \Leftrightarrow (N_2H_5)_{0.6}(H_3O)_{0.9}U_{0.5}Nd_{0.5}(C_2O_4)_{2.5}\cdot 7H_2O \quad (34)$$

and their solubility products can be expressed as follows:

$$K_1 = a_{N_2H_5^+,eq}^{0.6}\cdot a_{H_3O^+,eq}^{0.4}\cdot a_{U^{4+},eq}\cdot a_{C_2O_4^{2-},eq}^{2.5} \quad (35)$$

$$K_2 = a_{N_2H_5^+,eq}^{0.6}\cdot a_{H_3O^+,eq}^{0.9}\cdot a_{U^{4+},eq}^{0.5}\cdot a_{Nd^{3+},eq}^{0.5}\cdot a_{C_2O_4^{2-},eq}^{2.5} \quad (36)$$

In the case of a solid solution formed by the copre cipitation of the above endmembers, the equilibrium relations are:

$$K_1 = \frac{a_{N_2H_5^+,eq}^{0.6}\cdot a_{H_3O^+,eq}^{0.4}\cdot a_{U^{4+},eq}\cdot a_{C_2O_4^{2-},eq}^{2.5}}{\gamma_1 X_1} \quad (37)$$

$$K_2 = \frac{a_{N_2H_5^+,eq}^{0.6}\cdot a_{H_3O^+,eq}^{0.9}\cdot a_{U^{4+},eq}^{0.5}\cdot a_{Nd^{3+},eq}^{0.5}\cdot a_{C_2O_4^{2-},eq}^{2.5}}{\gamma_2 X_2} \quad (38)$$

As discussed before (Equation (20)), the total solubility term can be written as:

$$\sum \Pi_{eq}^{1/\sum\nu_i} = \left(a_{N_2H_5^+,eq}^{0.6}\cdot a_{H_3O^+,eq}^{0.4}\cdot a_{U^{4+},eq}\cdot a_{C_2O_4^{2-},eq}^{2.5}\right)^{1/4.5} + \left(a_{N_2H_5^+,eq}^{0.6}\cdot a_{H_3O^+,eq}^{0.9}\cdot a_{U^{4+},eq}^{0.5}\cdot a_{Nd^{3+},eq}^{0.5}\cdot a_{C_2O_4^{2-},eq}^{2.5}\right)^{1/5} = (K_1\gamma_1X_1)^{1/4.5} + (K_2\gamma_2X_2)^{1/5} \quad (39)$$

or:

$$\sum \Pi_{eq}^{1/\sum\nu_i} = \gamma_{\pm,1}\left([N_2H_5^+]_{eq}^{0.6}\cdot[H_3O^+]_{eq}^{0.4}\cdot[U^{4+}]_{eq}\cdot[C_2O_4^{2-}]_{eq}^{2.5}\right)^{1/4.5} + \gamma_{\pm,2}\left([N_2H_5^+]_{eq}^{0.6}\cdot[H_3O^+]_{eq}^{0.9}\cdot[U^{4+}]_{eq}^{0.5}\cdot[Nd^{3+}]_{eq}^{0.5}\cdot[C_2O_4^{2-}]_{eq}^{2.5}\right)^{1/5} = (K_1\gamma_1X_1)^{1/4.5} + (K_2\gamma_2X_2)^{1/5} \quad (40)$$

Based on these Equation (41), the supersaturation ratio during the coprecipitation is:

$$S(X) = \frac{\sum \Pi^{1/\sum\nu_i}}{\sum \Pi_{eq}^{1/\sum\nu_i}} = \frac{\gamma_{\pm,1}\left([N_2H_5^+]^{0.6}\cdot[H_3O^+]^{0.4}\cdot[U^{4+}]\cdot[C_2O_4^{2-}]^{2.5}\right)^{1/4.5}}{\gamma_{\pm,1}\left([N_2H_5^+]_{eq}^{0.6}\cdot[H_3O^+]_{eq}^{0.4}\cdot[U^{4+}]_{eq}\cdot[C_2O_4^{2-}]_{eq}^{2.5}\right)^{1/4.5}} + \frac{\gamma_{\pm,2}\left([N_2H_5^+]^{0.6}\cdot[H_3O^+]^{0.9}\cdot[U^{4+}]^{0.5}\cdot[Nd^{3+}]^{0.5}\cdot[C_2O_4^{2-}]^{2.5}\right)^{1/5}}{\gamma_{\pm,2}\left([N_2H_5^+]_{eq}^{0.6}\cdot[H_3O^+]_{eq}^{0.9}\cdot[U^{4+}]_{eq}^{0.5}\cdot[Nd^{3+}]_{eq}^{0.5}\cdot[C_2O_4^{2-}]_{eq}^{2.5}\right)^{1/5}}$$
$$= \frac{\gamma_{\pm,1}\left([N_2H_5^+]^{0.6}\cdot[H_3O^+]^{0.4}\cdot[U^{4+}]\cdot[C_2O_4^{2-}]^{2.5}\right)^{1/4.5}}{(K_1\gamma_1X)^{1/4.5} + [K_2\gamma_2(1-X)]^{1/5}} + \frac{\gamma_{\pm,2}\left([N_2H_5^+]^{0.6}\cdot[H_3O^+]^{0.9}\cdot[U^{4+}]^{0.5}\cdot[Nd^{3+}]^{0.5}\cdot[C_2O_4^{2-}]^{2.5}\right)^{1/5}}{(K_1\gamma_1X)^{1/4.5} + [K_2\gamma_2(1-X)]^{1/5}} \quad (41)$$

In Equation (41), the concentrations of all compounds in the liquid phase are known, while the solid phase composition X is not directly known, but must be determined by indirect complex calculations. As noted before, the supersaturation must be expressed only as a function of the liquid phase composition, since only this one isknown at any time of coprecipitation process. The compositions (molar fractions of endmembers) of the liquid and solid phases are defined as follows:

$$x_1 = x = \frac{C_{\text{endmember 1,solution}}}{C_{\text{endmember 1,solution}} + C_{\text{endmember 2,solution}}} = \frac{[U^{4+}]-[Nd^{3+}]}{([U^{4+}]-[Nd^{3+}]) + 2\cdot[Nd^{3+}]} = \frac{[U^{4+}]-[Nd^{3+}]}{[U^{4+}]+[Nd^{3+}]} \quad (42)$$

$$x_2 = 1 - x = \frac{C_{\text{endmember 2,solution}}}{C_{\text{endmember 1,solution}} + C_{\text{endmember 2,solution}}} = \frac{2\cdot\left[Nd^{3+}\right]}{\left(\left[U^{4+}\right]-\left[Nd^{3+}\right]\right)+2\cdot\left[Nd^{3+}\right]} = \frac{2\cdot\left[Nd^{3+}\right]}{\left[U^{4+}\right]+\left[Nd^{3+}\right]} \quad (43)$$

$$X_1 = X = \frac{\text{Number of moles of endmember 1 in the solid phase}}{\text{total number of moles of endmembers 1 and 2 in the solid phase}} \quad (44)$$

$$X_2 = 1 - X = \frac{\text{Number of moles of endmember 2 in the solid phase}}{\text{total number of moles of endmembers 1 and 2 in the solid phase}} \quad (45)$$

From Equations (6), (37) and (38) we can obtain:

$$\left[C_2O_4^{2-}\right]_{eq}^{2.5} = \frac{1}{\dfrac{\gamma_{\pm,1}^{4.5}\cdot\left[N_2H_5^+\right]_{eq}^{0.6}\cdot\left[H_3O^+\right]_{eq}^{0.4}\cdot\left[U^{4+}\right]_{eq}}{\gamma_1 K_1} + \dfrac{\gamma_{\pm,2}^{5}\cdot\left[N_2H_5^+\right]_{eq}^{0.6}\cdot\left[H_3O^+\right]_{eq}^{0.9}\cdot\left[U^{4+}\right]_{eq}^{0.5}\cdot\left[Nd^{3+}\right]_{eq}^{0.5}}{\gamma_2 K_2}} \quad (46)$$

Substituting Equation (46) in Equation (40) yields:

$$\sum \Pi_{eq}^{1/\Sigma\nu_i} = \gamma_{\pm,1}\left(\frac{\left[N_2H_5^+\right]_{eq}^{0.6}\cdot\left[H_3O^+\right]_{eq}^{0.4}\cdot\left[U^{4+}\right]_{eq}}{\dfrac{\gamma_{\pm,1}^{4.5}\cdot\left[N_2H_5^+\right]_{eq}^{0.6}\cdot\left[H_3O^+\right]_{eq}^{0.4}\cdot\left[U^{4+}\right]_{eq}}{\gamma_1 K_1} + \dfrac{\gamma_{\pm,2}^{5}\cdot\left[N_2H_5^+\right]_{eq}^{0.6}\cdot\left[H_3O^+\right]_{eq}^{0.9}\cdot\left[U^{4+}\right]_{eq}^{0.5}\cdot\left[Nd^{3+}\right]_{eq}^{0.5}}{\gamma_2 K_2}}\right)^{1/4.5}$$

$$+ \gamma_{\pm,2}\left(\frac{\left[N_2H_5^+\right]_{eq}^{0.6}\cdot\left[H_3O^+\right]_{eq}^{0.9}\cdot\left[U^{4+}\right]_{eq}^{0.5}\cdot\left[Nd^{3+}\right]_{eq}^{0.5}}{\dfrac{\gamma_{\pm,1}^{4.5}\cdot\left[N_2H_5^+\right]_{eq}^{0.6}\cdot\left[H_3O^+\right]_{eq}^{0.4}\cdot\left[U^{4+}\right]_{eq}}{\gamma_1 K_1} + \dfrac{\gamma_{\pm,2}^{5}\cdot\left[N_2H_5^+\right]_{eq}^{0.6}\cdot\left[H_3O^+\right]_{eq}^{0.9}\cdot\left[U^{4+}\right]_{eq}^{0.5}\cdot\left[Nd^{3+}\right]_{eq}^{0.5}}{\gamma_2 K_2}}\right)^{1/5}$$

$$= \gamma_{\pm,1}\left(\frac{\left[N_2H_5^+\right]_{eq}^{0.6}\cdot\left[H_3O^+\right]_{eq}^{0.4}}{\dfrac{\gamma_{\pm,1}^{4.5}\cdot\left[N_2H_5^+\right]_{eq}^{0.6}\cdot\left[H_3O^+\right]_{eq}^{0.4}}{\gamma_1 K_1} + \dfrac{\gamma_{\pm,2}^{5}\cdot\left[N_2H_5^+\right]_{eq}^{0.6}\cdot\left[H_3O^+\right]_{eq}^{0.9}}{\gamma_2 K_2}\sqrt{\dfrac{\left[Nd^{3+}\right]_{eq}}{\left[U^{4+}\right]_{eq}}}}\right)^{1/4.5} \quad (47)$$

$$+ \gamma_{\pm,2}\left(\frac{\left[N_2H_5^+\right]_{eq}^{0.6}\cdot\left[H_3O^+\right]_{eq}^{0.9}\sqrt{\dfrac{\left[Nd^{3+}\right]_{eq}}{\left[U^{4+}\right]_{eq}}}}{\dfrac{\gamma_{\pm,1}^{4.5}\cdot\left[N_2H_5^+\right]_{eq}^{0.6}\cdot\left[H_3O^+\right]_{eq}^{0.4}}{\gamma_1 K_1} + \dfrac{\gamma_{\pm,2}^{5}\cdot\left[N_2H_5^+\right]_{eq}^{0.6}\cdot\left[H_3O^+\right]_{eq}^{0.9}}{\gamma_2 K_2}\sqrt{\dfrac{\left[Nd^{3+}\right]_{eq}}{\left[U^{4+}\right]_{eq}}}}\right)^{1/5}$$

From the definition (42):

$$\frac{\left[Nd^{3+}\right]}{\left[U^{4+}\right]} = \frac{1-x}{1+x} \quad (48)$$

We can obtain:

$$\sum \Pi_{eq}^{1/\sum\nu_i} = \gamma_{\pm,1}\left(\frac{\left[N_2H_5^+\right]_{eq}^{0.6}\cdot\left[H_3O^+\right]_{eq}^{0.4}}{\frac{\gamma_{\pm,1}^{4.5}\cdot\left[N_2H_5^+\right]_{eq}^{0.6}\cdot\left[H_3O^+\right]_{eq}^{0.4}}{\gamma_1 K_1}+\frac{\gamma_{\pm,2}^{5}\cdot\left[N_2H_5^+\right]_{eq}^{0.6}\cdot\left[H_3O^+\right]_{eq}^{0.9}}{\gamma_2 K_2}\sqrt{\frac{1-x}{1+x}}}\right)^{1/4.5} + \gamma_{\pm,2}\left(\frac{\left[N_2H_5^+\right]_{eq}^{0.6}\cdot\left[H_3O^+\right]_{eq}^{0.9}\sqrt{\frac{1-x}{1+x}}}{\frac{\gamma_{\pm,1}^{4.5}\cdot\left[N_2H_5^+\right]_{eq}^{0.6}\cdot\left[H_3O^+\right]_{eq}^{0.4}}{\gamma_1 K_1}+\frac{\gamma_{\pm,2}^{5}\cdot\left[N_2H_5^+\right]_{eq}^{0.6}\cdot\left[H_3O^+\right]_{eq}^{0.9}}{\gamma_2 K_2}\sqrt{\frac{1-x}{1+x}}}\right)^{1/5} \quad (49)$$

and the supersaturation ratio as a function of the liquid phase composition only:

$$S(x) = \frac{\gamma_{\pm,1}\left(\left[N_2H_5^+\right]^{0.6}\cdot\left[H_3O^+\right]^{0.4}\cdot\left[U^{4+}\right]\cdot\left[C_2O_4^{2-}\right]^{2.5}\right)^{1/4.5}}{\sum\Pi_{eq}^{1/\sum\nu_i}\text{ (Equation (49))}} + \frac{\gamma_{\pm,2}\left(\left[N_2H_5^+\right]^{0.6}\cdot\left[H_3O^+\right]^{0.9}\cdot\left[U^{4+}\right]^{0.5}\cdot\left[Nd^{3+}\right]^{0.5}\cdot\left[C_2O_4^{2-}\right]^{2.5}\right)^{1/5}}{\sum\Pi_{eq}^{1/\sum\nu_i}\text{ (Equation (49))}} \quad (50)$$

In addition, we can notice that the equalization of Equation (50) with the right term of Equation (40) gives the cumbersome implicit equation for the determination of the solid phase composition at equilibrium with the liquid phase composition (X versus x).

3. Experimental

The nucleation kinetics is studied using a stopped flow apparatus patented by the CEA/AREVA shown in **Figure 1** [29].

This apparatus is equipped with a Hartridge-Roughton confined opposing jets mixing device. It is characterised by very high micromixing performances as micromixing times of the order of one millisecond can be achieved [30].

The apparatus, made of stainless steel material, is particularly appropriate to hostile environment, due to the simplicity and the experimental set up compactness. It requires a small solution volume and no unintentional release is possible. With all these safe use properties, this apparatus is well adapted to the radioactive material handling in glove boxes [31].

The block diagram of the stopped flow apparatus is represented in **Figure 2**. The thermostated tanks are supplied with reagents, 5 mL of each: uranium and neodymium nitrate solution and oxalic acid solution. The contact of the reagents is performed by two pistons, based on a crossbow mechanism. The pistons ensure a rapid and

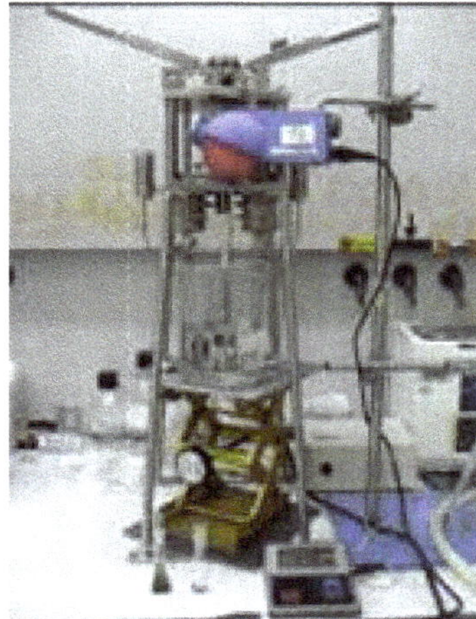

Figure 1. Patented apparatus for the nucleation kinetic. Patented apparatus for nucleation kinetic study [29].

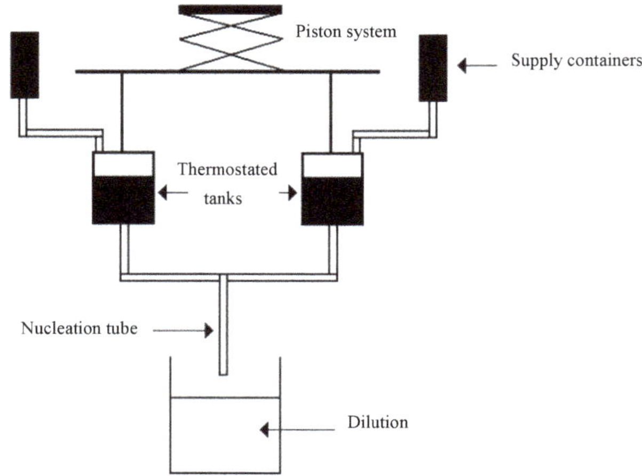

Figure 2. Principle of the experimental setup for the nucleation kinetic study.

simultaneous injection of the reagents in the nucleation tube. The nucleation occurs in the nucleation tube with a length of 10 cm and a diameter of 2 mm. The residence time is about 10 ms. Two optical detectors linked to an electronic system allows the nucleation process time in the tube to be measured.

The duration of the nucleation process is about 0.4 s. A dilution is performed at the tube outlet in a 500 mL vessel initially filled with nitric acid, oxalic acid and hydrazine solution. The role of this quench is to stop the nucleation process with a quick drop of the supersaturation. The dilution vessel is stirred by a 4-blade Rushton turbine at 1500 rpm with 4 baffles. As recommended by Nielsen [32], gelatine is introduced in a concentration of 2 $g \cdot L^{-1}$ in order to limit the agglomeration phenomena. The tube is placed in the discharging jet of the turbine to guarantee a quasi-instantaneous dilution. The solid particles formed can grow during 15 min to reach a detectable size range and the suspension is analysed by laser diffraction granulometry using a Coulter Counter apparatus. A preliminary study on the induction time [26] confirms that only nucleation takes place in the tube, as the residence time in the nucleation tube (10 ms) is lower than the induction period for all supersaturation values. The nuclei number retained corresponds to an average between 3 analysed samples. The nucleation rate is then determined using the total number of nuclei formed and the nucleation period measured with precision in each experiment. The method is well described in [31]. The crystals are observed by Scanning Electron Microscopy (SEM) and analysed by X-Ray diffractometry in order to be certain that the desired solid solution is formed.

A tetravalent uranium nitrate solution is purified on ion exchange resins in order to separate uranium(IV) from uranium(VI). In this way, pure solution containing 99% uranium(IV) is obtained and then characterized by UV-visible spectrophotometer (Shimadzu UV 2550). Uranium(IV) ions can be oxidized by nitrous acid (HNO_2) which is presented in small quantities in nitric acid media. For this reason, hydrazinium nitrate is added in order to stabilize the tetravalent uranium nitrate solution for several months. Neodymium nitrate hexahydrate (99.9%, Alfa Aesar) is dissolved in the uranium nitrate solution with a determined $\left[Nd^{3+}\right]/\left[U^{4+}\right]+\left[Nd^{3+}\right]$ ratio. Oxalic acid solution with a concentration of 0.7 $mol \cdot L^{-1}$ is prepared from oxalic acid dihydrate (Normapur, >99.5%, VWR Prolabo). Deionised water is used in all experiments. The experiments are performed at room temperature under experimental conditions shown in **Tables 1** and **2**.

4. Results and Discussion

4.1. Analysis of Mixed Oxalate Crystals

The SEM photographs validate the formation of mixed oxalates and the absence of agglomeration (see **Figure 3**).

The particles are hexagonal rods which are the characteristic morphology of the solid solutions formed by the precipitation Reaction (1) [23].

In addition, careful X-Ray diffractometric analyses are performed. As it is shown in **Figure 4**, the X-Ray spectrum confirms the formation of pure mixed oxalate as it is consistent with the PDF card 04-11-8368 [22].

The composition of the solid phase is also analysed on different individual particles. The results lead to a homogeneous composition of individual particles.

4.2. Endmembers Solubility

The solubility products of endmembers are experimentally determined under the experimental conditions re-

Table 2. Liquid phase compositions.

$\frac{\left[Nd^{3+}\right]}{\left[U^{4+}\right]+\left[Nd^{3+}\right]}$	10%	20%	30%
x (Equation (42))	0.80	0.60	0.40

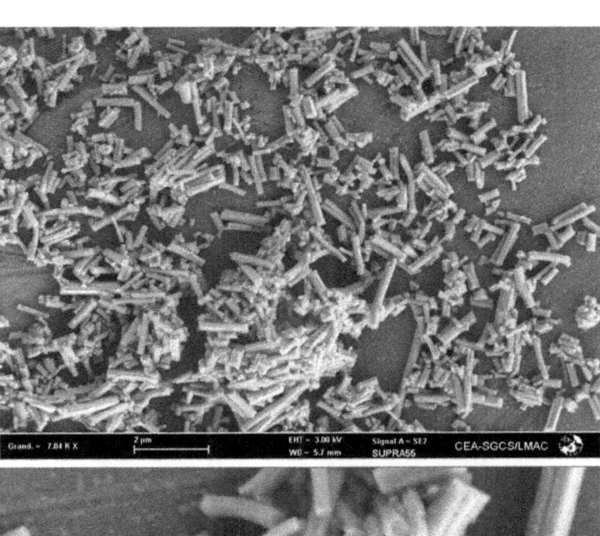

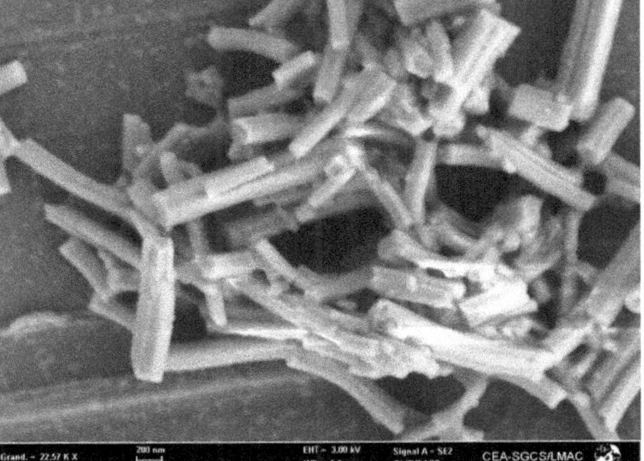

Figure 3. SEM photographs of uranium and neodymium mixed oxalates.

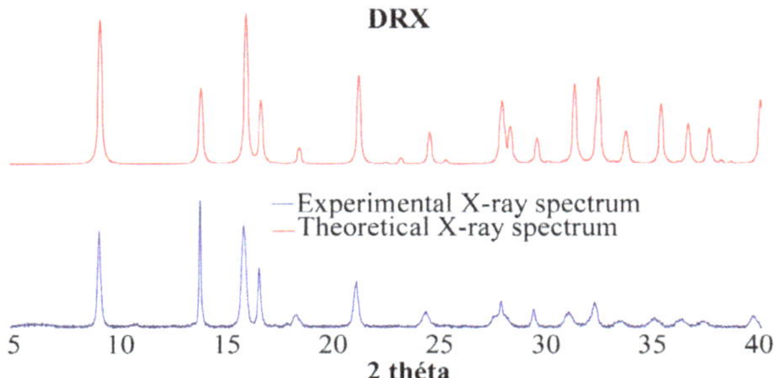

Figure 4. X-ray spectrum of uranium-neodymium oxalates.

ported in **Table 1**. The first step was to synthesize the two endmembers. Then solubility measurements are performed at room temperature in different vessels to study the dissolution of the co-precipitates as a function of time. When equilibrium is reached, the uranium(IV) and neodymium(III) concentrations are analysed by spectrophotometry and ICP-AES [28]. In his thesis, Costenoble remarked that the solid solutions of uranium-neodymium oxalates are the most probably ideal solid solutions due to their structural similarity, so $\gamma_1 = \gamma_2 = 1$. On the contrary, the activity coefficients in solution were impossible to be determined, as no data are available. For instance, the individual contributions of hydrazinium and oxalic ions are not known for the calculation of activity coefficients according to the Bromley method. From Equations (35) and (36), it can be written:

$$K_1 = a^{0.6}_{N_2H_5^+,eq} \cdot a^{0.4}_{H_3O^+,eq} \cdot a_{U^{4+},eq} \cdot a^{2.5}_{C_2O_4^{2-},eq} = \gamma^{4.5}_{\pm,1,eq}\left[N_2H_5^+\right]^{0.6}_{eq}\left[H_3O^+\right]^{0.4}_{eq}\left[U^{4+}\right]_{eq}\left[C_2O_4^{2-}\right]^{2.5}_{eq} \quad (51)$$

$$K_2 = a^{0.6}_{N_2H_5^+,eq} \cdot a^{0.9}_{H_3O^+,eq} \cdot a^{0.5}_{U^{4+},eq} \cdot a^{0.5}_{Nd^{3+},eq} \cdot a^{2.5}_{C_2O_4^{2-},eq} = \gamma^{5}_{\pm,2,eq}\left[N_2H_5^+\right]^{0.6}_{eq}\left[H_3O^+\right]^{0.9}_{eq}\left[U^{4+}\right]^{0.5}_{eq}\left[Nd^{3+}\right]^{0.5}_{eq}\left[C_2O_4^{2-}\right]^{2.5}_{eq} \quad (52)$$

Measuring the concentrations of all compounds in solution, Costenoble [24] has determined the values of the solubility apparent constants:

$$K_1' = \frac{K_1}{\gamma^{4.5}_{\pm,1,eq}} = \left[N_2H_5^+\right]^{0.6}_{eq}\left[H_3O^+\right]^{0.4}_{eq}\left[U^{4+}\right]_{eq}\left[C_2O_4^{2-}\right]^{2.5}_{eq} = 3.09\times10^{-7}\left(mol\cdot L^{-1}\right)^{4.5} \quad (53)$$

$$K_2' = \frac{K_2}{\gamma^{5}_{\pm,2,eq}} = \left[N_2H_5^+\right]^{0.6}_{eq}\left[H_3O^+\right]^{0.9}_{eq}\left[U^{4+}\right]^{0.5}_{eq}\left[Nd^{3+}\right]^{0.5}_{eq}\left[C_2O_4^{2-}\right]^{2.5}_{eq} = 4.37\times10^{-7}\left(mol\cdot L^{-1}\right)^{5} \quad (54)$$

where K_1' and K_2' are the solubility apparent constants.

The supersaturation ratio is calculated by Equation (50) using K_1' and K_2' values. In addition, the coprecipitation is performed in concentrated solutions of ions $N_2H_5^+$, H_3O^+, NO_3^- and $C_2O_4^{2-}$, so the ionic strength remains constant and then the activity coefficients in solution vary slightly during the precipitation process. For this reason, the supersaturation ratio can be calculated assuming $\gamma_{\pm1} \approx \gamma_{\pm2} \approx \gamma_{\pm1,eq} \approx \gamma_{\pm2,eq}$. So the activity coefficients can be simplified in Equation (50).

4.3. Determination of the Nucleation Kinetics

The primary nucleation kinetics of uranium and neodymium mixed oxalates is studied for different fixed $Nd/(U+Nd)$ ratio in solution (see **Table 2**) by varying their concentrations, so by varying the supersaturation ratio. In all published papers on this subject, the authors considered the primary nucleation kinetic law of solid solutions using the similar formula as for a single compound [33]:

$$R_N(x) = A(x)\cdot\exp\left\{-\frac{B(x)}{\left[\ln S(x)\right]^2}\right\} \text{ [33]} \quad (55)$$

All our experiments proved the validity of this equation. For illustration, the experimental data and Equation (55) are shown in **Figure 5** for $x_1 = 0.40$.

Using a non-linear regression technique on $R_N(x)-S(x)$ experimental data, it is found:

—primary homogeneous nucleation kinetics

$$R_{N,hom}(x_1 = 0.4) = 1.86\times10^{22}\times\exp\left\{-\frac{6.40}{\left[\ln S\right]^2}\right\} \quad (56)$$

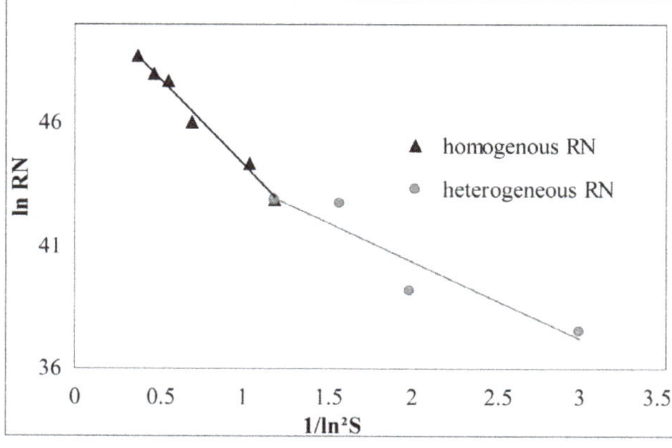

Figure 5. Primary nucleation kinetics for the $Nd/(U+Nd)$ ratio equals to 30% $(x_1 = 0.40)$.

Table 3. Nucleation kinetic parameters for the oxalate solid solutions and the simple oxalates.

	A_{hom} ($m^{-3}\cdot s^{-1}$)	B_{hom}	A_{het} ($m^{-3}\cdot s^{-1}$)	B_{het}
Oxalate solid solution $x_1 = 0.80$	$(6.78 \pm 2.2) \times 10^{21}$	4.79 ± 0.5	$(1.31 \pm 0.6) \times 10^{20}$	1.05 ± 0.3
Oxalate solid solution $x_1 = 0.60$	$(1.89 \pm 1.6) \times 10^{22}$	6.33 ± 1.6	$(6.86 \pm 2.6) \times 10^{20}$	1.63 ± 0.2
Oxalate solid solution $x_1 = 0.40$	$(1.86 \pm 1.2) \times 10^{22}$	6.40 ± 0.7	$(2.10 \pm 1.9) \times 10^{20}$	3.20 ± 1.4
Simple uranium oxalate	$(1.2 \pm 0.4) \times 10^{19}$	85 ± 14.2	$(2 \pm 0.6) \times 10^{18}$	33.4 ± 6.6
Simple neodymium oxalate	$(9.8 \pm 5) \times 10^{19}$	206 ± 25	-	-

—primary heterogeneous nucleation kinetics

$$R_{N,het}\left(x_1 = 0.4\right) = 2.10\times10^{20}\cdot\exp\left\{-\frac{3.20}{\left[\ln S\right]^2}\right\} \quad (57)$$

The method accuracy is estimated from five identical experimental runs performed under the same operating conditions. According to the Student distribution with a 95% confident level, the measure uncertainty is equal to about 30%, which is acceptable for the nucleation rate measurements. The same methodology is used for the other x_1 values and the results are gathered in **Table 3**.

4.4. Comparison with Simple Oxalates

The results obtained in this study can be compared with the results obtained for the nucleation rates of simple uranium oxalate and of simple neodymium oxalate [31]. The primary heterogeneous nucleation kinetic expression for neodymium oxalate was not determined. The data shown in **Table 3** prove that the corresponding pre-exponential factors A are higher, while the exponential coefficients B are lower for solid solutions compared with simple uranium and neodymium oxalates.

Particularly interesting is the fact that the energetic barrier (proportional to B) of solid solutions is much lower than the energetic barrier of simple oxalates. All these data point out that the oxalate solid solutions are kinetically favoured compared with simple oxalates. Generally, it also proves the easiness to obtain the oxalate solid solutions in nuclear industry when oxalic ions are added to a mixture of lanthanide and actinide ions in nitric acid media.

5. Conclusion

This paper describes how to obtain new expressions for the determination of equilibrium relation and supersaturation ratio for any solid solution-aqueous solution (SS-AS) system. An adequate nucleation rate expression is proposed in order to study the co-precipitation processes in nuclear industry. This approach is successfully applied to the uranium-neodymium oxalates co-precipitation. The experimental method developed to study the primary nucleation of solid solutions leads to satisfactory results although the expression of supersaturation ratio is complex. The experimental set up is simple and it is easy to carry out experiments, especially in the presence of dangerous and/or expensive elements. From all experimental results, the homogeneous and heterogeneous primary nucleation kinetic laws of solid solutions of uranium(IV)-neodymium(III) oxalate are obtained for different $\left[Nd^{3+}\right]/\left[U^{4+}\right]$ ratio $\left\{\left[Nd^{3+}\right]/\left[U^{4+}\right]=(1-x)/(1+x)\right\}$. The values of the kinetic parameters point out that in an aqueous mixture of Nd^{3+} and U^{4+}, the oxalate solid solutions are kinetically extremely favoured compared with their simple oxalates, which is a very interesting phenomenological result concerning the formation of oxalate solid solutions of lanthanides and actinides during the co-precipitation process. These kinetic results can be combined with a hydrodynamic model in the same way as it is done for the oxalic precipitation of uranium oxalate [34].

6. Acknowledgements

The authors gratefully acknowledge AREVA for its financial support. The authors are also thankful to the Metallographic and Chemical Analysis Laboratory from CEA for the SEM photos and the DRX spectrum.

REFERENCES

[1] S. Grandjean, A. Beres, J. Rousselle and C. Maillard, Patent No. FR/04510058, 2004.

[2] B. Arab-Chapelet, S. Grandjean, G. Nowogrocki and F. Abraham, "Synthesis of New Mixed Actinides Oxalates as Precursors of Actinides Oxide Solid Solutions," *Journal of Alloys and Compounds*, Vol. 444-445, 2007, pp. 387-390.

[3] F. Lippmann, "Phase Diagrams Depicting Aqueous Solubility of Binary Mineral Systems," *Neues Jahrbuch für Mineralogie Abhandlungen*, Vol. 139, No. 1, 1980, pp. 1-25.

[4] P. D. Glynn and E. J. Reardon, "Solid-Solution Aqueous-Solution Equilibria—Thermodynamic Theory and Representation," *American Journal of Science*, Vol. 290, No. 2, 1990, pp. 164-201.

[5] M. Prieto, A. Putnis and L. Fernandez Diaz, "Crystallization of Solid-Solutions from Aqueous-Solutions in a Porous-Medium—Zoning in $(Ba,Sr)SO_4$," *Geological Magazine*, Vol. 130, No. 3, 1993, pp. 289-299.

[6] H. Gamsjager, E. Konigsberger and W. Preis, "Lippmann Diagrams: Theory and Application to Carbonate Systems," *Aquatic Geochemistry*, Vol. 6, No. 2, 2000, pp. 119-132.

[7] C. M. Pina, M. Enders and A. Putnis, "The Composition of Solid Solutions Crystallising from Aqueous Solutions: The Influence of Supersaturation and Growth Mechanisms," *Chemical Geology*, Vol. 168, No. 3, 2000, pp 195-210.

[8] J. M. Astilleros, C. M. Pina, L. Fernandez Diaz and A. Putnis, "Supersaturation Functions in Binary Solid Solution-Aqueous Solution Systems," *Geochimica et Cosmochimica Acta*, Vol. 67, No. 9, 2003, pp. 1601-1608.

[9] C. M. Pina, A. Putnis and J. M. Astilleros, "The Growth Mechanisms of Solid Solutions Crystallising from Aqueous Solutions," *Chemical Geology*, Vol. 204, No. 1-2, 2004, pp. 145-161.

[10] A. J. Andara, D. M. Heasman, A. Fernandez Gonzales and M. Prieto, "Characterization and Crystallization of $Ba(SO_4,SeO_4)$ Solid Solution," *Crystal Growth & Design*, Vol. 5, No. 4, 2005, pp. 1371-1378.

[11] A. G. Shtukenberg, Y. O. Punin and P. Azimov, "Crystallization Kinetics in Binary Solid Solution-Aqueous Solution Systems," *American Journal of Science*, Vol. 306, No. 7, 2006, pp. 553-574.

[12] J. M. Astilleros, C. M. Pina, L. Fernandez Diaz, M. Prieto and A. Putnis, "Nanoscale Phenomena during the Growth of Solid Solutions on Calcite {101-4} Surfaces," *Chemical Geology*, Vol. 225, No. 3-4, 2006, pp. 322-335.

[13] C. A. YuHang, Asenjo, N. Sanchez Pastor, L. Fernandez Diaz, J. Gomez and C. M. Pina, "Growth of $BaxSr1\text{-}xSO_4$ Nano-Steps on Barite (0 0 1) Face," *Surface Science*, Vol. 601, No. 2, 2007, pp. 381-389.

[14] M. A. Prieto, F. Gonzalez, A. Putnis and L. F. Diaz, "Nucleation, Growth, and Zoning Phenomena in Crystallizing $(Ba, Sr)CO_3$, $Ba(SO_4, CrO_4)$, $(Ba, Sr)SO_4$, and $(Cd, Ca)CO_3$ Solid Solutions from Aqueous Solutions," *Geochimica Cosmochimica Acta*, Vol. 61, No. 16, 1997, pp. 3383-3397.

[15] C. Noguera, B. Fritz, A. Clement and Y. Amal, "Simulation of the Nucleation and Growth of Binary Solid Solutions in Aqueous Solutions," *Chemical Geology*, Vol. 269, No. 1-2, 2010, pp. 89-99.

[16] A. G. Shtukenberg, Y. O. Punin and P. Azimov, "Reply to Comment: Supersaturation in Binary Solid Solution—Aqueous Solution Systems by M. Prieto, J. M. Astilleros, C. M. Pina, L. Fernandez-Diaz and A. Putnis," *American Journal of Science*, Vol. 307, No. 7, 2007, pp. 1046-1050.

[17] M. Prieto, J. M. Astilleros, C. M. Pina, L. Fernandez Diaz and A. Putnis, "Comment: Supersaturation in Binary Solid Solution-Aqueous Solution Systems," *American Journal of Science*, Vol. 307, No. 7, 2007, pp. 1034-1045.

[18] D. C. Thorstenson and L. N. Plummer, "Equilibrium Criteria for 2-Component Solids Reacting with Fixed Composition in an Aqueous Phase Example—Magnesian Calcites," *American Journal of Science*, Vol. 277, No. 9, 1977, pp.1203-1223.

[19] C. M. Pina and A. Putnis, "The Kinetics of Nucleation of Solid Solutions from Aqueous Solutions: A New Model for Calculating Non-Equilibrium Distribution Coefficients," *Geochimica and Cosmochimica Acta*, Vol. 66, No. 2, 2002, pp. 185-192.

[20] A. Putnis, C. M. Pina, J. M. Astilleros, L. Fernandez Diaz and M. Prieto, "Nucleation of Solid Solutions Crystallizing from Aqueous Solutions," *Philosophical Transactions of the Royal Society of London Series A—Mathematical Physical and Engineering Sciences*, Vol. 361, No. 1804, 2003, pp 615-631.

[21] V. Pacary, Y. Barre and E. Plasari, "Modeling and Comparison of Continuous and Semicontinuous Process for Simulating Decontamination of Liquid Nuclear Wastes by Coprecipitation of Strontium Ions with Barium Sulphate," *International Journal of Chemical Reactor Engineering*, Vol. 6, No. A32, 2008.

[22] B. Chapelet-Arab, G. Nowogrocki, F. Abraham and S. Grandjean, "U(IV)/Ln(III) Unexpected Mixed Site in Polymetallic Oxalato Complexes. Part I. Substitution of Ln(III) for U(IV) from the New Oxalate $(NH_4)_2U_2(C_2O_4)_5 \cdot 0.7H_2O$," *Journal of Solid State Chemistry*, Vol. 178, No. 10, 2005, pp. 3046-3054.

[23] B. Arab-Chapelet, S. Grandjean, G. Nowogrocki and F. Abraham, "Synthesis and Characterization of Mixed An(IV) An(III) Oxalates (An(IV) = Th, Np, U or Pu and An(III) = Pu or Am)," *Journal of Nuclear Materials*, Vol. 373, No. 1-3, 2008, pp. 259-268.

[24] S. Costenoble, "Modélisation de la Coprécipitation d'Oxalates Mixtes d'Uranium et de Plutonium Dans le Cadre du Recyclage du Combustible Nucléaire: Solubilité des Solutions Solides Oxalate," Ph.D. Thesis, Lille Sciences and Technologies University, Lille, 2009.

[25] A. Putnis, M. Prieto and L. Fernandez Diaz, "Fluid Supersaturation and Crystallization in Porous-Media," *Geological Magazine*, Vol. 132, No. 1, 1995, pp. 1-13.

[26] D. Parmentier, M. Bertrand and E. Plasari, "Discussion sur l'Expression de la Sursaturation au Cours de la Coprécipitation d'Oxalates d'Actinides," *Proceedings of Cristal* 6, Marseille, 2010.

[27] B. Chapelet-Arab, G. Nowogrocki, F. Abraham and S. Grandjean, "U(IV)/LN(III) Unexpected Mixed Site in Polymetallic Oxalato Complexes. Part II. Substitution of U(IV) for Ln(III) in the New Oxalates $(N_2H_5)Ln(C_2O_4)2 \cdot nH_2O$ (Ln = Nd, Gd)," *Journal of Solid State Chemistry*, Vol. 178, No. 10, 2005, pp. 3055-3065.

[28] S. Costenoble, S. Grandjean, P. Pochin, S. Picart and F. Abraham, "Determination and Modelling of the Solubility of Mixed Actinide(IV)-actinide(III) oxalates (An(IV) = U, Np, Pu and An(III) = Pu, Am)," *Proceedings of Thermodynamics*, London, 23-25 September 2009.

[29] M. Bertrand, P. Baron, E. Plasari and R. Lorrain, Patent No. EN 04/50637, 2004

[30] M. Bertrand, E. Plasari and P. Baron, "Methods for Characterization and Comparison of Mixing Efficiency of Different Confined Opposing Jet Mixing Devices," *Proceedings of the 12th European Conference on Mixing*, Bologne, 2006.

[31] M. Bertrand-Andrieu, E. Plasari and P. Baron, "Determination of Nucleation and Crystal Growth Kinetics in Hostile Environment—Application to the Tetravalent Uranium Oxalate $U(C_2O_4)_2 \cdot 6H_2O$," *Canadian Journal of Chemical Engineering*, Vol. 82, No.5, 2004, pp. 930-938.

[32] A. E Nielsen, "Nucleation and Growth of Crystals at High Supersaturation," *Kristall und Technik*, Vol. 4, No. 1, 1969, pp. 17-38.

[33] M. Volmer and A. Weber, "Keimbildung in Tibersätting Gebilden," *Zeitschrift für physikalisch Chemie*, Vol. 119, 1926, pp. 277-301.

[34] M. Bertrand, E. Plasari, O. Lebaigue, P. Baron, N. Lamarque and F. Ducros, "Hybrid LES-Multizonal Modelling of the Uranium Oxalate Precipitation," *Chemical Engineering Science*, Vol. 77, 2012, pp 75-104.

Nomenclature

a_i: activity of ions in the aqueous solution, mol·m^{-3}
a_{BA}: activity of BA in the solid solution
A: pre-exponential kinetic parameter, m^{-3}·s^{-1}
$\prod a_i^{\upsilon_i}$: product of the ionic activity, mol·m^{-3}
B: kinetic parameter
C_i: $[i]$ concentration of ions i, mol·m^{-3}
K_{BA}: solubility product of BA
K_s: solubility product for simple salts
K' : solubility apparent constant
$\sum \prod_{eq}$: total solubility product
$\sum \prod_{eq}^{1/2}$: total solubility term ?
P_i: partial vapour pressure, Pa
P_{eq}: total vapour pressure of the liquid mixture, Pa
R_N: nucleation rate, m^{-3}·s^{-1}
s: absolute supersaturation ?
S: supersaturation ratio
x: mole fraction in the liquid phase
X: mole fraction in the solid phase

Greek letters

$\gamma_\pm$: mean activity coefficient in the liquid phase
γ: solid phase activity coefficients
υ_i: stoechiometric number of ion i

Subscripts

1: endmember BA
2: endmember CA
eq: at equilibrium
het: heterogenous
hom: homogeneous

13

Pseudosymmetric Features and Nonlinear Optical Properties of Potassium Titanyl Phosphate Crystals

Anastasia P. Gazhulina*, **Mikhail O. Marychev**
Lobachevsky State University of Nizhni Novgorod, Nizhni Novgorod, Russia

ABSTRACT

A number of publications containing structural data, characteristics of nonlinear optical properties of pure and doped crystals of potassium titanyl phosphate (KTP) family have been reviewed to analyze the structural and symmetry conditionality of nonlinear optical properties of these crystals. The pseudosymmetric features of KTP-type crystals with respect to inversion are investigated. Specifically, pseudoinversion distribution maps are calculated; pseudoinversion extrema and coordinates of pseudoinversion centres are found; and the distributions of pure and doped KTP-type structures and their individual atomic sublattices over the degree of pseudoinversion are analyzed. A correlation between the characteristics of nonlinear optical properties of a number of crystals belonging to the KTP family and the degree of pseudoinversion of their atomic structures is demonstrated.

Keywords: Potassium Titanyl Phosphate Family; Pseudosymmetry; Nonlinear Optical Properties

1. Introduction

Study of the relationship of structural and symmetric features of crystals with their physical properties is an urgent problem of condensed-matter physics. Point symmetry determines the set of possible physical properties of crystals, primarily, in correspondence with the Neumann principle. The symmetric features of atomic structures of crystals can be characterized more completely taking into account the phenomenon of pseudosymmetry, which makes it possible to establish finer relationships of the structure-property type. Fedorov pseudosymmetry of crystals [1] is the phenomenon of invariance of a considerable part of the crystal atomic structure (part of electron density and (or) subsystem of atomic nuclei) with respect to some group of symmetry operations compatible with the lattice (with respect to some supergroup of the symmetry space group of crystal).

The pseudosymmetry of a specific structure can quantitatively be characterized by the degree of invariance (degree of pseudosymmetry) of its total electron density $\rho(\mathbf{r})$ with respect to some isometric operation $\hat{g}$ [1,2]:

$$\eta_g\left[\rho(\mathbf{r})\right] = \frac{\int_V \rho(\mathbf{r})\rho(\hat{g}\mathbf{r})\,dV}{\int_V \rho^2(\mathbf{r})\,dV}. \quad (1)$$

Integration in Equation (1) is performed over the volume V of crystal unit cell. If $\hat{g}$ is not a symmetry operation for the function $\rho(\mathbf{r})$, the degree of pseudosymmetry $\eta_g\left[\rho(\mathbf{r})\right] < 1$; however, if $\rho(\mathbf{r})$ is symmetric with respect to the operation $\hat{g}$, $\eta_g\left[\rho(\mathbf{r})\right]$ reaches a maximally possible value: unity.

The second-order susceptibility of crystal determines the intensity of generation of the second optical harmonic and is a structure- and symmetry-sensitive property of crystal. For centrosymmetric crystals, the second-order susceptibility should be zero. One might suggest that reduction of symmetry will lead to some dependence of the second-order susceptibility of crystal on the degree of invariance of crystal structure with respect to inversion.

2. Nonlinear Optical Properties of Potassium Titanyl Phosphate Crystals: A Review

The family of crystals with potassium titanyl phosphate (KTP) structure includes more than 100 compounds [3,4]. Their general formula can be written as $MM'OXO_4$, where M = K, Rb, Na, Cs, Tl, NH_4; M' = Ti, Sn, Sb, Zr,

*Corresponding author.

Ge, Al, Cr, Fe, V, Nb, Ta, Ga; X = P, As, Si, Ge. Inter-octahedral (M'-O-M') oxygen atoms can be replaced with OH- and F-; the resulting compounds with the general formula MM'(F,OH)XO4 also belong to the KTP family. The structure of KTP crystals is described by the space group $Pna2_1$.

We considered 118 crystals belonging to the KTP family, including 29 pure and 89 doped ones. Information about the nonlinear optical characteristics of 108 crystals was found in the corresponding publications. All crystals under consideration were separated into three groups with respect to the available data on their structure and nonlinear optical properties; the relationship between these groups is clearly shown in **Figure 1**.

The characteristics of nonlinear optical properties of KTP crystals are listed in **Table 1**, where the parameter $I/I_{reference}$ is the ratio of second harmonic intensities from a sample under study studied and a powder sample of reference crystal.

The characteristics of nonlinear optical properties were determined in [8,10,11,24] by the Kurtz-Perry method [26] and in [3,5-7,12,14,19,22] by the method described in [27]. The components of the second-order susceptibility tensor were found in [16,17,20] using the Maker fringe technique [28,29].

With allowance for the results of our analysis of the corresponding publications, we can select crystals whose characteristics of nonlinear optical properties are comparable with those for $KTiOPO_4$ crystal ($K_{0.5}Rb_{0.5}TiOPO_4$, RTA, $K_{0.966}Ti_{0.966}Nb_{0.034}OPO_4$, $K_{0.921}Ti_{0.921}Nb_{0.079}OPO_4$, RTP, $K_{0.99}Ti_{0.99}Sb_{0.01}OPO_4$, $KTi_{0.96}Zr_{0.04}OPO_4$, TTP, $KTi_{0.9975}V_{0.0025}OPO_4$, $K_{0.5}Ti_{0.5}Nb_{0.5}OPO_4$, CTA, $K_{0.5}Ta_{0.5}Ti_{0.5}OPO_4$, $KTiO(PO_4)_{0.5}(AsO_4)_{0.5}$, TTA, $KTi_{0.7}Nb_{0.3}OP_{0.7}Si_{0.3}O_4$, $KTi_{0.65}Nb_{0.35}OP_{0.65}Si_{0.35}O_4$, $KTi_{0.6}Nb_{0.4}OP_{0.6}Si_{0.4}O_4$, $RbTi_{0.98}Nb_{0.02}OPO_4$, $Na_{0.87}K_{0.13}TiOAsO_4$, $KTi_{0.7}Nb_{0.3}OAs_{0.7}Si_{0.3}O_4$, $KTi_{0.6}Nb_{0.4}OAs_{0.6}Si_{0.4}O_4$ $RbTi_{0.927}Nb_{0.056}Er_{0.017}OPO_4$) and crystals with characteristics of nonlinear optical properties exceeding those of $KTiOPO_4$ crystal (KTA, $K_{0.98}Ti_{0.98}Nb_{0.02}OPO_4$, $K_{0.96}Ti_{0.96}Nb_{0.04}OPO_4$, $K_{0.97}Ti_{0.97}Sb_{0.07}OPO_4$, $K_{0.88}Ti_{0.98}Zr_{0.06}OP_{0.99}O_4$, $K_{0.88}Ti_{0.93}Zr_{0.11}OP_{0.99}O_4$, $K_{0.97}Ti_{0.99}OAs_{0.53}P_{0.49}O_4$, $KTi_{0.9}Nb_{0.1}OP_{0.9}Si_{0.1}O_4$, $K_{0.80}Ti_{0.26}Zr_{0.78}OAs_{1.01}O_4$, $KTi_{0.9}Nb_{0.1}OAs_{0.9}Si_{0.1}O_4$, $K_{0.94}Nb_{0.12}Ti_{0.91}OAs_{0.89}Ge_{0.09}O_4$, $KTi_{0.8}Nb_{0.2}OAs_{0.8}Si_{0.2}O_4$, $K_{1.02}Nb_{0.25}Ti_{0.76}OAs_{0.75}Ge_{0.23}O_4$, $K_{0.68}Rb_{0.32}TiOPO_4$, $Cs_{0.5}K_{0.5}TiOAsO_4$, $KTi_{0.97}Zr_{0.03}OPO_4$, $K_{0.54}Li_{0.46}TiOAsO_4$, $K_{1.03}Nb_{0.52}Ti_{0.48}OAs_{0.48}Ge_{0.51}O_4$, $Rb_{0.855}Ti_{0.955}Nb_{0.045}OPO_4$, $KNb_{0.52}Ti_{0.48}OAs_{0.48}Ge_{0.51}O_4$, $RbTi_{0.96}Nb_{0.04}OPO_4$, $K_{0.98}Nb_{0.46}Ti_{0.56}OAs_{0.58}Ge_{0.39}O_4$).

There are data in the literature on $KTiOPO_4$ crystals doped with Nb [30-36], Ge [37], Sn [33,38-40], Zr [41-43], Sb [35,44,45], Ta [35], Fe [46], Hf [47], and Zn [48] and RTP crystals doped with Cs [49] and Zr [50,51]. It was indicated in [39] that an increase in the Sn content leads to a dramatic decrease in the output second-harmonic intensity to zero. The intensity of second-harmonic generation (SHG) for $KTi_{1-x}Zr_xOPO_4$ crystals reaches a maximum at x = 0.28, where it is more than doubled in comparison with the $KTiOPO_4$ sample [41]. The SHG intensity increases with an increase in the zirconium content in $RbTi_{1-x}Zr_xOPO_4$ crystals; at x = 0.034, it rises by 40% [50]. The SHG intensity increases by approximately 35% in comparison with pure $KTiOPO_4$ samples after replacement of 6% titanium atoms with hafnium [47]. $RbTi_{1-x}Ta_xOPO_4$ and $RbTi_{1-x}Nb_xOPO_4$ crystals were investigated in [52], as well as Yb-doped $RbTi_{1-x}Ta_xOPO_4$ crystals and $RbTi_{1-x}Nb_xOPO_4$ crystals doped with Yb, Ln and Er. $KTiOPO_4$ crystals doped with transition metals and RTA crystals doped with lanthanides were studied in [53]. A number of compounds ($RbTi_{0.98}Er_{0.01}Nb_{0.01}OAsO_4$, $RbTi_{0.96}Er_{0.02}Nb_{0.02}OAsO_4$, and $KTi_{0.98}Cr_{0.02}O_{0.98}F_{0.02}PO_4$, $KTi_{0.99}Fe_{0.01}O_{0.99}F_{0.01}PO_4$ $RbTi_{0.98}Er_{0.02}O(AsO_4)_{0.98}(SO_4)_{0.02}$) exhibited an increase in the SHG intensity in comparison with RTA and KTP crystals, respectively.

3. Analysis of the Degree of Invariance of the Structure of KTP Crystals with Respect to Inversion

The complete characteristic of pseudosymmetry of any crystal under study with respect to inversion is a three-dimensional distribution map of the degree of structural invariance (electron density) with respect to this operation (hereinafter, pseudoinversion), calculated for different positions of inversion points within their unit cell. These maps were obtained with scanning steps over the unit-cell axes a, b, and c chosen to be 0.05 of the cor responding unit-cell parameters. For 118 crystals (**Figure 1**) with known structure, we calculated three-dimensional pseudoinversion maps using Equation (1). The calculations were performed using the computer program and technique described in [54]. Within this approach the electron density function is expanded in a Fourier series in structural amplitudes ([1], see Formulas (5) and (6)).

Figure 2 presents typical examples of cross sections of three-dimensional distribution maps of the degree of pseudoinversion for the structures of $KTiOPO_4$, $KSnOPO_4$, $KTiOAsO_4$, $Cs_{0.625}K_{0.375}TiOAsO_4$ crystals (cuts by the plane z = 0.25). For the structures presented in **Figure 2**, the origin of coordinates is chosen on the two fold screw axis, and the coordinates of pseudoinversion peaks on the x and y axes are 0.25. We chose cuts by the plane z = 0.25 in **Figure 2** because the z coordinate of the pseudoinversion peaks for the structures of the aforementioned crystals is also 0.25.

This situation is typical of most structures under study; in KTP crystals is accompanied by a phase transition to the centrosymmetric space group $Pnan$. Indeed, having

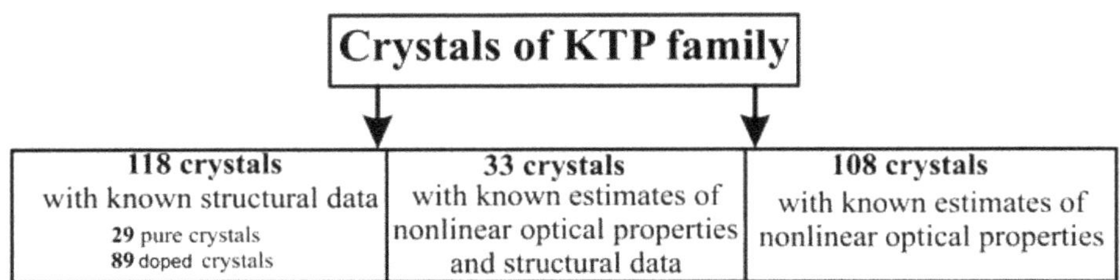

Figure 1. Groups of KTP crystals under consideration.

Table 1. Characteristics of nonlinear optical properties of KTP crystals.

№	Crystal	Characteristics of nonlinear optical properties	References
1	$KTiOPO_4$ (KTP)	6000 (I/I_{SiO2})	[3,5]
		910 (I/I_{SiO2})	[6,7]
		4.24 ± 0.17 ($\chi_{2\omega}$, relative to KDP)	[8]
		d_{15} (0.852 μm) = 1.9 ± 0.1 pm/V d_{24} (0.852 μm) = 3.9 ± 0.2 pm/V d_{33} (0.852 μm) = 16.6 ± 0.8 pm/V d_{15} (1.064 μm) = 1.9 ± 0.1 pm/V d_{24} (1.064 μm) = 3.7 ± 0.2 pm/V d_{31} (1.064 μm) = 2.2 ± 0.1 pm/V d_{32} (1.064 μm) = 3.7 ± 0.2 pm/V d_{33} (1.064 μm) = 14.6 ± 0.7 pm/V d_{15} (1.313 μm) = 1.4 ± 0.1 pm/V d_{24} (1.313 μm) = 2.6 ± 0.1 pm/V d_{33} (1.313 μm) = 11.1 ± 0.6 pm/V	[9]
2	$RbTiOPO_4$ (RTP), rubidium titanyl phosphate	6000 (I/I_{SiO2})	[3,5]
		0.73 (I/I_{KTP})	[10]
		0.7 (I/I_{KTP})	[11]
		d_{31} (1.064 μm) = 3.3 ± 0.6 pm/V d_{32} (1.064 μm) = 4.1 ± 0.8 pm/V d_{33} (1.064 μm) = 17.1 ± 3.4 pm/V	[9]
3	$TlTiOPO_4$ (TTP), thallium titanyl phosphate	6000 (I/I_{SiO2})	[3,5]
4	$NaTiOPO_4$	160 (I/I_{SiO2})	[3]
5	$AgTiOPO_4$	5 (I/I_{SiO2})	[3,5,7]
6	$(NH_4)TiOPO_4$ (NTP)*, ammonium titanyl phosphate	2400 (I/I_{SiO2})	[5,12]
7	$KSnOPO_4$	0.50 (I/I_{SiO2})	[13]
8	$KGeOPO_4$	3.3 (I/I_{SiO2})	
9	$NaGeOPO_4$	4 (I/I_{SiO2})	[3,5]
10	$KVOPO_4$	opaque	
11	$KTiOAsO_4$ (KTA), potassium titanyl arsenate	6000 (I/I_{SiO2})	[3]
		990 (I/I_{SiO2})	[6,7]
		1.09 (I/I_{KTP})	[14]
		1.01 (I/I_{KTP})	[15]
		d_{15} (1.064 μm) = 1.3 × d_{15}(KTP) d_{24} (1.064 μm) = (1.8 ± 0.1) × d_{15}(KTA) d_{31} (1.064 μm) = 2.8 ± 0.3 pm/V d_{31} (1.064 μm) = (1.3 ± 0.1) × d_{31}(KTP) d_{32} (1.064 μm) = 4.2 ± 0.4 pm/V d_{32} (1.064 μm) = (1.8 ± 0.1) × d_{31}(KTA) d_{33} (1.064 μm) = 16.2 ± 1.0 pm/V d_{15} (1.32 μm) = 1.2 × d_{15}(KTP) d_{24}(1.32 μm) = 1.7 × d_{15}(KTP)	[9,16]

Continued

		6000 (I/I_{SiO2})	[3]
12	$RbTiOAsO_4$ (**RTA**), rubidium titanyl arsenate	d_{31} (1.064 μm) = 2.3 ± 0.5 pm/V d_{31} (1.064 μm) = 3.55 × d_{36}(KDP) d_{32} (1.064 μm) = 3.8 ± 0.7 pm/V d_{32} (1.064 μm) = 11.71 × d_{36}(KDP) d_{33} (1.064 μm) = 15.8 ± 1.6 pm/V d_{33} (1.064 μm) = 31.05 × d_{36}(KDP)	[9]
13	$CsTiOAsO_4$ (CTA), cesium titanyl arsenate	d_{31} (1.064 μm) = 2.1 ± 0.4 pm/V d_{32} (1.064 μm) = 3.4 ± 0.7 pm/V d_{33} (1.064 μm) = 18.1 ± 1.8 pm/V d_{31} (1.32 μm) = 1.1 ± 0.1 pm/V d_{32} (1.32 μm) = 1.7 ± 0.6 pm/V	[9,17]
14	$TlTiOAsO_4$ (TTA), thallium titanyl arsenate	6000 (I/I_{SiO2})	
15	$NH_4TiOAsO_4$	100 (I/I_{SiO2})	
16	$KGeOAsO_4$	0.03 (I/I_{SiO2})	
17	$KSnOAsO_4$	0.53 (I/I_{SiO2})	[3]
18	$RbZrOAsO_4$	3 (I/I_{SiO2})	
19	$CsZrOAsO_4$	2 (I/I_{SiO2})	
20	$NH_4ZrOAsO_4$	1 (I/I_{SiO2})	
21	$KSbOSiO_4$	0.5 (I/I_{SiO2})	[18]
22	$NaSbOSiO_4$	0.4 (I/I_{SiO2})	
23	$AgSbOSiO_4$	1.1 (I/I_{SiO2})	[18]
24	$KSbOGeO_4$	0.95 (I/I_{SiO2})	
25	$NaSbOGeO_4$	0.8 (I/I_{SiO2})	
26	$AgSbOGeO_4$	1.5 (I/I_{SiO2})	
27	$KFePO_4F$	2.66 (I/I_{SiO2})	[3,5]
28	$KGaAsO_4F$	0.02 (I/I_{SiO2})	[3]
29	$KFeAsO_4F$	1 (I/I_{SiO2})	
30	$K_2FeNb(PO_5)_2$	1 (I/I_{SiO2})	[5]
31	$RbScFAsO_4$	0.5 (I/I_{SiO2})	[19]
32	$CsScFAsO_4$	1.2 (I/I_{SiO2})	
33	$Ag_{0.5}K_{0.5}TiOPO_4$	130 (I/I_{SiO2}) 135 (I/I_{SiO2})	[3] [7]
34	$Ag_{0.85}K_{0.15}TiOPO_4$	7 (I/I_{SiO2}) 0.01 (I/I_{KTP}) ***	[3,6,7] [14]
35	$(NH_4)_{0.5}K_{0.5}TiOPO_4$	1100 (I/I_{SiO2})	[3,5]
36	$K_{0.5}Rb_{0.5}TiOPO_4$	6000 (I/I_{SiO2})	[3]
37	$K_{0.68}Rb_{0.32}TiOPO_4$	d_{31} (1.06 μm) = 6.5 pm/V d_{32} (1.06 μm) = 5.0 pm/V d_{33} (1.06 μm) = 13.7 pm/V d_{24} (1.06 μm) = 7.6 pm/V d_{15} (1.06 μm) = 6.1 pm/V	[20]
38	$Na_{0.2}K_{0.8}TiOPO_4$	675 (I/I_{SiO2})	[7]
39	$Na_{0.4}K_{0.6}TiOPO_4$	620 (I/I_{SiO2})	
40	$Na_{0.55}K_{0.45}TiOPO_4$	570 (I/I_{SiO2})	
41	$Na_{0.65}K_{0.35}TiOPO_4$	590 (I/I_{SiO2})	
42	$Na_{0.95}K_{0.05}TiOPO_4$	100 (I/I_{SiO2}) 90 (I/I_{SiO2}) 0.11 (I/I_{KTP}) ***	[6] [7] [14]
43	$K_{0.55}Li_{0.45}TiOPO_4$	620 (I/I_{SiO2}) 0.68 (I/I_{KTP}) ***	[6] [14]

Continued

44	$K_{0.966}Ti_{0.966}Nb_{0.034}OPO_4$	d_{15} (1.064 μm) = (0.8 ± 0.1) × d_{15}(KTP) d_{24} (1.064 μm) = (2.2 ± 0.1) × d_{15}(Nb: KTP)	[9]
45	$K_{0.921}Ti_{0.921}Nb_{0.079}OPO_4$	d_{15} (1.064 μm) = 0.75 × d_{15}(KTP) ± 10% d_{24} (1.064 μm) = 1.13 × d_{24}(KTP) ± 10% d_{33} (1.064 μm) = 0.9 × d_{33}(KTP) ± 10%	
46	$K_{0.98}Ti_{0.98}Nb_{0.02}OPO_4$	4.56 ± 0.18	
47	$K_{0.96}Ti_{0.96}Nb_{0.04}OPO_4$	4.97 ± 0.18	
48	$K_{0.89}Ti_{0.89}Nb_{0.11}OPO_4$	2.39 ± 0.25	
49	$K_{0.99}Ti_{0.99}Sb_{0.01}OPO_4$	4.18 ± 0.22 ($\chi_{2\omega(\text{relative to KDP})}$)	[8]
50	$K_{0.97}Ti_{0.97}Sb_{0.07}OPO_4$	4.50 ± 0.18	
51	$K_{0.83}Ti_{0.83}Sb_{0.17}OPO_4$	1.02 ± 0.05	
52	$KTi_{0.97}Zr_{0.03}OPO_4$	4.58 ± 0.21	
53	$KTi_{0.96}Zr_{0.04}OPO_4$	4.33 ± 0.2	
54	$K_{0.88}Ti_{0.98}Zr_{0.06}OP_{0.99}O_4$	1.8 (I/I_{KTA})	[21]
55	$K_{0.88}Ti_{0.93}Zr_{0.11}OP_{0.99}O_4$	1.7 (I/I_{KTA})	
56	$KTi_{0.5}V_{0.5}OPO_4$	0.0008 (I/I_{KTP})	[22]
57	$KTi_{0.75}V_{0.25}OPO_4$	0.05 (I/I_{KTP})	[5]
58	$KTi_{0.85}V_{0.15}OPO_4$	0.1 (I/I_{KTP})	
59	$KTi_{0.95}V_{0.05}OPO_4$	0.13 (I/I_{KTP})	[22]
60	$KTi_{0.9975}V_{0.0025}OPO_4$	1 (I/I_{KTP})	
61	$K_{0.67}Ti_{0.5}V_{0.5}OPO_4$	0.20 (I/I_{KTP})	[5]
62	$K_{0.75}Ti_{0.75}V_{0.25}OPO_4$	0.24 (I/I_{KTP})	
63	$K_{0.85}Ti_{0.85}V_{0.15}OPO_4$	0.36 (I/I_{KTP})	
64	$K_{0.5}Ti_{0.5}Nb_{0.5}OPO_4$	0.9 (I/I_{KTP})	[23]
65	$K_{0.5}Ta_{0.5}Ti_{0.5}OPO_4$	0.8 (I/I_{KTP})	
66	$KGa_{0.5}Nb_{0.5}OPO_4$	1 (I/I_{SiO2})	[3]
67	$KFe_{0.5}Nb_{0.5}OPO_4$	2.7 (I/I_{SiO2})	
68	$K_{0.5}Nb_{0.5}V_{0.5}OPO_4$	0.5 (I/I_{KTP})	[23]
69	$K_{0.5}Ta_{0.5}V_{0.5}OPO_4$	0.4 (I/I_{KTP})	
70	$KTiO(PO_4)_{0.5}(AsO_4)_{0.5}$	6000 (I/I_{SiO2})	[3]
71	$K_{0.97}Ti_{0.99}OAs_{0.53}P_{0.49}O_4$	1.6 (I/I_{KTA})	[21]
72	$KTi_{0.9}Nb_{0.1}OP_{0.9}Si_{0.1}O_4$	1.05 (I/I_{KTP})	[15]
73	$KTi_{0.8}Nb_{0.2}OP_{0.8}Si_{0.2}O_4$	0.96 (I/I_{KTP})	
74	$KTi_{0.7}Nb_{0.3}OP_{0.7}Si_{0.3}O_4$	0.84 (I/I_{KTP})	
75	$KTi_{0.65}Nb_{0.35}OP_{0.65}Si_{0.35}O_4$	0.81(I/I_{KTP})	
76	$KTi_{0.6}Nb_{0.4}OP_{0.6}Si_{0.4}O_4$	0.72 (I/I_{KTP})	
77	$K_2GaGeP_2O_9(F, OH)$	10 (I/I_{SiO2})	[5]
78	$KTi_{0.5}Ga_{0.5}O_{0.5}PO_4F_{0.35}(OH)_{0.15}$	200 (I/I_{SiO2})	[3]
79	$KGaPO_4F_{0.7}(OH)_{0.3}$	0.72 (I/I_{SiO2})	[3, 5]
80	$RbTi_{0.98}Nb_{0.02}OPO_4$	0.97 (I/I_{KTP})	[10]
81	$RbTi_{0.96}Nb_{0.04}OPO_4$	1.23 (I/I_{KTP})	
82	$RbTi_{0.93}Nb_{0.07}OPO_4$	0.73 (I/I_{KTP})	
83	$Rb_{0.855}Ti_{0.955}Nb_{0.045}OPO_4$	1.2 (I/I_{KTP})	[11]
84	$RbTi_{0.927}Nb_{0.056}Er_{0.017}OPO_4$	0.7 (I/I_{KTP})	
85	$Rb_{0.855}Ti_{0.95}Ta_{0.04}OPO_4$	0.95 (I/I_{KTP})	[24]
86	$RbTi_{0.95}Ta_{0.03}Y_{0.02}OPO_4$	0.80 (I/I_{KTP})	
87	$RbGa_{0.5}Nb_{0.5}OPO_4$	1 (I/I_{SiO2})	[3]
88	$(NH_4)_{0.5}H_{0.5}TiOPO_4$ **	60 (I/I_{SiO2}) 40 (I/I_{SiO2})	[5] [12]

Continued

89	$(NH_4)_{0.5}(H_3O)_{0.5}TiOPO_4$	700 (I/I_{SiO2})	[3,5]
		650 (I/I_{SiO2})	[5,12]
90	$Cs_{0.5}K_{0.5}TiOAsO_4$	6700 (I/I_{SiO2})	[3]
91	$Na_{0.87}K_{0.13}TiOAsO_4$	790 (I/I_{SiO2})	[5,6]
		0.87 (I/I_{KTP}) ***	[14]
92	$Na_{0.98}K_{0.02}TiOAsO_4$	0.01 (I/I_{KTP})	
93	$K_{0.54}Li_{0.46}TiOAsO_4$	970 (I/I_{SiO2})	[6]
		1.07 (I/I_{KTP}) ***	[14]
94	$Ag_{0.98}K_{0.02}TiOAsO_4$	10 (I/I_{SiO2})	[6]
95	$(NH_4)_{0.5}K_{0.5}TiOAsO_4$	100 (I/I_{SiO2})	[3]
96	Sc: KTA (0.22 % dopant)	d_{24} (1.32 μm) = 1.4 × d_{15}(KTP)	[16]
97	$K_{0.80}Ti_{0.26}Zr_{0.78}OAs_{1.01}O_4$	1.2 (I/I_{KTA})	[21]
98	$KTi_{0.9}Nb_{0.1}OAs_{0.9}Si_{0.1}O_4$	1.04 (I/I_{KTP})	[15]
99	$KTi_{0.8}Nb_{0.2}OAs_{0.8}Si_{0.2}O_4$	1.03 (I/I_{KTP})	
100	$KTi_{0.7}Nb_{0.3}OAs_{0.7}Si_{0.3}O_4$	0.98 (I/I_{KTP})	
101	$KTi_{0.6}Nb_{0.4}OAs_{0.6}Si_{0.4}O_4$	0.90 (I/I_{KTP})	
102	$K_{0.94}Nb_{0.12}Ti_{0.91}OAs_{0.89}Ge_{0.09}O_4$	1.3 (I/I_{KTA})	[21]
103	$K_{1.02}Nb_{0.25}Ti_{0.76}OAs_{0.75}Ge_{0.23}O_4$	1.1 (I/I_{KTA})	
104	$K_{1.03}Nb_{0.52}Ti_{0.48}OAs_{0.48}Ge_{0.51}O_4$	1.1 (I/I_{KTA})	
105	$KNb_{0.52}Ti_{0.48}OAs_{0.48}Ge_{0.51}O_4$	1.3 (I/I_{KTA})	
106	$K_{0.98}Nb_{0.46}Ti_{0.56}OAs_{0.58}Ge_{0.39}O_4$	1.2 (I/I_{KTA})	
107	$KGa_{0.5}Nb_{0.5}OAsO_4$	1 (I/I_{SiO2})	[3]
108	$RbGa_{0.5}Nb_{0.5}OAsO_4$	5.5 (I/I_{SiO2})	

*A value of 1100 was indicated in [3], with reference to [5], where a value of 2400 was reported. **A value of 140 was indicated in [25], with reference to [12], where a value of 40 was reported, and a value of 6 was indicated in [3], with reference to [5], where the corresponding value was found to be 60. ***Values of second-harmonic generation intensity for KTiOPO4 crystal were reported in [14] with reference to [6], where the corresponding values were given for quartz crystal. The values of [14] correspond to those of [6], when divided by I/I_{SiO2} value for KTiOPO4 crystal (also taken from [6]).

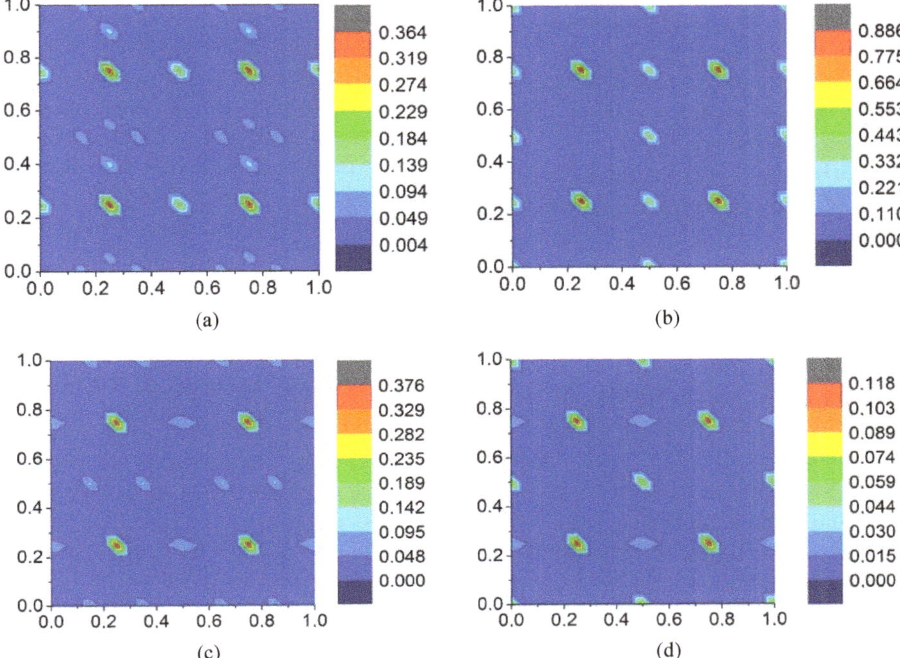

Figure 2. Cut of three-dimensional distribution maps of the degree of pseudoinversion of crystal structure by the plane z = 0.25: (a) $KTiOPO_4$ (CSD-№ 20970); (b) $KSnOPO_4$ (CSD-№ 68706); (c) $KTiOAsO_4$ (CSD-№ 202158); and (d) $Cs_{0.625}K_{0.375}TiOAsO_4$ (CSD-№ 74595).

added inversion to the set of symmetry operations of the space group $Pna2_1$, which describes the structure of KTP crystals at room temperature, we obtain the group *Pnan*, where the inversion centre with respect to the twofold screw axis has coordinates (0.25, 0.25).

Thus, in the polar phase of KTP structures, the pseudoinversion peaks are located specifically at inversion centres of these crystals in their high-symmetry nonpolar phase. With allowance for this circumstance, we will characterize the pseudosymmetry of the electron density in each crystal with a known structure by the maximum value of pseudoinversion η_{max} in the three-dimensional map, and the point with the coordinates corresponding to the found η_{max} values will be referred to as pseudoinversion centres.

Since the origin of coordinates is arbitrarily chosen in X-ray diffraction analysis, the coordinates of the pseudoinversion centres may differ from 0.25. In the $Pna2_1$ group [55], the origin of coordinates on the z axis can be chosen at any point, while in the directions of the x and y axes it may lie either on the twofold axis or at the intersection of mirror planes; therefore, the x and y coordinates of pseudoinversion centres can be either (0.25, 0.25) or (0, 0).

To refine the coordinates of pseudoinversion centres and η_{max} values, we additionally calculated the distribution of the degree of pseudoinversion with a relative scanning step of 0.025 over the unit cell axes. Fixed refined x and y coordinates of pseudoinversion centres were used for repeated calculation of pseudoinversion distribution along the z axis with a relative scanning step of 0.001.

Table 2 contains the maximum pseudoinversion values η_{max} and coordinates of pseudoinversion centres $z(\eta_{max})$ for a number of KTP structures.

Figures 3(a) and **3(b)** show the distribution histograms for the degree of pseudoinversion η_{max} for pure and doped KTP crystals.

The distribution of pure KTP crystals over pseudoinversion is fairly uniform. As is indicated in **Table 2**, the mean value $< \eta_{max} >$ for them is 0.606. The situation for doped crystals is different: the pronounced maximum in the histogram in **Figure 3(b)**, which amounts to 31%, lies in the range of pseudoinversion values of 0.4 - 0.5, which is followed by a sharp falloff. Therefore, the fraction of pseudo-centrosymmetric structures among doped KTP crystals is very small. The mean value $< \eta_{max} >$ for doped crystals is 0.490. Thus, doped KTP crystals are "less symmetric" with respect to inversion than pure compositions.

For the crystals listed in **Table 2**, along with the calculations of the pseudoinversion of their structures as a whole, pseudoinversion extrema for sublattices of individual types of atoms (η_{max}(sublattices)) were also calculated. Pseudoinversion was calculated for the pure sublattices of all 118 crystals in **Table 2**; the distribution histogram of the corresponding extrema is shown in **Figure 3(c)**. For 89 doped crystals in **Table 2**, the results of similar calculations for M- and M'-type sublattices containing doped atoms are presented as histograms in **Figure 3(d)**. The histograms in **Figure 3(c)** indicate that the sublattices of X, O, and F atoms are most pseudo-centrosymmetric, sublattices of M-type atoms are least pseudo-centrosymmetric, and the pseudoinversion of the M' sublattice is intermediate ($< \eta_{max}(X) \geq 0.857$, $< \eta_{max}(O) \geq 0.720$, $< \eta_{max}(F) \geq 0.870$, $< \eta_{max}(M) \geq 0.395$, $< \eta_{max}(M') \geq 0.661$).

In the presence of impurities, the general view of the histogram for the M' sublattice (**Figure 3(d)**) barely differs from that in **Figure 3(c)**; its characteristic maximum shifts to higher pseudoinversion values and the mean value $< \eta_{max}(M') >$ becomes 0.700.

The pseudoinversion histogram for the M-type sublettice changes more radically: the pronounced peak in the range of 0.3 - 0.4 in **Figure 3(c)** disappears in **Figure 3(d)**, and the distribution becomes more uniform in a wider pseudoinversion range; the fraction of crystals with ultimately acentric M sublattices increases. The mean pseudo inversion $< \eta_{max}(M) >$ becomes 0.384; *i.e.*, it barely changes in comparison with $< \eta_{max}(M) >$ for M sublattices without impurities. Thus, the analysis of the pseudo inversion of individual sublattices suggests that the reductions of pseudoinversion of structures as a whole at a transition to doped KTP compositions, which is noted in **Table 2** and **Figure 3(b)**, is related to a great extent to the higher sensitivity of the pseudoinversion of M-type sublattice to the presence of doped atoms. Note that pseudosymmetry was previously studied [58] by the atomic displacement method [1] for 11 KTP-type structures. In particular, it was established that the potassium sublattice is less centrosymmetric in comparison with the TiO_6-PO_4 subsystem, and its pseudosymmetry relative to inversion is more sensitive to introduction of impurities.

4. Comparison of the Nonlinear Optical Characteristics of KTP Crystals and the Pseudoinversion of Their Structures

A model was proposed in [8], according to which the second-order susceptibility of crystals is related to the symmetry of KTP-type structures and their pseudoinversion as follows:

$$\chi_{2\omega} \sim \sqrt{1-\eta} \quad (2)$$

As can be seen in **Table 1**, the SHG data with respect to the reference sample (powder of pure SiO_2 crystal) differ by an order of magnitude in different studies for KTP [3,5-7], and KTA [3,6,7] crystals. Based on this fact, we illustrated Equation (2) by selecting a group of com-

Table 2. Maximum pseudoinversion values η_{max} and z coordinates of pseudoinversion centres $z(\eta_{max})$ for a number of KTP structures.

№	Crystal	CSD-№ [56]	$\eta_{max} \pm 0.005$	$z(\eta_{max})$
1	$KTiOPO_4$	20970	0.363	0.254
2	$RbTiOPO_4$	281379	0.350	0.451
3	$TlTiOPO_4$	81436	0.534	0.205
4	$KSnOPO_4$	68706	0.886	0.250
5	$KGeOPO_4$	39735	0.812	0.251
6	$KVOPO_4$	79651	0.314	0.254
7	$KTiOAsO_4$	202158	0.375	0.258
8	$RbTiOAsO_4$	71907	0.276	0.243
9	$CsTiOAsO_4$	280315	0.539	0.252
10	$KSnOAsO_4$	80976	0.846	0.247
11	$RbSnOAsO_4$	80977	0.714	0.234
12	$KSbOSiO_4$	69429	0.884	0.250
13	$NaSbOSiO_4$	66354	0.474	0.250
14	$KSbOGeO_4$	39463	0.634	0.252
15	$RbSbOGeO_4$	71933	0.557	0.248
16	$NaSbOGeO_4$	39788	0.408	0.251
17	$TlSbOGeO_4$	84128	0.449	0.252
18	$KTaOGeO_4$	39585	0.686	0.250
19	$AgSbOSiO_4$	39789	0.644	0.250
20	$BiCdOVO_4$	91474	0.580	0.133
21	$KFeFPO_4$	39560	0.702	0.250
22	NH_4FeAsO_4F	170672	0.880	0.208
23	NH_4FePO_4F	75110	0.826	0.251
24	NH_4GaPO_4F	89953	0.920	0.251
25	$CsScFAsO_4$	87817	0.355	0.309
26	$KAlFPO_4$	39445	0.612	0.250
27	$KCrPO_4F$	39440	0.687	0.498
28	$KGaFPO_4$	80893	0.771	0.264
29	$RbScFAsO_4$	87816	0.485	0.233
30	$Ag_{0.85}K_{0.15}TiOPO_4$	67540	0.442	0.053
31	$Ba_{0.06}K_{0.88}TiOPO_4$	280413	0.426	0.254
32	$K_{0.981}Cr_{0.019}TiOPO_4$	98245	0.410	0.245
33	$K_{0.565}Li_{0.34}TiOPO_4^*$	83482	0.755	0.259
34	$Na_{0.95}K_{0.05}TiOPO_4$	67539	0.371	0.260
35	$K_{0.845}Na_{0.155}TiOPO_4$	85092	0.406	0.254
36	$Na_{0.114}K_{0.886}K(TiO)_2(PO_4)_2$	281363	0.400	0.246
37	$Na_{0.48}K_{0.52}TiOPO_4$	71239	0.378	0.255
38	$K_{0.42}Na_{0.58}TiOPO_4$	71928	0.407	0.265
39	$K_{0.433}Na_{0.567}TiOPO_4$	71929	0.406	0.252
40	$Na_{0.992}K_{0.008}TiOPO_4$	59284	0.377	0.251
41	$K_{0.5}Rb_{0.5}TiOPO_4$	71243	0.363	0.325
42	$K_{0.84}Rb_{0.16}TiOPO_4$	81251	0.378	0.244
43	$K_{0.88}Rb_{0.12}TiOPO_4$	88030	0.377	0.255
44	$K_{1.14}Rb_{0.86}TiOPO_4$	400849	0.271	0.257
45	$K_{0.535}Rb_{0.465}TiOPO_4$	71905	0.270	0.246
46	$K_{0.857}Rb_{0.143}TiOPO_4$	81250	0.365	0.245
47	$Sr_{0.06}Cr_{0.05}K_{0.87}Ti_{0.95}OPO_4$	280412	0.514	0.248

Continued

48	$K_{0.59}Tl_{0.41}TiOPO_4$	39777	0.190	0.199
49	$K_{0.812}Tl_{0.188}TiOPO_4$	85099	0.217	0.255
50	$KGe_{0.042}Ti_{0.958}OPO_4$	39950	0.439	0.254
51	$KGe_{0.063}Ti_{0.937}OPO_4$	39882	0.467	0.253
52	$KGe_{0.184}Ti_{0.816}OPO_4$	39951	0.568	0.252
53	$K_{0.84}Ti_{0.92}Nb_{0.08}OPO_4$	67120	0.546	0.254
54	$K_{0.89}Nb_{0.11}Ti_{0.89}OPO_4$	250046	0.822	0.252
55	$K_{0.93}Nb_{0.07}Ti_{0.93}OPO_4$	250016	0.593	0.252
56	$K_{0.96}Nb_{0.04}Ti_{0.96}OPO_4$	91556	0.480	0.253
57	$K_{0.97}Nb_{0.03}Ti_{0.97}OPO_4$	54149	0.439	0.253
58	$K_{0.99}Ti_{0.988}Sb_{0.0125}OPO_4^*$	250298	0.430	0.254
59	$K_{0.874}Ti_{0.927}Sb_{0.074}OPO_4^*$	250299	0.587	0.254
60	$K_{0.893}Ti_{0.833}Sb_{0.166}OPO_4^*$	250300	0.956	0.250
61	$KSn_{0.53}Ti_{0.47}OPO_4$	250087	0.705	0.249
62	$KSn_{0.064}Ti_{0.934}OPO_4$	91534	0.461	0.253
63	$KSn_{0.75}Ti_{0.25}OPO_4$	250088	0.840	0.250
64	$KSn_{0.504}Ti_{0.496}OPO_4$	72720	0.769	0.241
65	$K_{0.998}Ti_{0.998}W_{0.002}OPO_4$	82601	0.394	0.254
66	$KTi_{0.99}Zr_{0.01}OPO_4$	418713	0.408	0.068
67	$KTi_{0.975}Zr_{0.025}OPO_4$	418715	0.415	0.068
68	$KTi_{0.981}Zr_{0.019}OPO_4$	418714	0.425	0.068
69	$KTi_{0.97}Zr_{0.03}OPO_4$	173235	0.404	0.254
70	$KTi_{0.96}Zr_{0.04}OPO_4$	173233	0.414	0.254
71	$KTi_{0.88}Hf_{0.12}OPO_4$	421394	0.473	0.001
72	$KTi_{0.97}Hf_{0.03}OPO_4$	421393	0.432	0.253
73	$KTi_{0.99}Hf_{0.01}OPO_4$	421392	0.410	0.254
74	$KTiOP_{0.5}As_{0.5}O_4$	72051	0.585	0.255
75	$KTiOP_{0.38}As_{0.62}O_4$	80024	0.546	0.259
76	$KTiOP_{0.56}As_{0.44}O_4$	80023	0.473	0.259
77	$KTiOP_{0.58}As_{0.42}O_4$	71904	0.485	0.242
78	$KTiOP_{0.75}As_{0.25}O_4$	80022	0.440	0.257
79	$KTiOP_{0.57}As_{0.43}O_4$	400850	0.520	0.261
80	$K_{0.5}Na_{0.5}Sn_{0.5}Ti_{0.5}OPO_4$	67585	0.695	0.269
81	$K_{0.5}Rb_{0.5}Sn_{0.5}Ti_{0.5}OPO_4$	67587	0.648	0.257
82	$K_2(Cr_{0.63}Ti_{0.37})(Cr_{0.43}Ti_{0.57})(PO_4)_2(F_{0.65}O_{0.35})(F_{0.41}O_{0.59})$	87835	0.776	0.257
83	$Na_{0.505}Rb_{0.495}TiOPO_4$	71240	0.505	0.325
84	$Tl_{0.23}Rb_{0.77}TiOPO_4$	81438	0.362	0.201
85	$Rb_{0.766}Tl_{0.234}TiOPO_4$	85100	0.363	0.201
86	$RbTi_{0.927}Nb_{0.056}Er_{0.017}OPO_4$	96408	0.335	0.015
87	$Rb_{0.98}Ti_{0.99}Nb_{0.01}OPO_4$	250274	0.303	0.198
88	$Rb_{0.855}Ti_{0.955}Nb_{0.045}OPO_4$	[11]	0.261	0.197
89	$RbTi_{0.97}Zr_{0.03}OPO_4$	417985	0.311	0.200
90	$RbTi_{0.98}Zr_{0.02}OPO_4$	418599	0.254	0.197
91	$RbTi_{0.98}Zr_{0.016}OPO_4$	418598	0.317	0.201
92	$Rb_2TiGe_{0.121}Ti_{0.879}O_2(PO_4)_2$	281380	0.342	0.198
93	$Na_{0.5}Rb_{0.5}Sn_{0.5}Ti_{0.5}OPO_4$	67586	0.452	0.495
94	$KNb_{0.5}V_{0.5}OPO_4$	86787	0.730	0.250
95	$KGa_{0.5}Ge_{0.5}F_{0.5}O_{0.5}PO_4$	80894	0.881	0.262
96	$K_{0.5}Rb_{0.5}SnOPO_4$	67584	0.638	0.253
97	$Cs_{0.6}K_{0.4}TiOAsO_4$	74597	0.389	0.263

Continued

98	$Cs_{0.61}K_{0.39}TiOAsO_4$	74596	0.464	0.151
99	$Cs_{0.595}K_{0.405}TiOAsO_4$	74598	0.638	0.253
100	$Cs_{0.625}K_{0.375}TiOAsO_4$	74595	0.265	0.239
101	$K_{0.534}Li_{0.34}TiOAsO_4$ *	83483	0.385	0.263
102	$Na_{0.87}K_{0.13}TiOAsO_4$	67541	0.436	0.259
103	$K_{1.65}V(V_{0.78}W_{0.22})O_2(AsO_4)_2$	260558	0.807	0.253
104	$KAlNbO_2((As_{0.8}Nb_{0.2})O_4)_2$	[57]	0.881	0.251
105	$Cs_{0.068}Rb_{0.95}TiOAsO_4$	280501	0.376	0.243
106	$Cs_{0.62}Rb_{1.38}TiO_2(AsO_4)_2$	280502	0.331	0.255
107	$Cs_{1.12}Rb_{0.85}(TiO)_2(AsO_4)_2$	280503	0.286	0.252
108	$Cs_{1.43}Rb_{0.57}(TiO)_2(AsO_4)_2$	280504	0.372	0.344
109	$Cs_{1.73}Rb_{0.27}(TiO)_2(AsO_4)_2$	280505	0.371	0.345
110	$Cs_{1.4}Rb_{0.6}(TiO)_2(AsO_4)_2$	280506	0.369	0.159
111	$Cs_{1.72}Rb_{0.28}(TiO)_2(AsO_4)_2$	280507	0.373	0.340
112	$Cs_{0.9}Rb_{0.1}TiOAsO_4$	280508	0.330	0.109
113	$NH_4Fe(AsO_4)_{0.19}(PO_4)_{0.81}F$	420019	0.877	0.208
114	$NH_4Fe(AsO_4)_{0.37}(PO_4)_{0.63}F$	420020	0.865	0.207
115	$NH_4Fe(AsO_4)_{0.74}(PO_4)_{0.26}F$	420021	0.898	0.207
116	$NH_4VAsO_4F_{0.8}O_{0.2}$	419640	0.852	0.207
117	$(NH_4)_2Ga_2(PO_4)(HPO_4)F_3$	89952	0.429	0.418
118	$(NH_4)_{0.875}K_{0.125}FePO_4F$	260152	0.772	0.167
	Mean value $<\eta_{max}>$ for pure crystals (29 structures)			0.606
	Mean value $<\eta_{max}>$ for doped crystals (89 structures)			0.490

For crystals with numbers 2, 13, 16, 20, 25, 30, 34, 40, 41, 71, 83, 86, 92, 93, 98, 101, 102, 108-112, and 117, the (x, y) coordinates of pseudoinversion centres are (0, 0); for other crystals they are (0.25, 0.25). The numbers of crystals with known estimated characteristics of nonlinear optical properties are bolded. *The chemical composition of the crystals is given in correspondence with the CIF files indicated here; it somewhat differs from the corresponding chemical formulas in **Table 1**, which are given in correspondence with the references to original studies.

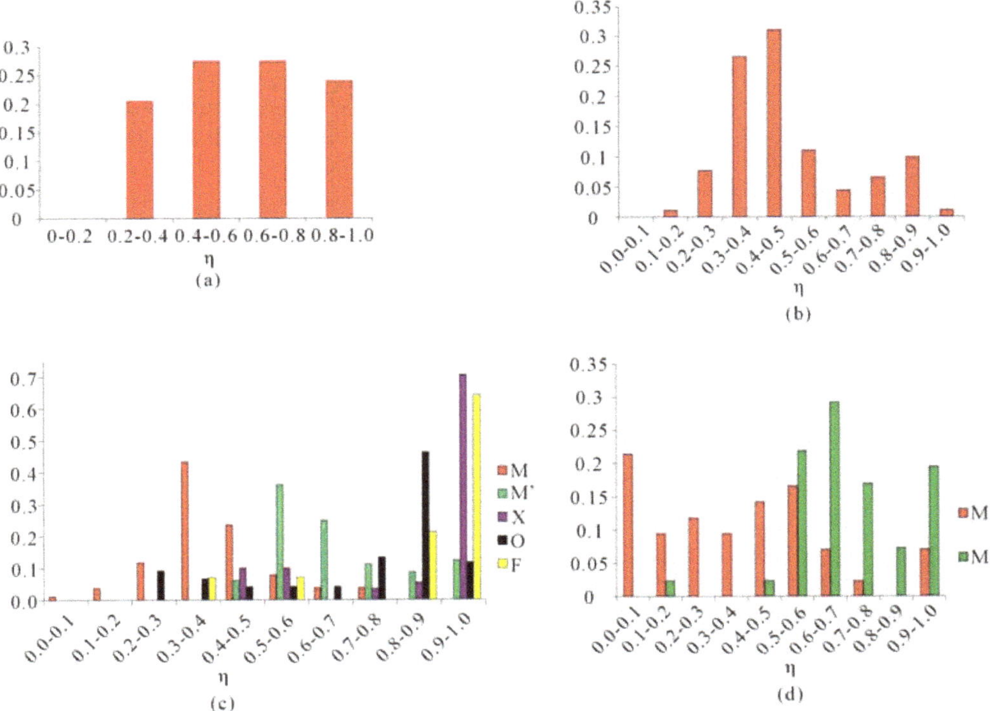

Figure 3. Distribution histogram of the degree of pseudoinversion for (a) pure KTP crystals (29 structures), (b) doped KTP crystals (89 structures), (c) pure atomic sublattices of individual types for 118 KTP crystals from Table 2, and (d) doped sublattices of individual types of atoms for 89 KTP crystals from Table 2.

positions (from the aforementioned set of crystals) for which experimental SHG data were obtained either by the powder method [26], or directly with respect to a powder of pure $KTiOPO_4$ crystal, or the data can be recalculated with respect to it based on a specific publication. In addition, since most sources yield data on the ratio of second-harmonic intensities for the studied and reference samples ($I/I_{KTP} = I_{2\omega}/I_{2\omega}(KTP)$), they were additionally recalculated into estimated values of the relative effective second-order susceptibility (it will be denoted as $\chi_{2\omega}/\chi_{2\omega}(KTP)$). In the first approximation, one can assume that

$$I_{2\omega} \sim I_{\omega}^2 \cdot \chi_{2\omega}^2,$$

where $I_{2\omega}$ is the second harmonic intensity and I_{ω} is the primary radiation intensity. Therefore, the desired ratio $\chi_{2\omega}/\chi_{2\omega}(KTP)$ was estimated to be

$$\chi_{2\omega}/\chi_{2\omega}(\mathrm{KTP}) \sim I_{2\omega}/I_{2\omega}(\mathrm{KTP}).$$

Figure 4 shows the dependence of the set of $\chi_{2\omega}/\chi_{2\omega}$ (KTP) values for $KTiOPO_4$ (CSD-№ 20970), $RbTiOPO_4$ (CSD-№ 281379, [10, 11]), $KTiOAsO_4$ (CSD-№ 202158, [14, 15]), $K_{0.565}Li_{0.34}TiOPO_4$ (CSD-№ 83482, [14]), $RbTi_{0.927}Nb_{0.056}Er_{0.017}OPO_4$ (CSD-№ 96408, [11]), $Rb_{0.855}Ti_{0.955}Nb_{0.045}OPO_4$ ([11]), $K_{0.534}Li_{0.34}TiOAsO_4$ (CSD-№ 83483, [14]), $Na_{0.87}K_{0.13}TiOAsO_4$ (CSD-№ 67541, [14]), $K_{0.89}Nb_{0.11}Ti_{0.89}OPO_4$ (CSD-№ 250046, [8]), $K_{0.96}Nb_{0.04}Ti_{0.96}OPO_4$ (CSD-№ 91556, [8]), $K_{0.99}Ti_{0.988}Sb_{0.0125}OPO_4$ (CSD-№ 250298, [8]), $K_{0.874}Ti_{0.927}Sb_{0.074}OPO_4$ (CSD-№ 250299, [8]), $K_{0.893}Ti_{0.833}Sb_{0.166}OPO_4$ (CSD-№ 250300, [8]), $KTi_{0.97}Zr_{0.03}OPO_4$ (CSD-№ 173235, [8]), and $KTi_{0.96}Zr_{0.04}OPO_4$ (CSD-№ 173233, [8]) crystals on the pseudoinversion $\eta = \eta_{max}$ of their atomic structures in the

$$\left(\sqrt{1-\eta},\ \chi_{2\omega}/\chi_{2\omega}(\mathrm{KTP})\right)$$

coordinates. The linear approximation of the dependence presented in **Figure 4** within the model described in [8], is characterized by a correlation coefficient of 0.91, and the confidence interval boundaries are (0.76, 0.97) at a confidence probability of 0.95.

Equation (2) can be more pronounced within the concentration series of samples of the same qualitative composition. For example, the SHG intensity decreases with an increase in the tin fraction in the $KTi_{1-x}Sn_xOPO_4$ series, and the calculation of pseudoinversion for a series of known structures of this composition indicates a monotonic increase in the latter (**Figure 5**).

The boundary-composition crystal $KSnOPO_4$ has almost zero SHG intensity and the largest (in the $KTi_{1-x}Sn_xOPO_4$ series) pseudoinversion: 0.886 (**Table 1**, no. 7; **Table 2**, no. 4). This fact is in agreement with the

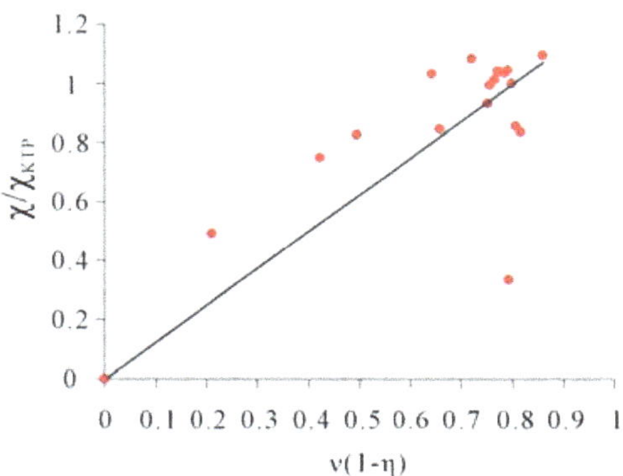

Figure 4. Correlation between the relative effective second-order susceptibility χ2ω/χ2ω(KTP) for a number of KTP crystals and their pseudoinversion in the

$$\left(\sqrt{1-\eta},\ \chi_{2\omega}/\chi_{2\omega}(\mathrm{KTP})\right)$$

coordinates (see [8] for the approximation model).

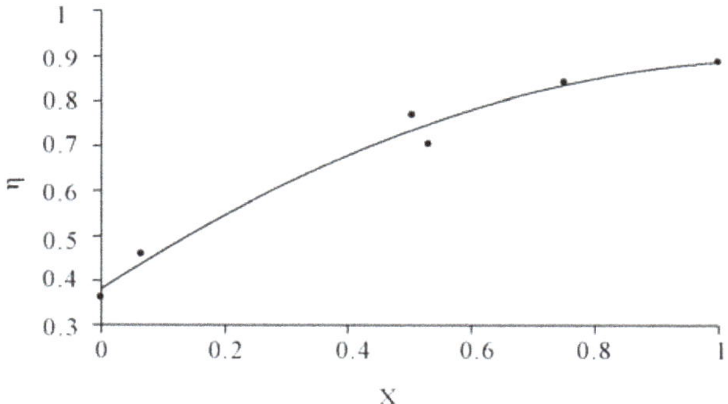

Figure 5. Dependence of the pseudoinversion on the tin content in KTi1-xSnxOPO4 crystals (calculation based on the structural data CSD-№ 20970, 91534, 72720, 250087, 250088, 68706).

data of Godfrey *et al.* [58], who established that the $KSnOPO_4$ structure can be partially described (in good approximation) by the *Pnan* group; exact description is obtained within the $Pna2_1$ group.

5. Conclusions

To date, despite the numerous publications on the structure and properties of KTP crystals, the question of the structural conditionality of the behavior of their nonlinear optical properties has not been completely clarified.

In this paper, we reported the results of studying the pseudosymmetric features of known structures of KTP crystals with respect to inversion and tried to analyze the entire set of known nonlinear optical parameters of these crystals in view of the obtained pseudosymmetric characteristics. In particular, it was shown that doped structures of KTP crystals have on average a lower degree of pseudoinversion than "pure" compositions; in some cases this feature adequately correlates with the increase in the relative intensity of the second optical harmonic. This correlation may manifest itself within the concentration series samples of the same qualitative composition.

We believe that, in order to establish the fundamental correlations between the structural and symmetric features of crystals (in particular, those belonging to the KTP family) and their nonlinear optical properties, for example, having the degree of pseudoinversion as a symmetric characteristic, it is necessary to primarily calculate this characteristic for the entire structure. This thesis is justified by the fundamental principles of symmetry in physical crystallography. The Neumann princeple, which sets a relationship between the symmetry of a medium (crystal) and the set of physical properties that are forbidden or allowed in this medium, deals with specifically the symmetry of the medium as a whole rather than the symmetry of its individual structural fragments within the unit cell. This approach was applied both in [8] and in this study. However, this does not depreciate the validity of the analysis of the characteristics of sublattices of individual types of atoms. Due to this analysis one can find sublattices with pseudosymmetric characteristics exhibiting a more significant sensitivity, for example, at a transition to doped compositions, and therefore, can determine to greater extent the behavior of the pseudosymmetric characteristics of crystal structures, as whole and physical properties of crystals.

6. Acknowledgements

This work was supported financially by the Ministry of Education and Science of the Russian Federation, project 14.B37.21.1158.

REFERENCES

[1] C. Lee and E. V. Chuprunov, "Fedorov Pseudosymmetry of Crystals: Review," *Crystallography Reports*, Vol. 52, No. 1, 2007, pp. 1-11.

[2] E. V. Chuprunov and E. A. Soldatov and T. N. Tarkhova, "The Quantitative Estimation of the Symmetry of the Crystal Structures," *Kristallografiya*, Vol. 33, No. 3, 1988, pp. 759-761.

[3] G. D. Stucky, M. L. F. Phillips and T. E. Gier, "The Potassium Titanyl Phosphate Structure Field: A Model for New Nonlinear Optical Materials," *Chemistry of Materials*, Vol. 1, No. 5, 1989, pp. 492-509.

[4] N. I. Sorokina and V. I. Voronkova, "Structure and Properties of Crystals in the Potassium Titanyl Phosphate Family: A Review," *Crystallography Reports*, Vol. 52, No. 1, 2007, pp. 80-93.

[5] M. L. F. Phillips, T. E. Gier, M. M. Eddy, N. L. Keder and G. D. Stucky, "Inclusion Tuning of Nonlinear Optical Materials: KTP Isomorphs," *Solid State Ionics*, Vol. 32-33, Part 1, 1989, pp. 147-153.

[6] M. L. F. Phillips, W. T. A. Harrison and G. D. Stucky, "Nonlinear Optical Properties of New $KTiOPO_4$ Isostructures," *SPIE Inorganic Crystals for Optics, Electro-Optics, and Frequency Conversion*, Vol. 1561, 1991, pp. 84-92.

[7] M. L. F. Phillips, W. T. A. Harrison, G. D. Stucky, E. M. McCarron, J. C. Calabrese and T. E. Gier, "Effects of Substitution Chemistry in the Potassium Titanyl Phosphate ($KTiOPO_4$) Structure Field," *Chemistry of Materials*, Vol. 4, No. 1, 1992, pp. 222-233.

[8] V. A. Ivanov, V. A. Burdov, M. O. Marychev, D. N. Titaev, M. A. Faddeev and E. V. Chuprunov, "On the Effect of Structural and Symmetrical Features of Potassium Titanyl Phosphate Crystals with Different Contents of Niobium, Antimony, and Zirconium on the Second-Harmonic Intensity," *Crystallography Reports*, Vol. 53, No. 4, 2008, pp. 678-682.

[9] D. N. Nikogosyan, "Nonlinear Optical Crystals: A Complete Survey," Springer Science+Business Media, USA, 2005.

[10] J. J. Carvajal, V. Nikolov, R. Sole, J. Gavalda, J. Massons, M.Rico, C. Zaldo, M. Aguilo and F. Díaz, "Enhancement of the Erbium Concentration in $RbTiOPO_4$ by Codoping with Niobium," *Chemistry of Materials*, Vol. 12, No. 10, 2000, pp. 3171-3180.

[11] J. J. Carvajal, J. L. Garcıa-Mun, R. Sole, J. Gavalda, J. Massons, X. Solans, F. Díaz and M. Aguilo, "Charge Self-Compensation in the Nonlinear Optical Crystals $Rb_{0.855}Ti_{0.955}Nb_{0.045}OPO_4$ and $RbTi_{0.927}Nb_{0.056}Er_{0.017}OPO_4$," *Chemistry of Materials* , Vol. 1, No. 10, 2003, pp. 2338-2345.

[12] M. M. Eddy, T. E. Gier, N. L. Keder, G. D. Stucky, D. E. Cox, J. D. Bierlein and G. Jones, "Inclusion Tuning of Nonlinear Optical Materials: Sorbates on the KTP Structure," *Inorganic Chemistry*, 1988. Vol. 27, No. 11, 1988, pp. 1856-1858.

[13] M. L. F. Phillips, W. T. A. Harrison and G. D. Stucky, "Influence of Electronic Configuration on the Structure and Optical Properties of Potassium Tin Oxide Phosphate," *Inorganic Chemistry*, Vol. 29, No. 17, 1990, pp. 3245-3247.

[14] M. E. Hagerman and K. R. Poeppelmeier, "Review of the Structure and Processing-Defect-Property Relationships of Potassium Titanyl Phosphate: A Strategy for Novel Thin-Film Photonic Devices," *Chemistry of Materials*, Vol. 7, No. 4, 1995, pp. 602-621.

[15] K. K. Rangan, B. R. Prasad, C. K. Subramanian and J. Gopalakrishnan, "Coupled Substitution of Niobium and Silicon in Potassium Titanyl Phosphate and Arsenate ($KTiOPO_4$ and $KTiOAsO_4$. Synthesis and Nonlinear Optical Properties of $KTi_{1-x}Nb_xOX_{1-x}Si_xO_4$ (X = P, As)," *Inorganic Chemistry*, Vol. 32, No. 20, 1993, pp. 4291-4293.

[16] L. K. Cheng, L. T. Cheng, J. D. Bierlein and F. C. Zumsteg, "Properties of Doped and Undoped Crystals of Single Domain $KTiOAsO_4$," *Applied Physics Letters*, Vol.

62, No. 4, 1993, pp 346-348.

[17] L. T. Cheng, L. K. Cheng, J. D. Bierlein and F. C. Zumsteg, "Nonlinear Optical and Electro—Optical Properties of Single Crystal $CsTiOAsO_4$," *Applied Physics Letters*, Vol. 63, No. 19, 1993, pp. 2618-2620.

[18] B. V. Mill', A. V. Butashin and S. Yu. Stefanovich, "New Compounds with $KTiOPO_4$-$NaSbOGeO_4$, $AgSbOGeO_4$ and $AgSbOSiO_4$ Structure," *Zhurnal Neorganicheskoi Khimii*, Vol. 38, No. 6, 1993, pp. 947-949.

[19] W. T. A. Harrison and M. L. F. Phillips, "Syntheses, Structures, and Properties of $RbScFAsO_4$ and $CsScFAsO_4$: Scandium-Containing Analogues of Potassium Titanyl Phosphate ($KTiOPO_4$)," *Chemistry of Materials*, Vol. 11, No. 12, 1999, pp. 3555-3560.

[20] F. C. Zumsteg, J. D. Bierlein and T. E. Gier, "$K_xRb_{1-x}TiOPO_4$: A New Nonlinear Optical Material," *Journal of Applied Physics*, Vol. 47, No. 11, 1976, pp. 4980-4985.

[21] V. I. Chani, K. Shimamura, S. Endo and T. Fukuda, "Growth of Mixed Crystals of the $KTiOPO_4$ (KTP) Family," *Journal of Crystal Growth*, Vol. 171, No. 3-4, 1997, pp. 472-476.

[22] M. L. F. Phillips, W. T. A. Harrison, T. E. Giers, G. D. Stucky, G. V. Kulkarni, J. K. Burdett, "Electronic Effects of Substitution Chemistry in the Potassium Titanyl Phosphate ($KTiOPO_4$) Structure Field: Structure and Optical Properties of Potassium Vanadyl Phosphate," *Inorganic Chemistry*, Vol. 29, No. 11, 1990, pp. 2158-2163.

[23] J. Gopalakrishnan, K. K. Rangan, B. R. Prasad and C. K. Subramanian, "New Transition Metal Phosphates Related to $KTiOPO_4$. Synthesis of $K_{0.5}M_{0.5}M'_{0.5}OPO_4$ (M = Nb, Ta; M' = Ti, V) and $K_{1-x}Ti_{1-x}V_xOPO_4$ Exhibiting Nonlinear Optical Behavior," *Journal of Solid State Chemistry*, Vol. 111, No. 1, 1994, pp. 41-47.

[24] A. Peña, J. J. Carvajal, J. Massons, Jna. Gavaldà, F. Díaz and M. Aguiló, "Yb: Ta:$RbTiOPO_4$, A New Strategy to Further Increase the Lanthanide Concentration in Crystals of the $KTiOPO_4$ Family," *Chemistry of Materials*, Vol. 19, No. 16, 2007, pp. 4069-4076.

[25] T. Hikita, "35B-1 (A, B)(C, D)OPO_4 (A, B = H, NH_4, Na, K, Rb, Ag, Cs, Tl; C, D = Ti, V, Nb)," In: Y. Shiozaki, E. Nakamura, T. Mitsui Eds., *SpringerMaterials—The Landolt-Börnstein Database*, Vol. 36B1, 2004, pp 1-28, 9e.

[26] S. K. Kurtz and T. T. Perry, "A Powder Technique for the Evaluation of Nonlinear Optical Materials," *Journal of Applied Physics*, 1968. Vol. 39, No. 8, 1968, pp. 3798-3813.

[27] J. P. Dougherty and S. K. L. Kurtz, "A Second Harmonic Analyzer for the Detection of Non-Centrosymmetry," *Journal of Applied Crystallography*, Vol. 9, Part 2, 1976, pp. 145-158.

[28] P. D. Maker, R. W. Terhune, M. Nisenoff and C. M. Savage, "Effects of Dispersion and Focusing on the Production of Optical Harmonics," *Physical Review Letters*, Vol. 8, No. 1, 1962, pp. 21-22.

[29] J. Jerphagnon and S. K. Kurtz, "Optical Nonlinear Susceptibilities: Accurate Relative Values for Quartz, Ammonium Dihydrogen Phosphate, and Potassium Dihydrogen Phosphate," *Physical Review B*, Vol. 1, No. 4, 1970, pp. 1739-1744.

[30] P. A. Thomas and B. E. Watts, "An Nb-doped Analogue of $KTiOPO_4$; Structural and Nonlinear Optical Properties," *Solid State Communications*, Vol. 73, No. 2, 1990, pp. 97-100.

[31] T. Yu. Losevskaya, O. A. Alekseeva, V. K. Yanovskii, V. I. Voronkova, N. I. Sorokina, V. I. Simonov, S. Yu. Stefanovich, S. A. Ivanov, S. Eriksson and S. A. Zverkov, "Structure and Properties of Niobium-Doped Potassium Titanyl Phosphate Crystals," *Crystallography Reports*, Vol. 45, No. 5, 2000, pp. 739-743.

[32] O. A. Alekseeva, M. K. Blomberg, V. N. Molchanov, I. A. Verin, N. I. Sorokina, T. Yu. Losevskaya, V. I. Voronkova and V. K. Yanovskii, "Refinement of the $K_{0.96}Ti_{0.96}Nb_{0.04}OPO_4$ Structure," *Crystallography Reports*, Vol. 46, No. 4, 2001, pp. 642-646.

[33] V. K. Yanovski, V. I. Voronkova, T. Yu. Losevskaya, S. Yu. Stefanovich, S. A. Ivanov, V. I. Simonov and N. I. Sorokina, "Growth and Properties of Nb-or Sn-Doped $KTiOPO_4$ Crystals," *Crystallography Reports*, Vol. 47, Suppl. 1, 2002, pp. S99-S104.

[34] O. A. Alekseeva, N. I. Sorokina, I. A. Verin, T. Yu. Losevskaya, V. I. Voronkova, V. K. Yanovskii and V. I. Simonov, "Structure and Properties of Potassium Titanyl Phosphate Single Crystals with 7 and 11 at. % Nb," *Crystallography Reports*, Vol. 48, No. 2, 2003, pp. 205-211.

[35] V. I. Voronkova, V. K. Yanovskii, T. Yu. Losevskaya, S. Yu. Stefanovich, S. A. Zver'kov, O. A. Alekseeva and N. I. Sorokina, "Electrical and Nonlinear Optical Properties of $KTiOPO_4$ Single Crystals Doped with Niobium, Antimony, and Tantalum," *Crystallography Reports*, Vol. 49, No. 1, 2004, pp. 123-129.

[36] A. P. Dudka, I. A. Verin, V. N. Molchanov, M. K. Blomberg, O. A. Alekseeva, N. I. Sorokina, N. E. Novikova and V. I. Simonov, "Structural Study of $K_{0.93}Ti_{0.93}Nb_{0.07}OPO_4$ Single Crystals at 30 K," *Crystallography Reports*, Vol. 50, No. 1, 2005, pp. 36-41.

[37] N. I. Sorokina, V. I. Voronkova, V. K. Yanovskii, I. A. Verin and V. I. Simonov, "Crystal Structures of Compounds in the $KTiOPO_4$-$KGeOPO_4$ System," *Crystallography Reports*, Vol. 41, No. 3, 1996, pp. 435-435.

[38] L. Wen, N. I. Sorokina, V. I. Voronkova, V. K. Yanovskii, I. A. Verin, A. G. Vigdorchik and V. I. Simonov, "Crystal Structure of $KTi_{0.93}Sn_{0.07}OPO_4$," *Crystallography Reports*, Vol. 45, No. 3, 2000, pp. 386-388.

[39] L. Wen, V. I. Voronkova, V. K. Yanovskii, S. Y. Stefanovich, N. I. Sorokina and I. A. Verin, "Growth and Properties of $KTi_{1-x}Sn_xOPO_4$ Crystals," *Inorganic Materials*, Vol. 37, No. 3, 2001, pp. 290-293.

[40] O. D. Krotova, N. I. Sorokina, I. A. Verin, V. I. Voronkova, V. K. Yanovskii and V. I. Simonov, "Structure

and Properties of Single Crystals of Tin-Doped Potassium Titanyl Phosphate," *Crystallography Reports*, Vol. 48, No. 6, 2003, pp. 925-932.

[41] V. I. Voronkova, V. K. Yanovskii, I. N. Leont'eva, E. I. Agapova, E. P. Kharitonova, S. Y. Stefanovich and S. A. Zver'kov, "Growth and Properties of Zr-Doped $KTiOPO_4$ Crystals," *Inorganic Materials*, Vol. 40, No. 12, 2004, pp. 1321-1323.

[42] O. A. Alekseeva, A. P. Dudka, N. I. Sorokina, A. Pietraszko, M. Kh. Rabadanov, E. I. Agapova, V. I. Voronkova and V. I. Simonov, "Crystal Structure of Potassium Titanyl Phosphate Doped with Zirconium," *Crystallography Reports*, Vol. 52, No. 4, 2007, pp. 659-667.

[43] N. E. Novikova, I. A. Verin, N. I. Sorokina, O. A. Alekseeva, E. I. Agapova and V. I. Voronkova, "Structural Reasons for the Nonlinear Optical Properties of $KTi_{0.96}Zr_{0.04}OPO_4$ Single Crystals," *Crystallography Reports*, Vol. 54, No. 2, 2009, pp. 219-227.

[44] O. A. Alekseeva, O. D. Krotova, N. I. Sorokina, I. A. Verin, T. Y. Losevskaya, V. I. Voronkova, V. K. Yanovskii and V. I. Simonov, "Structure and Properties of Antimony-Doped Potassium Titanyl Phosphate Single Crystals," *Crystallography Reports*, Vol. 50, No. 4, 2005, pp. 554-565.

[45] V. V. Atuchin, O. A. Alekseeva, V. G. Kesler, L. D. Pokrovsky, N. I. Sorokina and V. I. Voronkova, "Chemical Shifts of Atomic Core Levels and Structure of $K_{1-x}Ti_{1-x}Sb_xOPO_4$, x=0-0.23, Solid Solutions," *Journal of Solid State Chemistry*, Vol. 179, No. 8, 2006, pp. 2349-2355.

[46] V. I. Voronkova, I. N. Leont'eva, N. I. Sorokina, T. I. Ovsetsina and I. A. Verin, "Growth, Structure, and Properties of $KTiOPO_4$ Crystals Doped with Iron," *Crystallography Reports*, Vol. 51, No. 6, 2006, pp. 977-981.

[47] E. I. Orlova, E. P. Kharitonova, N. E. Novikova, I. A. Verin, O. A. Alekseeva, N. I. Sorokina and V. I. Voronkova, "Synthesis, Properties, and Structure of Potassium Titanyl Phosphate Single Crystals Doped with Hafnium," *Crystallography Reports*, Vol. 55, No. 3, 2010, pp. 404-411.

[48] E. I. Orlova, N. E. Novikova, A. Gagor, I. A. Verin, A. Pietraszko, D. A. Belov and V. I. Voronkova, "Growth of $KTiOPO_4$ Crystals Doped with Zinc and Studies of Their Physical Properties and Specific Structural Features," *Crystallography Reports*, Vol. 55, No. 4, 2010, pp. 594-601.

[49] L. Wen, V. I. Voronkova, V. K. Yanovskii, N. I. Sorokina, I. A. Verin and V. I. Simonov, "Synthesis, Atomic Structure, and Properties of Crystals in the $RbTiOPO_4$-$CsTiPO_5$ System," *Crystallography Reports*, Vol. 45, No. 3, 2000, pp. 380-385.

[50] E. I. Agapova, V. I. Voronkova, E. P. Kharitonova, I. N. Leont'eva, S. Y. Stefanovich, N. I. Sorokina, A. P. Dudka, O. A. Alekseeva and N. N. Kononkova, "Synthesis and Properties of Zirconium-Doped $RbTiOPO_4$ Single Crystals," *Crystallography Reports*, Vol. 53, No. 2, 2008, pp. 285-290.

[51] O. A. Alekseeva, A. P. Dudka, N. E. Novikova, N. I. Sorokina, E. I. Agapova and V. I. Voronkova, "Structure of the $RbTi_{0.98}Zr_{0.02}OPO_4$ Single Crystal at Temperatures of 293 and 105 K," *Crystallography Reports*, Vol. 53, No. 4, 2008, pp. 557-564.

[52] A. R. Peña, "Ytterbium and Erbium Doped $RbTi_{1-x}M_xOPO_4$ (M=Nb or Ta) Crystals. New Laser and Nonlinear Bifunctional Materials," Doctoral Thesis, Física I Cristallografia de Materials (FiCMA), Tarragona, 2007. http://www.tesisenred.net/handle/10803/9087

[53] M. T. Anderson, M. L. F. Phillips, M. B. Sinclair and G. D. Stucky, "Synthesis of Transition-Metal-Doped $KTiOPO_4$ and Lanthanide-Doped $RbTiOAsO_4$ Isomorphs that Absorb Visible Light," *Chemistry of Materials*, Vol. 8, No. 1, 1996, pp. 248-256.

[54] N. V. Somov and E. V. Chuprunov, "Pseudosymmetry of Atomic Crystal Structures," In: E. V. Chuprunov, Ed., *Crystallography: Laboratory Workshop*, Fizmatlit, Moskow, 2005, pp. 80-90.

[55] T. Hahn, "International Tables for Crystallography. Vol. A: Space-Group Symmetry," Springer, Berlin, 2005.

[56] Inorganic Crystal Structure Data-base. Version 2010-1. http://www.fiz-karlsruhe.de/icsd.html/

[57] T. Nakagawa, T. Matsumoto, V. I. Chani and T. Fukuda, "$K_2NbAlO_2[(As,Nb)O_4]_2$, Isostructural with $KTiOPO_4$," *Acta Crystallographica Section C*, Vol. 55, 1999, pp. 1391-1393.

[58] K. W. Godfrey, P. A. Thomas and B. E. Watts, "The Structural and Optical Properties of Potassium Titanyl Phosphate and Its Analogues," *Materials Science and Engineering B*, Vol. 9, No. 4, 1991, pp. 479-483.

X-Ray Crysatllographic and Vibrational Spectroscopic Studies of Thorium Bromate Hydrate

M. Junaid Bushiri[1*], T. C. Kochuthresia[2], S. Athimoolam[3], V. Ramakrishnan[4], V. K. Vaidyan[2]
[1]Department of Physics, Cochin University of Science and Technology, Kochi, India
[2]Department of Physics, University of Kerala, Thiruvananthapuram, India
[3]Department of Physics,University College of Engineering, Anna University Tirunelveli Region, Nagercoil, India
[4]School of Physics, Madurai Kamaraj University, Madurai, India

ABSTRACT

$Th(BrO_3)_3{\cdot}H_2O$ single crystals were grown from its aqueous solution at room temperature. Single crystal XRD, Raman and FTIR techniques were used to investigate the crystal structure. The crystal structure was solved by Patterson method. The as grown crystals are in monoclinic system with space group P21/c. The unit cell parameters are a = 12.8555(18) Å, b = 7.8970(11) Å, c = 9.0716(10) Å, $\alpha = 90°$, $\beta = 131.568°$ and $\gamma = 90°$ and unit cell volume is 689.1(2) $Å^3$. Z = 8, R factor is 5.9. The Raman and FTIR studies indicate the lowering of symmetry of bromate anion from C_{3V} to C_1. Hydrogen bonds with varying strengths are present in the crystal. The centrosymmetric space group P21/c of the crystal is confirmed by the non-coincidence of majority of Raman and IR bands.

Keywords: Thorium Bromate; X-Ray Diffraction; IR Spectroscopy; Raman; Crystal Structure; BrO_3^- Anion

1. Introduction

Hydrated bromates, coordinated to rare earth metals are an interesting group of compounds due its magnetic and optical properties [1,2]. Further, Bromates are attained considerable interest from chemists because its role in chemical oscillator systems and chemiluminescence systems and as an oxidizing agent for the fast oxidation of secondary alcohols [3-7]. Recently, Thionine-Bromate is identified as a new reaction system for kinetic spectrophotometric determination of hydrazine in cooling tower water samples [8]. However a limited attention is paid by researchers to understand structural and physical properties of bromate compounds. Single crystal XRD and vibrational spectroscopic techniques are being used by different authors for elucidating structure of hydrated metallic bromates [9-15]. Interestingly, most of these compounds having hydrogen bonded crystalline structure and oxyhalogen anions in these compounds posses distorted pyramidal structure. Thorium is one of the important actinide elements, actinide bromate compounds are rarely reported in the literature. In the present work we are reporting structural studies of hydrated rare earth thorium bromate by using X-ray diffraction, Raman and infrared spectroscopy for the first time.

2. Experimental

Crystals used in the present investigation were grown from aqueous solution of thorium bromate by slow evaporation at room temperature (32°C) over a period of one month. Single crystal X-ray diffraction data were collected by using single crystal X-ray diffractometer sealed tube CAD4/MACH3, with Mo$K\alpha$ radiation (λ = 0.71073 Å). CAD4 Software. Version 5.0. Enraf-Nonius were used for data collection [16,17]. The cell data were obtained from the least-squares refinement. Structure was solved by using ShellX97 program. Raman spectrum of the sample was recorded using Bruker RFS 100/s Raman spectrometer equipped with Nd:YAG laser (Wavelength 1064 nm at a Laser power of 150 mW) in the stokes region from 50 - 4000 cm^{-1} with a spectral resolution better than 4 cm^{-1}. Standard Ge detector was used for detecting the Raman signals. FTIR spectrum of the sample in the range 400 - 4000 cm^{-1} was obtained from Shimadzu spectrometer by KBr pellet method and in the region 50 -

*Corresponding author.

400 cm^{-1} with polyethylene pellet method by using MAGNA 550 FTIR spectrometer.

3. Results and Discussion

3.1. X-Ray Crystal Structural Studies

The structure of the title compound is solved by Patterson method [16]. Thorium atoms are coordinated to oxygen atoms and some of these are coordinated to bromine and H_2O (**Figure 1**). The average bond length of water (oxygen) coordinated Th(1)-O1w is of the order of 2.6972(0.0232) Å which is comparable to Th-O (water) distance (2 - 46 Å) reported in thorium nitrate pentahydrate [18]. The thorium atom coordinated to the oxygen atoms of bromate group (Th(1)-O (of BrO_3) is of the order of 2.7766(0.0125) Å. Generally lanthanide hydrated bromates form disordered bromate anions at room temperature [15]. In the actinide bromate also disordered bromate anion is seen which is formed by the bonding of oxygen atoms with bromine and Thorium atoms is shown in the ORTEP diagram (**Figure 2**). The average Br(1)-O(1) distance is 1.6504(0.0133) Å, Br(1)-O(3) is 1.6573(0.0129) Å and Br(1)-O(2) 1.6712(0.0128) Å (**Table 1**). These distances are consistent with those of previously reported one in the literature [12-14].

3.2. Raman and FTIR Spectral Studies

$Th(BrO_3)_3{\cdot}H_2O$ crystallizes in the monoclinic crystal system (space group P21/c, Z = 8) and all the atoms are occupying in the C_1 sites. The factor group analysis was carried out by the correlation method developed by Fateley *et al.* [19]. The total irreducible representation, excluding the acoustic modes, are distributed as

$$\Gamma 189 = 48A_g + 48B_g + 47A_u + 46B_u .$$

The BrO_3^- anion with C_{3V} symmetry has four fundamental modes of vibrations, symmetric stretching ν_1—805 (A_1), asymmetric stretching ν_3—805 (E), symmetric bending ν_2—418 (A_1) and asymmetric bending ν_4—358 (E) cm^{-1} [13,14,16-20]. For XY_3 pyramidal (C_{3v}) molecules $\Gamma = 2A_1+2E$, and all the modes are active both in the Raman and IR spectra [20]. In Raman spectrum, a strong band is observed at 820 cm^{-1} followed by a shoulder at 796 cm^{-1} and another strong band is appeared at 784 cm^{-1} (**Figure 3**). The strongest band observed in the IR spectrum is at 789 followed by a strong shoulder at 821 cm^{-1}. These bands correspond to ν_1-ν_3 modes of BrO^{3-} anion. These spectral band positions are in the same region as reported previously in the literature which is having comparable Br-O distance [21-25]. The sym-me-tric stretching frequency of BrO_3^- anion observed in this crystal is almost similar (784 cm^{-1}) to that of reported in crystals like $Cd(BrO_3)_2{\cdot}2H_2O$, $Nd(BrO_3)_3{\cdot}9H_2O$, $Gd(BrO_3)_3{\cdot}9H_2O$ and $La(BrO_3)_3{\cdot}9H_2O$ [23-26]. This

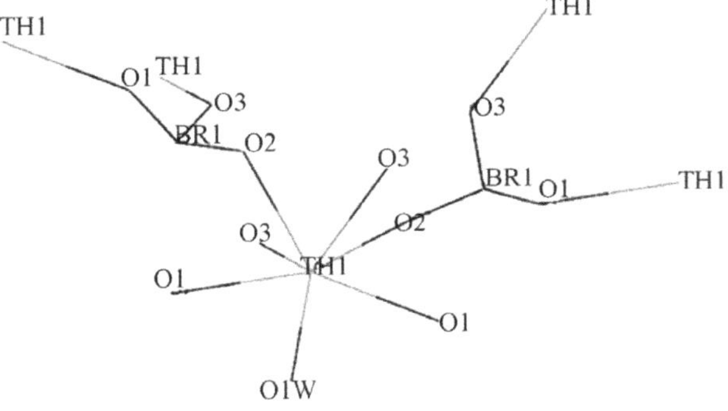

Figure 1. ORTEP diagram of hydrated thorium bromate.

Figure 2. BrO_3^- anion and thorium water coordination in hydrated thorium bromate.

Table 1. Selected bond distances of thorium bromate hydrate .

Br(1)-O(1)	1.6504(0.0133) Å
Br(1)-O(2)	1.6712(0.0128) Å
Br(1)-O(3)	1.6573(0.0129) Å
Th(1)-O(1)w	2.6972(0.0232) Å
Th(1)-O (of BrO_3)	2.7766(0.0125) Å

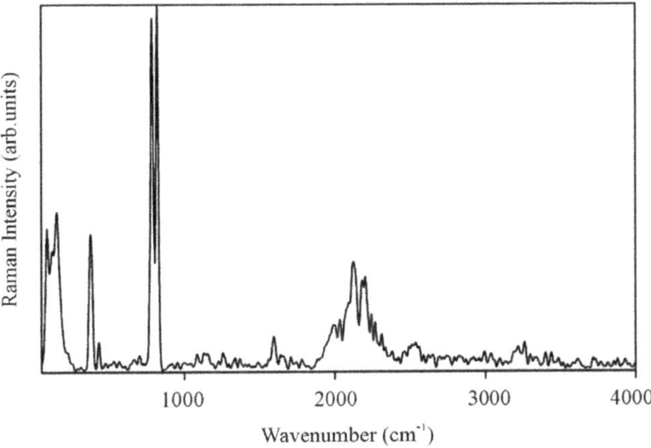

Figure 3. FT Raman spectrum of hydrated thorium bromate in the region 50 - 4000 cm^{-1}.

downward shift is attributed to the presence of hydrogen bonding, which is attached to water molecules.

Symmetric bending mode ν_2 is observed at 429 cm^{-1} as a weak band in the Raman spectrum similar to that in the case of [Cu $(H_2O)_6$] $(BrO_3)_2$ (**Table 2**) [25]. The degeneracy of the ν_4 mode of BrO_3^- anion is lifted in the Raman spectrum and two bands are observed at 371 and at 383 cm^{-1} (**Figure 3**). Lifting of degeneracy is also observed in the IR spectrum and two bands are observed at 363 and 389 cm^{-1}. The lifting of degeneracy of ν_4 mode both in the Raman and IR spectra indicate the lowering of symmetry of BrO_3^- anion from C_{3v} pyramidal to lower one at C_1. A moderately intense broad band is observed at 152 cm^{-1} in Raman spectra is assigned to the BrO_3^- rotational mode [27].

Table 2. Raman and IR Spectral data of thorium bromate hydrate (cm^{-1}).

Raman	IR	Assignments
	66 vvw	
85 ms	86 vvw	libr. BrO_3^-
	113 w	
125 w	126 ms	rot. BrO_3^-
	146 ms	
152 ms	156 m	
	163 m	
	199 m	
	218 m	
200 vw	236 m	
	251 m	
	260 m	
	295 m	
371 ms	363 vvs	ν_4 BrO_3^-
383 sh	389 s	
429 vw	522 w	ν_2 BrO_3^-
760 sh		
784 vs	789 vs	ν_1 BrO_3^-
796 sh		
820 vs	821 sh	ν_3 BrO_3^-
	1026 sh	
	1597 s	ν_2 H_2O
2050 m		
2122 ms	2075 w	Res.Raman
2204 m		
	2857 vvw	
	2992 vvw	
	3155 sh	ν_1-ν_3 H_2O
	3458 vvs br	
	3521 sbr	

vvw, very very weak; vw, very weak; w, weak; wsh, weak shoulder; sh, shoulder; br, broad; vs, very strong; s, strong; vvs, very very strong; ms, moderately strong; m, moderately; sbr, strong and broad; rot., rotational; libr., librational.

In the stretching mode region of H_2O a broad band extending from 3000 - 3708 cm^{-1} is obtained in the IR spectrum (**Figure 4**) with shoulder at 3155 and followed by in tense peaks 3458 and 3521 cm^{-1}. In the bending mode region, a moderately intense band is observed at 1597 cm^{-1} in the IR spectrum. The appearance of OH stretching modes at lower wavenumber than those of free state values of H_2O indicate the presence of hydrogen bonds in the crystal [24]. Due to hydrogen bonds, OH bending modes are usually expected to shift upwards. But the coordination of Thorium atoms to the water Oxygen atoms causes a downward shift. Therefore, this almost equally energetic upward and downward pulling of O atoms causes the intense OH bending modes to appear without any shift in wavenumber values with respect to its free state value (1595-ν_2 H_2O). The appearance of additional bands in the OH stretching regions and broad spectral profile of both bending and stretching regions confirms the existence of hydrogen bonds of different strengths in the crystal [24,25]. The bands observed below 250 cm^{-1} are mostly of lattice modes and are assigned in **Table 2**. Interestingly, majority of the Raman and IR bands are not coinciding each other which indicate that the hydrated thorium bromate is crystallized in centrosymmetric space group which is in agreement with X-ray crystallographic information.

The strong Raman bands observed at 2130 and 2206 cm^{-1} probably attributed to resonance Raman scattering process since there is no vibrational bands are expected in this region for the title compound. Thorium is having closely lying "f" electronic energy levels, the electronic transitions between these levels initiate the formation of bands in the region 1925 - 2280 cm^{-1} similar to that reported previously in several U(IV) and Zr(IV) bis (ketimide) complexes, and in a few Th(IV) compounds [28,29].

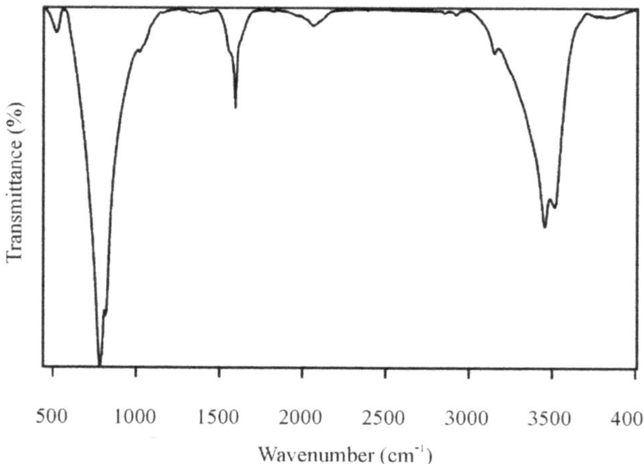

Figure 4. FTIR spectrum of the hydrated thorium bromate in the region 500 - 4000 cm^{-1}.

4. Conclusion

The crystal structure of hydrated thorium bromate shows the Br(1)-O(1) distance is 1.6504(0.0133) Å, Br(1)-O(2) 1.6712(0.0128) Å and Br(1)-O(3) is 1.6573(0.0129) Å. Additional bands obtained in the vibrational spectra in the region of ν_1 modes in the thorium bromate hydrate are due to the lifting of degeneracy of BrO_3^- ion, which occupies a site of lower symmetry than its free state at C_{3v}. The lifting of degeneracy of ν_4 mode both in the Raman and IR spectra indicate the lowering of symmetry of BrO_3^- anion from C_{3v} pyramidal to lower one. The shifting of stretching modes of water towards low wavenumber region as compared to that of its free state value and the broadening of the stretching bands of H_2O indicate the presence of hydrogen bonds of varying strengths in the crystal. Non-coincidence of majority of Raman and IR bands confirms the centrosymmetric space group P21/c of the crystal. Resonance Raman scattering process is seen in addition to vibrational bands attributed to closely lying "f" electronic energy levels. The present compound may be useful to study as a chemical oscillator and chemiluminescence systems in future.

REFERENCES

[1] S. K. Misra and X. Li, "Electron Paramagnetic Resonance of Gd^{3+}-Doped $Dy(BrO_3)_3{\cdot}9H_2O$ and $Eu(BrO_3)_3{\cdot}9H_2O$ Single Crystals: Structural Phase Transitions and Spin-Hamiltonian Parameters," *Journal of Physics: Condensed Matter*, Vol. 4, No. 13, 1992, pp. 3559-3670.

[2] D. Neogy and T. Purohit, "Magnetic Behavior and Crystal Field of $Dy(BrO_3)_3{\cdot}9H_2O$," *Physical Review B*, Vol. 35, No. 11, 1987, pp. 5849-5855.

[3] N. Matsuyama, N. Okazaki, Y. Tanimoto and I. Hanazaki, "Photo-Response of the Bromate-Sulfite Chemical Oscillator with Tris-(bipyridine)ruthenium(II) as a Catalyst," *Chemical Physics Letters*, Vol. 323, No. 3, 2000, pp. 372-376.

[4] K. Kurin-Csörgei, I. R. Epstein and M. Orbán, "Systematic Design of Chemical Oscillators Using Complexation and Precipitation Equilibria," *Nature*, Vol. 433, No. 7022, 2005, pp. 139-142.

[5] M. Iranifam, M. A. Segundo, J. L. M. Santos, J. L. F. C. Lima and M. H. Sorouraddin, "Oscillating Chemiluminescence Systems: State of the Art," *Luminescence*, Vol. 25, No. 6, 2010, pp. 409-418.

[6] S. Pääkkönen, J. Pursiainen and M. Lajunen, "Fast Oxidation of Secondary Alcohols by the Bromate-Bromide System Using Cyclic Microwave Heating in Acidic Water," *Tetrahedron Letters*, Vol. 51, No. 51, 2010, pp. 6695-6699.

[7] D. Deepa and G. Chandramohan, "Kinetic and Mechanistic Study on the Oxidation of Indole-3-Propionic Acid in Acetic Acid Medium," *Research Journal of Chemical Sciences*, Vol. 2, No. 10, 2012, pp. 70-74.

[8] M. R. Shishehbore, A. Sheibani and M. Eslami, "Thionine-Bromate as a New Reaction System for Kinetic Spectrophotometric Determination of Hydrazine in Cooling Tower Water Samples," *Journal of Chemistry*, Vol. 2013, 2013, pp. 1-5.

[9] J. Albertsson and I. Elding, "The Geometry of the Nonaaqualanthanoid (3+) Complex in the Solid Bromates and Ethyl Sulphates," *Acta Crystallographica*, Vol. B33, 1977, pp. 1460-1469.

[10] H. Poulet, J. P. Mathieu, D. Vergnat, B. Vergnat, A. Hadni and X. Gerbaux, "Vibrational Spectra, Structure, and Phase Transition in Neodymium and Gadolinium Bromate Enneahydrates," *Physica Status Solidi*, Vol. 32, No. 2, 1975, pp. 509-520.

[11] R. E. Gerkin and W. J. Reppart, "The Structures of the Lanthanide Ethyl Sulfate Enneahydrates, $M(C_2H_5SO_4)_3.9H_2O$ [M = La-Lu (except Pm)], at 171 K," *Acta Crystallographica*, Vol. C40, No. 5, 1984, pp. 781-786.

[12] R. E. Gerkin and W. J. Reppart, "Structures of Holmium Bromate Enneahydrate at 168 and 294 K and Their Implications for the Isomorphic Series of Rare-Earth Bromate Enneahydrates," *Acta Crystallographica*, Vol. C43, No. 4, 1987, pp. 623-631.

[13] A. C. Blackburn, J. C. Gallucci and R. E. Gerkin, "The Structure of Hexaaquaaluminium(III) Bromatetrihydrate, $[Al(H_2O)_6](BrO_3)_3{\cdot}3H_2O$," *Acta Crystallographica*, Vol. C48, No. 7, 1992, pp. 1185-1188.

[14] K. Lieselotte, L. K. Templeton and D. H. Templeton, "Structure of Barium Bromate Monohydrate," *Acta Crystallographica*, Vol. C45, No. 4, 1989, pp. 672-673.

[15] A. Abbasi and L. Eriksson, "Nonaaquayttrium(III) Tris(Bromate)," *Acta Crystallographica*, Vol. E62, No. 5, 2006, pp. 126-128.

[16] A. C. T North, D. C. Philips and F. S. Mathews, "A Semi-Empirical Method of Absorption Correction," *Acta Crystallographica*, Vol. A24, No. 3, 1968, pp. 351-359.

[17] Enraf-Nonius, "CAD-4 Software. Version 5.0.," Enraf-Nonius, Delft, 1994.

[18] T. Ueki, A. Zalkin and D. H. Templeton, "Crystal Structure of Thorium Nitrate Pentahydrate by X-Ray Diffraction," *Acta Crystallographica*, Vol. 20, No. 6, 1966, pp. 836-841.

[19] W. G. Fateley, F. R. Dollish, N. T. McDevitt and F. F. Bentley, "Infrared and Raman Selection Rules for Molecular and Lattice Vibrations—The Correlation Method," Wiley-Interscience, New York, 1972.

[20] K. Nakamoto, "Infrared and Raman Spectra of Inorganic and Coordination Compounds—Part A," 5th Edition Wiley-Interscience, New York, 1997.

[21] D. M. Adams, J. Barlow, H. Tan and M. J. Taylor, "The Vibrational Spectra of Mercury (I) Bromate, Sulphate, and Nitrate Dihydrate," *Journal of Raman Spectroscopy*, Vol. 5, No. 1, 1976, pp. 63-73.

[22] R. S. Jayasree, M. J. Bushiri, A. John and V. U. Nayar, "Temperature Dependent Polarized Raman Spectra of Nonaaqualanthanoid (Pr) Single Crystal," *Spectrochimica Acta*, Vol. A64, No. 2, 2006, pp. 518-525.

[23] T. Devanathan and T. K. K. Srinivasan, "Raman Spectra of Single-Crystal $Cd(BrO_3)_2 \cdot 2H_2O$ and Polycrystalline $Cd(BrO_3)_2 \cdot 2D_2O$," *Journal of Raman Spectroscopy*, Vol. 18, No. 7, 1987, pp. 525-531.

[24] T. K. K. Srinivasan and T. Devanathan, "Raman Spectral Studies of Dehydration of $Sr(BrO_3)_2 \cdot H_2O$," *Journal of Raman Spectroscopy*, Vol. 21, No. 2, 1990, pp. 99-102.

[25] M. J. Bushiri and V. U. Nayar, "Raman and FTIR Spectra of $[Cu(H_2O)_6](BrO_3)_2$ and $[Al(H_2O)_6](BrO_3)_3 \times 3H_2O$," *Spectrochimica Acta*, Vol. A58, No. 5, 2002, pp. 899-909.

[26] M. J. Bushiri and V. U. Nayar, "Raman and FTIR Spectra of $RE(BrO_3)_3 \cdot 9H_2O$ (RE = Eu, Tb) and Electronic Transitions in $Eu(BrO_3)_3 \cdot 9H_2O$," *International Journal of Modern Physics*, Vol. B15, No. 18, 2001, pp. 2499-2507.

[27] A. Viste and D. E. Irish, "Raman and Infrared Spectral Studies of Polycrystalline Thallium(I) Halates, $TlXO_3$," *Canadian Journal of Chemistry*, Vol. 55, No. 18, 1977, pp. 3218-3227.

[28] E. J. Schelter, P. Yang, B. L. Scott, R. E. Da Re, K. C. Jantunen, R. L. Martin, P. J. Hay, D. E. Morris and J. L. Kiplinger, "Systematic Studies of Early Actinide Complexes: Thorium(IV) Fluoroketimides," *Journal of the American Chemical Society*, Vol. 129, No. 16, 2007, pp. 5139-5152.

[29] D. E. Morris, R. E. Da Re, K. C. Jantunen, I. C. Rodriguez and J. L. Kiplinger, "Trends in Electronic Structure and Redox Energetics for Early-Actinide Pentamethylcyclopentadienyl Complexes," *Organometallics*, Vol. 23, No. 22, 2004, pp. 5142-5153.

Influence of Oxygen to Argon Ratio on the Structural and Morphological Properties of Nb-Doped $SrTiO_3$ Epitaxial Films Grown by Reactive Ion Beam Sputter Deposition

Gasidit Panomsuwan[1], Nagahiro Saito[1,2,3]
[1]Department of Materials, Physics and Energy Engineering, Graduate School of Engineering, Nagoya, Japan
[2]EcoTopia Science Institute, Nagoya University, Nagoya, Japan
[3]Green Mobility Collaborative Research Center, Nagoya, Japan

ABSTRACT

Nb-doped $SrTiO_3$ (STNO) films were grown on (001)-oriented $LaAlO_3$ substrates by a reactive ion beam sputter deposition at various $O_2/(Ar+O_2)$ mixing ratios (OMRs) with a substrate temperature of 800˚C. The STNO films exhibited good crystallinity with an epitaxial orientation as characterized by high-resolution X-ray diffraction, grazing-incidence X-ray diffraction, and in-plane pole figure analysis. A decrease of out-of-plane and in-plane lattice constants was observed with an increase of OMR. The surface morphology of the STNO films showed a very dense fine-grain structure. The root-mean-square roughness was found to be increased as the OMR increased. Moreover, the elemental compositions of the STNO films were examined by X-ray photoelectron spectroscopy.

Keywords: Nb-Doped $SrTiO_3$; Epitaxial Films; Crystal Structure; Ion Beam Sputter Deposition.

1. Introduction

$SrTiO_3$ (STO) has received much attention in the past decade as a potential material for a wide range of electronic devices owing to its excellent dielectric properties [1-3]. Another interesting property of STO is the ability to adjust the electrical conductivity, which is achievable with an n-type semiconductor and metallic behaviors by doping an appropriate level of impurity atoms (e.g. La^{2+}, Nb^{5+}, etc.) [4,5]. Very recently, n-type semiconductor Nb-doped STO (STNO) has become an attractive material for thermoelectric applications, and has also served as the bottom electrodes for perovskite ferroelectric thin-film capacitors [6,7]. Most studies extensively used pulsed laser deposition (PLD) and molecular beam epitaxy (MBE) to grow the epitaxial STNO films with control at an atomic level. However, they are not suitable for large area deposition in the viewpoint of industrial processes. Electron-cyclotron-resonance ion beam sputter deposition (ECR-IBSD) has been demonstrated as a potential technique for fabricating complex oxide films with high crystalline quality and flat surface. It also provides several advantages, such as well-controlled ion energy and growth rate, low operating pressure, (~5×10^{-5} - 5×10^{-4} Torr), good film adhesion, and ease of fabrication of thin film over a large area [8,9]. The ECR ion source is a non-filament type, leading to a long lifetime and allowing for the use of various kinds of gas species [10]. Not only selecting a suitable growth technique, but the effect of growth conditions on the physical properties of the film is also very important to consider. Oxygen pressure has been found to be a critical parameter for the growth of oxide films because it significantly affects the crystal structure, surface morphology, stoichiometry, and electrical properties [11-14]. Therefore, careful control of the oxygen pressure during growth is very important in order to obtain films with desirable properties.

In the present study, we have focused on the growth of epitaxial STNO films using the reactive ECR-IBSD technique. The effects of the $O_2/(Ar+O_2)$ mixing ratio (OMR) to their structural and morphological properties were studied and discussed.

2. Experimental

STNO films were grown on (001)-oriented $LaAlO_3$ (LAO) single crystal substrates using the ECR-IBSD technique. 20 mol% Nb-doped STO was used as a target. To prepare the substrate with atomically flat surface, the LAO substrates (MTI Corporation) were ultrasonically cleaned with acetone and ethanol (purity 99.0%, Wako Pure Chemical Industries) followed with etching in concentrated hydrochloric acid (37 wt%, Sigma-Aldrich) under ultrasonic agitation at room temperature [15]. The chamber was initially evacuated to a base pressure of <2 × 10^{-6} Torr. The cleaned LAO substrates were placed above the target in parallel at a distance of 40 mm. The STNO films were grown on the LAO substrates at various OMRs (12.5%, 25.0%, 37.5% and 50.0%), while total pressure of the Ar + O_2 gas mixtures was maintained at 6 × 10^{-4} Torr. These were the optimal condition to obtain the stable beam during film growth. The gas mixtures were discharged by a microwave ECR ion source $(f = 2.45\ \text{GHz})$ at a fixed power output of 180 W. Then the ions were extracted from the source at a voltage of 1800 V and delivered to the target at an incident angle of 45˚. All films were grown at a substrate temperature of 800˚C.

Structural properties of the STNO films were characterized by x-ray diffraction (XRD, Rigaku SmartLab) with CuK_α radiation (λ = 1.5418 Å). The X-ray power was 9 kW. Surface morphology of the films was observed by atomic force microscopy (AFM, Seiko Instrument, SPA-300HV). To analyze elemental composition, X-ray photoelectron spectroscopy (XPS) measurement was performed with an Omicron ESCA Probe (Omicron, Nanotechnology). Monochromatic MgK_α radiation (photon energy of 1256.6 eV) was used as an excitation source.

3. Results and Discussion

3.1. Structural Properties

Figure 1 shows a grazing-incidence X-ray reflection (GIXRR) spectrum of the STNO film grown on the LAO substrate. The oscillation pattern began to be observed at an angle greater than the critical angle $(\alpha_C \approx 0.315^\circ)$. The film thickness could be determined from the period of this oscillation. The inset of **Figure 1** demonstrates average growth rate as a function of OMR. By increasing OMR, the average growth rate decreased from 0.56 to 0.17 nm/min due to its low sputtering yield. This growth rate was calculated by dividing the film thickness (as extracted from GIXRR) by the growth time. A high-resolution X-ray diffraction (HRXRD) ω–2θ scan in **Figure 2(a)** revealed only 00l reflection peaks corresponding to the STNO film and the LAO substrate without the reflection peaks from randomly oriented grains and impurity phases, such as with Nb_2O_3, $SrNb_2O_7$, $SrNb_2O_6$ or Nb_2TiO_7. This indicated that the STNO films were single phase, and preferred (001) orientation in a direction normal to the substrate surface. A grazing incidence x-ray diffraction (GIXRD) $2\theta\chi$ scan was also measured and shown in **Figure 2(b)**. An incident angle was fixed at ω = 0.3˚ for the measurement, which provided the penetration depth near the film surface. The measurement was fixed at a scattering vector normal to the 200_{LAO} plane. Only the 200_{STNO} peak together with the 200_{LAO} peak along the $2\theta\chi$ scan was detected, indicating good in-plane alignment. The same HRXRD and GIXRD patterns were found for all STNO films.

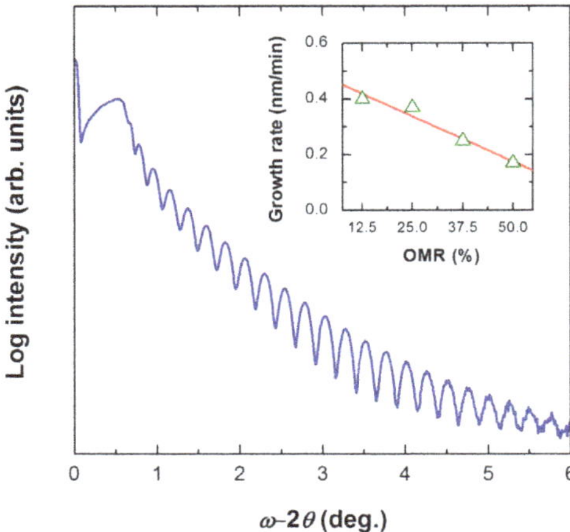

Figure 1. GIXRR spectrum of the STNO film grown on LAO substrate. The inset shows growth rate as a function of OMR.

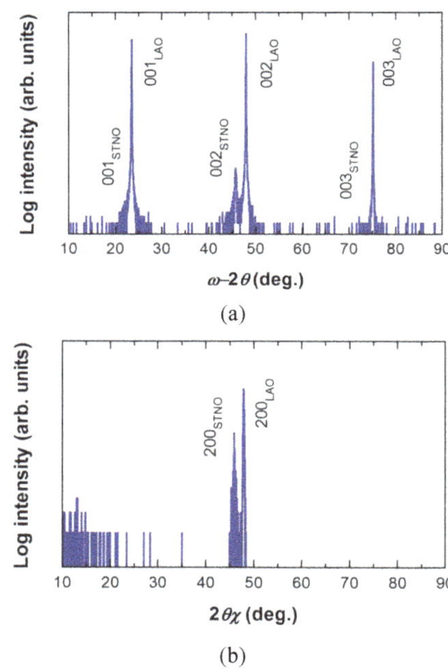

Figure 2. (a) HRXRD ω–2θ scan and (b) GIXRD $2\theta\chi$ scan of the STNO film grown on LAO substrate at 25% OMR.

Figures 3(a) and **(b)** present the HRXRD ω–2θ scans and GIXRD $2\theta\chi$ scans around the 002_{STNO} and 200_{STNO} reflections, respectively. Pendellösung fringes were clearly seen around the 002_{STNO} reflection from the HRXRD spectra, which provided evidence that the grown films had good crystallinity, smooth surfaces, and well-defined film/substrate interfaces. Another observation was a slight shift of the 002_{STNO} and 200_{STNO} reflections toward a higher angle, suggesting a change of the crystal lattice structure. Out-of-plane and in-plane lattice constants calculated from the angle position of the 002_{STNO} and 200_{STNO} reflections, respectively, using Bragg's law were plotted as a function of OMR in **Figure 4**. The out-of-plane lattice constants were larger than in-plane lattice constants for all films. This implied that the STNO films grew on the LAO substrate with a tetragonal structure due to an in-plane compressive stress from the LAO substrate. With an increase of OMR, both out-of-plane and in-plane lattice constants decreased, while tetragonality (c/a) increased from 1.001 to 1.003. The larger lattice constants at low OMR were attributed to Ti^{3+} and Nb^{4+} ions induced by oxygen vacancies. The ion radii of Ti^{3+} (67 pm) and Nb^{4+} (68 pm) are much larger than those of Ti^{4+} (60.5 pm) and Nb^{5+} ions (64 pm), resulting in lattice expansion [16].

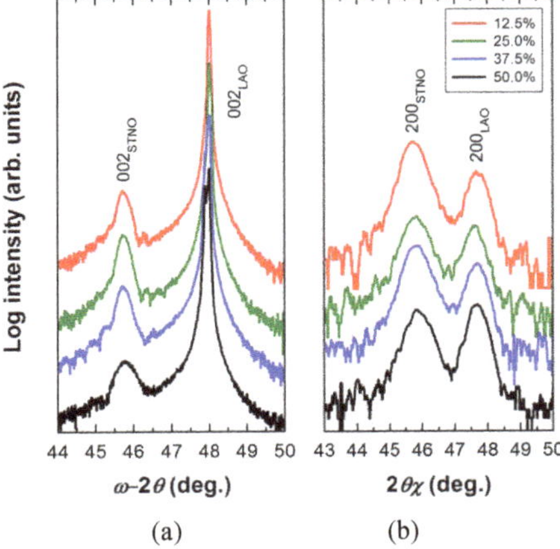

Figure 3. (a) HRXRD $\omega-2\theta$ scans around 002_{STNO}; and (b) GIXRD $2\theta\chi$ scans around the 200_{STNO} reflections of the STNO films grown at various OMRs.

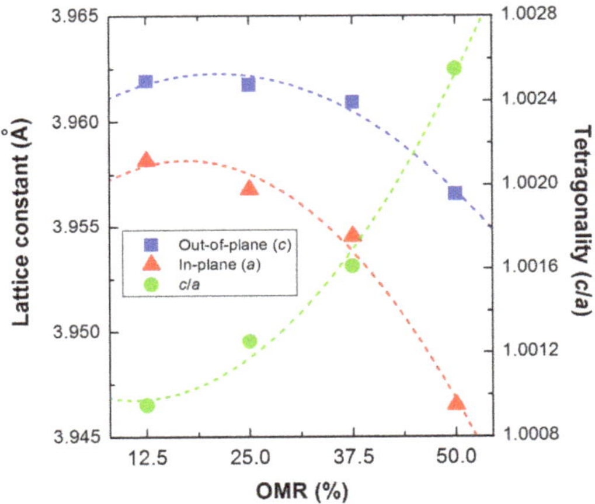

Figure 4. Lattice constants and tetragonality (c/a) of the STNO films as a function of OMR.

Orientation quality of the STNO films was confirmed by out-of-plane and in-plane rocking curve measurements on the 002_{STNO} reflection (ω scan) and the 200_{STNO} reflection (ϕ scan), respectively. The out-of-plane rocking curve consisted of two intensity components: a narrow peak corresponding to the reflection from a good alignment of crystal, and a broad peak corresponding to local atomic displacement. This behavior is normally found in epitaxial layers due to the presence of misfit dislocations lying at the film/substrate interface. Full-width at half-maximum (FWHM) of the out-of-plane rocking curve was found to be the lowest value for the film grown at 25% OMR, which was about 0.1° and 1° for the narrow and broad peaks, respectively. A gradual increase of the FWHM of out-of-plane rocking curve was investigated when the OMR was greater than 25%. This suggested that the orientation quality of the films dropped when they grew at over 25% OMR. On the other hand, in-plane rocking curve revealed only one peak with the FWHM of about 1° and remained almost constant with increasing OMR. The FWHMs of out-of-plane and in-plane rocking curves plotted versus the OMR are shown in **Figure 5**.

To confirm in-plane orientation of the STNO films on the LAO substrates, an in-plane pole figure analysis was examined on the $\{011\}_{LAO}$ and $\{011\}_{STNO}$ planes by fixing the 2θ at ~33.4° and ~32.2°, respectively. By varying $\beta\left(0<\beta<360^\circ\right)$ and $\alpha\left(0<\alpha<90^\circ\right)$, the samples were rotated relative to the scattering vector. Both in-plane pole figures revealed four intense spots at $\alpha \approx 45^\circ$ with an equal space of 90° at the same β-angle position $\left(\beta = 0^\circ, 90^\circ, 180^\circ \text{ and } 270^\circ\right)$, as depicted in **Figure 6**. This result confirmed that the films exhibited a cube-on-cube orientation and a good in-plane alignment. The orientation relationship could be given as:

$$(001)_{STNO} \| (001)_{LAO} \text{ and } [110]_{STNO} \| [110]_{LAO}$$

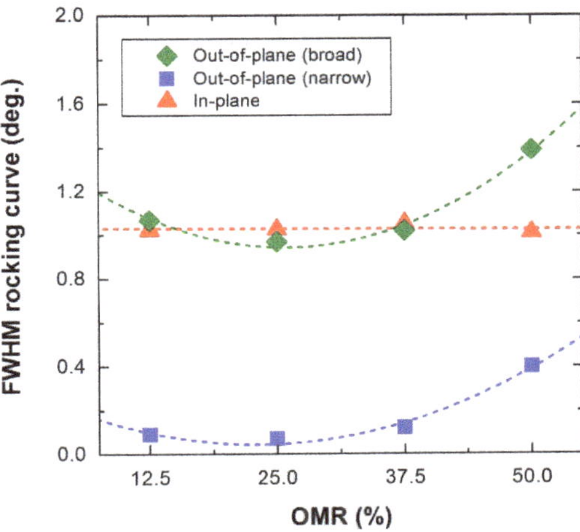

Figure 5. FWHMs of out-of-plane and in-plane rocking curves of the STNO films grown at various OMRs.

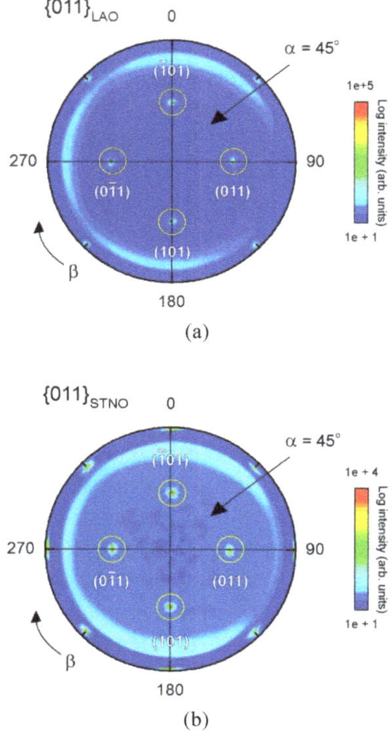

Figure 6. In-plane pole figures measured on (a) the $\{011\}_{LAO}$ and (b) the $\{011\}_{STNO}$ planes.

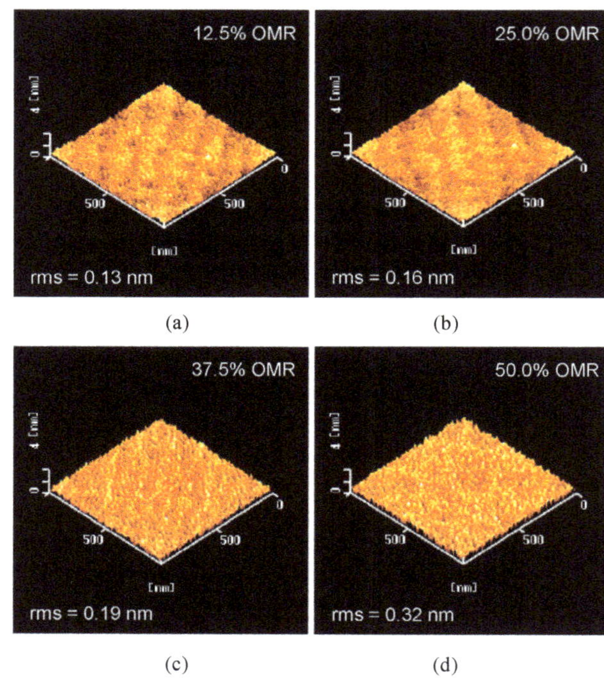

Figure 7. AFM topography images (1 × 1 μm^2 scan area) of the STNO films grown at various OMRs: (a) 12.5%; (b) 25.0%; (c) 37.5%; and (d) 50.0%.

3.2. Surface Morphology

AFM topography images of the STNO film surfaces grown at various OMRs are illustrated in **Figure 7**. Three-dimensional islands covered on the substrate terrace were clearly observed. Moreover, the film surfaces became rougher as the OMR increased. The rms roughness was increased up to about 0.32 nm when the OMR reached 50%. This might be attributed to the different physical and chemical properties of the vapor species. At higher OMR, these species lost their kinetic energy due to the scattering effect of oxygen molecules, leading to a decrease of surface mobility and lateral growth inhibition.

3.3. Elemental Composition Analysis

All binding energies detected with XPS wide scans showed that the STNO films were composed of Sr, Ti, Nb, and O elements. Elemental compositions near the film surface were obtained by quantitative analysis using Sr3d, Ti3p, Nb3d, and O1s peaks. It was found that the compositional ratio of $Sr/(Ti+Nb)$ was close to unity. **Figure 8** shows a narrow scan of O1s, which consisted of two overlapping peaks. One peak at a lower binding energy $(E_B \approx 530\ \text{eV})$ was attributed to the lattice oxygen (O_L) in the film. Another peak at a higher binding energy $(E_B \approx 532\ \text{eV})$ arose from the chemically adsorbed oxygen on the film surface (O_C). The peak of

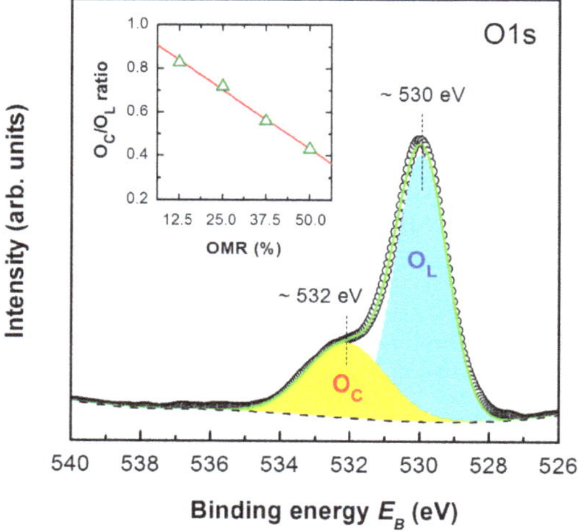

Figure 8. Narrow scan XPS spectrum of the O1s peak of STNO film. The inset shows the relative ratio of O_C/O_L as a function of OMR.

chemically adsorbed oxygen was much weaker than that of lattice oxygen in the XPS spectrum. The relative ratio of O_C/O_L calculated from peak areas was found to decrease with an increase of OMR, as demonstrated in the inset of **Figure 8**. A decrease of O_C indicated that oxygen vacancies decreased when the films were grown at higher OMR. This might be explained that there were only a few atoms on the film surface that allowed for the chemical adsorption of oxygen.

4. Conclusion

STNO films were epitaxially grown on the LAO substrates by reactive ECR-IBSD at various OMRs. The OMR was found to significantly influence the growth rate, crystal structure and film quality. Moreover, the rms surface roughness of the films became rougher when the OMR increased. The present study showed that an optimum value for the growth of STNO thin films with good crystallinity and a relatively smooth surface was at 25% OMR. The XPS result showed that the ratio of O_L/O_C was also related to the OMR. We expect that the ECR-IBSD can be a capable method for fabricating high quality films in the research and development of production technology. The obtained results are also useful for selecting optimum growth conditions for further experiments.

5. Acknowledgements

This work has been partially supported by Micro-Nano Global Center of Excellence (G-COE), Nagoya University.

REFERENCES

[1] C. W. Schneider, R. Schneider and H. Rietschel, "High Dielectric Constant and Tunability of Epitaxial $SrTiO_3$ Thin Film Capacitors," *Journal of Applied Physics*, Vol. 85, No. 10, 1999, pp. 7362-7369.

[2] K. Eisenbeiser, J. M. Finder, Z. Yu, J. Ramdani, J. A. Curless, J. A. Hallmark, R. Droopad, W. J. Ooms, L. Salem, S. Bradshaw and C. D. Overgaard, "Field Effect Transistors with $SrTiO_3$ Gate Dielectric on Si," *Applied Physics Letters*, Vol. 76, No. 10, 2000, pp, 1324-1326.

[3] J. Robertson, "High Dielectric Constant Gate Oxides for Metal Oxide Si Transistors," *Reports on Progress in Physics*, Vol. 69, No. 2, 2006, pp. 327-396.

[4] T. Tomio, H. Miki, H. Tabata, T. Kawai and S. Kawai, "Control of Electrical Conductivity in Laser Deposited $SrTiO_3$ Thin Films with Nb Doping," *Journal of Applied Physics*, Vol. 76, No. 10, 1994, pp. 5886-5890.

[5] D. Olaya, F. Pan, C. T. Roger and J. C. Price, "Electrical Properties of La-Doped Strontium Titanate Thin Films," *Applied Physics Letters*, Vol. 80, No. 16, 2002, pp. 2928-2930.

[6] S. Y. Wang, B. L. Cheng, C. Wang, S. Y. Dai, H. B. Lu, Y. L. Zhou, Z. H. Chen and G. Z. Yang, "Reduction of Leakage Current by Co Doping in $Pt/Ba_{0.5}Sr_{0.5}TiO_3/$ Nb–$SrTiO_3$ Capacitor," *Applied Physics Letters*, Vol. 84, 2004, pp. 4116-4118.

[7] S. Ohta, T. Nomura, H. Ohta, M. Hirano, H. Hosono and K. Koumoto, "Large Thermoelectric Performance of Heavily Nb-Doped $SrTiO_3$ Epitaxial Film at High Temperature," *Applied Physics Letters*, Vol. 87, No. 9, 2005, Article ID: 092108.

[8] A. Biswas, A. K. Poswal, R. B. Tokas and D. Bhattacharyya, "Characterization of Ion Beam Sputter Deposited W and Si Flms and W/Si Interfaces by Grazing Incidence X-Ray Reflectivity, Atomic Force Microscopy and Spectroscopic Ellipsometry," *Applied Surface Science*, Vol. 254, No. 11, 2008, pp. 3347-3356.

[9] G. Panomsuwan, O. Takai and N. Saito, "Fabrication and Characterization of Epitaxial $SrTiO_3$/Nb-Doped $SrTiO_3$ Superlattices by Double ECR Ion Beam Sputter Deposition," *Vacuum*, Vol. 89, 2013, pp. 35-39.

[10] N. Sakudo, K. Tokiguchi, H. Koike and I. Kanomata, "Microwave Ion Source," *Review of Scientific Instruments*, Vol. 48, No. 7, 1977, pp. 462-466.

[11] K. Fukushima and S. Shibagaki, "Nb Doped $SrTiO_3$ Thin Films Deposited by Pulsed Laser Ablation," *Thin Solid Films*, Vol. 315, No. 1-2, 1998, pp. 238-243.

[12] E. J. Tarsa, E. A. Hachfeld, F. T. Quinlan and J. S. Speck, "Growth-Related Stress and Surface Morphology in Homoepitaxial $SrTiO_3$ Films," *Applied Physics Letters*, Vol. 68, No. 4, 1996, pp. 490-492.

[13] H. L. Cai, X. S. Wu and J. Gao, "Effect of Oxygen Content on Structural and Transport Properties in $SrTiO_{3-x}$ Thin Films," *Chemical Physics Letters*, Vol. 467, No. 4-6, 2009, pp. 313-317.

[14] R. Reshmi, M. K. Jayaraj and M. T. Sebastian, "Influence of Oxygen to Argon Ratio on the Properties of RF Magnetron Sputtered $Ba_{0.7}Sr_{0.3}TiO_3$ Thin Films," *Journal of the Electrochemical Society*, Vol. 158, No. 5, 2011, pp. G124-G127.

[15] T. Ohnishi, K. Takahashi, M. Nakamura, M. Kawasaki, M. Yoshimoto and H. Koinuma, "A-Site Layer Terminated Perovskite Substrate: $NdGaO_3$," *Applied Physics Letters*, Vol. 74, No. 17, 1999, pp. 2531-2533.

[16] W. Martienssen and H. Warlimont, "Springer Handbook of Condensed Matter and Materials Data," Springer, Berlin, 2005.

16

Ce-SAD Phasing of Glucose Isomerase and Thermolysin Using Cu *Kα* Radiation

Sibi Narayanan, Devadasan Velmurugan*
Centre of Advanced Study in Crystallography and Biophysics, University of Madras,
Maraimalai (Guindy) Campus, Chennai, India

ABSTRACT

Current structural genomics projects aim to solve a large number of selected protein structures as fast as possible. High degree of automation and standardization is required at every step of the whole process to speed up protein structure determination. Phase problem is a bottleneck in macromolecular structure determination and also in model building which is a time-consuming task. The simplest approach to phasing macromolecular crystal structures is the use of a SAD signal. SAD data can be collected using the in-house copper (1.54 Å) wavelength source. Data collected using copper wavelength with the incorporation of anomalously scattering heavy metal atoms may serve as a powerful tool for structural biologists to solve novel protein structures as well where synchrotron beam line is not available. A short soak of protein crystals in heavy metal solution or by incorporating heavy atoms into the protein drop while crystallizing the protein (co-crystallization) leads to incorporation of these heavy metal ions into the ordered solvent shell around the protein surface. The present work aims to determine whether cerium ion can be successfully incorporated into the protein crystal through quick-soaking method while maintaining the isomorphism. The study also aims in understanding whether this metal ion can be used for phasing purpose. The intensity data are collected and analyzed for anomalous signal, substructure solution and the binding sites.

Keywords: Anomalous Scattering; SAD phasing; Ce; Glucose Isomerase; Thermolysin

1. Introduction

More than 85% of the structures in the Protein Data Bank [1] have been solved by using the versatile tool, X-ray crystallography. Automation in X-ray crystallography is necessary to speed up the structure determination process. Phase problem is rate limiting and a bottleneck in macromolecular crystallography [2]. Phases can be derived using Molecular Replacement (MR) method using the atomic coordinates of a structurally similar protein. MR is widely used, when appropriate models are available [3]. Phase problem can also be solved by Multiple Isomorphous Replacement (MIR) method for novel proteins. But in this approach, protein crystals are derivatized by soaking in solutions of organometallic compounds or metal salts and the choice of proper derivatives is very much dependent on the nature of the protein under study and therefore not easy to generalize. Non-isomorphism that exists between the derivative and native crystals also creates problems [4]. MIR has largely been replaced with the Multi-wavelength Anomalous Diffraction (MAD) and Single-wavelength Anomalous Diffraction (SAD) methods fuelled by progress in the technology of X-ray sources, detectors as well as in the methodology of diffraction data-acquisition and phasing algorithms [5,6]. Anomalous scattering methods are widely used to solve the phase problem in macromolecular crystallography where molecular replacement fails [7,8]. For MAD data, selenomethionine has to be introduced during expression itself instead of methionine present in the protein and the data has to be collected at varied wavelengths at a synchrotron data source near the Selenium absorbtion edge. Sometimes MAD may fail due to disorder in methionine residues [9], and producing selenomethionine labeled proteins is not amenable for all expression systems. The simplest approach to phasing macromolecular crystal structures is the use of a SAD signal. SAD data can be collected using the in-house copper (1.54 Å) and chromium (2.29 Å) wavelength sources [10]. A single

*Corresponding author.

wavelength is sufficient for the collection of a SAD data [11]. The first step in solving macromolecular crystal structures by SAD is the location of anomalous scatterers. Currently sophisticated algorithms are very much available for data processing, substructure determination, phasing, density modification and model building. Recent studies show the great success of in-house SAD data even at low redundancy and low anomalous signal to noise ratio case [12].

Success in the location of the substructures and subsequent phasing depend critically on the quality of data and on the extent of the anomalous signal [13]. Exploiting the anomalous signal already present in the native protein or in the solvent would eliminate the extra experimental work in derivatization [14]. In the absence of such anomalous scatterers, the classic derivatization approach involves prolonged soaking of the native crystals in diluted solutions of various heavy metal salts [15].

A short soak of protein crystals in heavy metal solution or by incorporating heavy atoms into the protein drop while crystallizing the protein (co-crystallization) leads to incorporation of these heavy metal ions into the ordered solvent shell around the protein surface. In addition, incorporation of the heavy atoms also improves the nucleation process in the crystallization of proteins. Recently many novel protein structures have been solved using cobalt and cadmium incorporated as anomalous markers [16]. The use of heavy metal ion concentration and the soaking time vary from the different experiments carried out so far. It also depends on various factors such as protein crystal stability, molecular weight, pH, cryoprotectant etc. Heavy atoms can exist as individual ions when they are used in solution. Heavy atoms can constitute a small molecule (a metal complex) or a small peptide such that, the isomorphism of the protein crystal does not get disturbed [17]. Heavy atoms have a property of dissolving into charged ions when they are used as their salt solution, which may later bind with the oppositely charged amino acid residues present in the proteins. There are also cases, where the heavy metal ion used may replace the similarly charged metal ions that are naturally present in the proteins [18].

Cerium ion is a heavy metal ion that has not been explored much. It exists as a trivalent ion in solution and has its L_1 absorption edge at 1.8932 Å. The anomalous scattering coefficients are $f' = -1.88$ and $f'' = 9.74$ ē, at copper wavelength (1.54 Å). Cerium has higher anomalous signal at copper wavelength than chromium source [19]. Cerium exists as a trivalent ion in $CeCl_3$ solution. The van der Waals radius of cerium ion is 1.81 Å. Cerium has been derivatized using quick-soaking method into proteins *viz.*, Glucose Isomerase (GI) and Thermolysin (TL). Sufficient occupation of sites and successful phasing has been achieved with the solution concentration of the metal ion concentration ranging from 0.1 M to 0.3 M. The present work aims to determine whether cerium ion can be successfully incorporated into the protein crystal through quick-soaking method while maintaining the isomorphism. The study also aims in understanding whether this metal ion can be used for phasing purpose. The intensity data were collected and analyzed for anomalous signal, substructure solution and the binding sites.

2. Methods

2.1. Crystallization

GI and TL were purchased from Sigma-Aldrich, USA [20] and used without further purification. The GI protein solution containing 33 mg/ml was mixed with the well solution consisting of 200 mM Magnesium chloride and 100 mM Tris, pH 4.7 at 2:1 ratio (2 µl protein solution + 1 µl well solution). TL protein solution containing 25 mg/ml was mixed with the well solution consisting of 1.4 mM Calcium Acetate, 10 mM Zinc Acetate, 1 mM Sodium Nitrate and 50 mM Tris; pH 7.3 at 1:1 ratio (1 µl protein solution + 1 µl well solution).

Cerium chloride used was of analytical grade (Sigma-Aldrich, USA). Cerium derivatives were obtained by co-crystallizing or soaking cerium chloride in the range of 100 mM to 300 mM. Well diffracting crystals were obtained in the above mentioned ranges. The crystal was transferred for a short period (120 seconds) to mother liquor supplemented with 25 % MPD and 25 % ethylene glycol for cryo-protection for GI and TL, respectively and then flash-freezed in nitrogen gas stream at 100 K. Protein crystals were obtained using hanging-drop vapour-diffusion method. Crystallization, cryoprotectant and soaking conditions are shown in **Table 1**.

2.2. Data Collection and Processing

Data sets were collected using mar 345 dtb Image Plate Detector equipped with in-house Microstar Cu *Kα* rotating-anode X-ray generator [21] operated at 40 kV and

Table 1. Crystallization, soaking concentration and cryoprotectant details.

	Native GI	Native TL
Crystallization Condition	200 mM Magnesium Chloride and 100 mM Tris, pH 4.7	1.4 mM Calcium Acetate, 10 mM Zinc Acetate, 1 mM Sodium Nitrate and 50 mM Tris; pH 7.3
Soaking concentration of $CeCl_3$	300 mM	100 mM
Cryoprotectant	25% MPD	25% Ethylene Glycol

60 mA with Helios focusing mirrors. The slit size was set to 0.5/0.6, the crystal was mounted arbitrarily and no inverse beam technique was used to collect *Bijvoet* pairs. The detector was placed at a distance of 200 mm (GI) and 180 mm (TL), respectively for data collection. The diffraction images were collected over a total angular range of 360° with an oscillation angle of 1°. The time per frame varied from 120 sec to 180 sec. Data sets were integrated and scaled using *automar* [22]. The anomalous signals of the data sets were analyzed using *Phenix. xtriage* [23].

Anomalous scatterers were found using the dual-space recycling algorithm enabled in *SHELXD* [24]. The density modified phases obtained from *SHELXE* were used for model building using the *ARP/wARP* program web server [25] and refined using *REFMAC*5.0 [26] of *CCP4i* software [27]. Clearly interpretable maps were obtained after automated model building. Totally, two datasets were collected (300 mM Ce for GI; 100 mM Ce for TL). The crystallographic data are summarized in **Table 2**.

Table 2. Crystal Data Statistics, phasing and model building details.

	Ce-soaked GI dataset	Ce-soaked TL dataset
Cell Parameters	a = 78.34 Å b = 98.20 Å c = 129.43 Å	a = b = 92.61 Å c = 129.07 Å Γ = 120 °
Space Group	$P2_12_12$	$P6_122$
Mosaicity	0.49	0.52
Resolution range	27.08 - 2.21 Å	25.23 - 2.03 Å
Solvent content (%)	56	46
Redundancy	14.7 (12.8)	40.40 (38.23)
Completeness (%)	100.0	98.6 (98.3)
$I/\sigma(I)$	19.4 (16.7)	16.7 (4.48)
Anomalous signal (%)	1.88	1.54
FOM	0.57	0.63
Map Correlation	0.78	0.76
B_{Wilson}	14.8	34.50
Total Residues Built	772	316
Side Chains	761	315
R_{work} (%)	17.3	18.2
R_{free} (%)	21.6	20.3
Number of solvent molecules	207	171

3. Results and Discussion

3.1. Data Analysis

The data was processed with *automar* by separately merging the anomalous pairs as I^+ and I^-. The first five frames of 360 intensity images collected were used for peaksearch by the same program. The R_{merge} value was found to be 5.3% and 6.8% for GI and TL data sets, respectively. Mosaicity for both the datasets was 0.49 and 0.52. The overall redundancy of the data collected to 2 Å at 100 K for the complete dataset were 14.7 and 40.40 for GI and TL datasets, respectively at high resolution bin.

The full resolution range of the data had anomalous signal measurability greater than 10%. The mean signal to noise ratio for all the datasets were found between 16.5 and 19.5. Both the datasets consist of significant anomalous signal measurability. Solvent contents of both the datasets were 56% and 46%, respectively.

Completeness for GI dataset was 100% and TL dataset was above 98%. Both the redundancy and completeness play a major role in model building after substructure determination. The B-factor (thermal factor) in both the datasets were 14.8 and 34.5, respectively. The final model of all the datasets were refined to reasonable R_{work} and R_{free} values. The substructure solution, phasing and refinement statistics are shown in **Table 2**.

3.1.1. Ce Soaked GI Data

Ce soaked GI crystallized in *orthorhombic* space group, $P2_12_12$. GI crystallized as two molecules in the asymmetric unit. The structure is that of a homo-dimer. For Ce-soaked GI dataset during the substructure solution, ten anomalous scatterer peaks were located. f″ for manganese ion and magnesium ion at copper wavelength is 2.8 ē and 1.75 ē as estimated from program *CROSSEC* [28]. The resolution cutoff for finding heavy atom site was left to *SHELXC* [29]. Peaks for eight cerium ions, one manganese ion and one magnesium ion were picked or determined using *SHELXD*. The resolution cutoff used for the substructure determination was 3.0 Å. The electron density map correlation was above 75% for the dataset. The best solution obtained from *SHELXD* run was used directly for phase calculation and further improvement was done by density modification with program *SHELXE*. The mean Figure of Merit (FOM) was found to be 0.57, indicating good phasing.

The poly ala or preliminary model building was carried out using *SHELXE beta* version and the output model was fed as input to *ARP/wARP* for automated building. 772 residues (dimer) were built out of a total of 776 amino acids (388 amino acids in each molecule) using *ARP/wARP* for GI data, which clearly shows the good phasing and model building. Clearly interpretable electron density maps were obtained after the automated model building. More than 98% of the residues were able to be built using *ARP/wARP* program. The redundancy of data around 14 collected with three minute exposure time was sufficient to solve the structure at 2.21 Å resolution data by in-house Ce-SAD. Manual rebuilding was carried out using COOT [30] and the model was finally refined to R_{work} and R_{free} of 17.3% and 21.6%, respectively. Anomalous map of monomeric GI with eight cerium ions, one manganese ion and one magnesium ion at 5σ with water molecules is shown in **Figure 1**.

3.1.2. Cerium Soaked TL Data

Ce soaked TL crystallized in *hexagonal* space group, $P6_122$. TL was crystallized as a monomer in the asymmetric unit. At the copper wavelength of 1.54 Å, f″ for S, Ca and Zn are 0.56 ē, 1.28 ē and 0.68 ē, respectively as estimated from program *CROSSEC*. The resolution cut-off for finding heavy atom site was left to *SHELXC*. For the cerium soaked TL dataset, *SHELXD* located eight peaks as anomalous scatterers. The peaks corresponded to three cerium ions, three calcium ions, one zinc ion and one sulfur atom. Beta-test version of *SHELXE* was used for phasing, density modification and main-chain auto-tracing.

The poly ala residues obtained as output from *SHELXE* were given to *ARP/wARP* as input for model building. The automated solution and the model building by *ARP/wARP* built 316 amino acid residues totally. *ARP/wARP* built 99% of the residues for the dataset. Clearly interpretable maps were obtained after automated model building. The electron density map correlation was above 75% for the dataset. The mean Figure of Merit (FOM) was found to be 0.63, indicating good phasing. The model obtained was manually corrected using *COOT* and refined using *REFMAC*5.0. The final atomic model has crystallographic R_{work} and R_{free} values of 18.2% and 20.3%, respectively. Anomalous difference Fourier map of monomeric TL with three cerium ions, three calcium ions, one zinc ion and one sulfur atom at 5σ with water molecules is shown in **Figure 2.**

3.2. Cerium Ion Binding Sites

f″ for cerium at copper wavelength is 9.76 ē. For Ce-soaked GI dataset, substructure solution resulted in eight high peaks corresponding to cerium (III) ions with data limited up to 3.0 Å. The molecular structure remains the same after the binding of cerium ion to the protein. All the cerium peaks showed good spherical density. The occupancies of the cerium ions in GI were (0.42, 0.46, 0.45, 0.53, 0.56, 0.61, 0.62, 0.64). Metal ions have the nature of coordinating with the atoms of amino acid residues. This knowledge allowed us in assigning the peaks as cerium ions. The positions of the cerium ions were validated by the anomalous map shown in **Figure 1**. Manganese and Manganese ion peaks also had spherical density in the anomalous map. The positions of the manganese ion and magnesium ion peaks were identified by comparing the manganese and magnesium ion positions or occupied sites in the earlier reported GI structure [31]. This structure is a dimer with 776 amino acids.

Coordination distances for cerium ions w.r.t protein atoms and water molecules varied from 2.6 Å to 4.3 Å. The amine group present in the *Arginine residue* (ARG 266) interacts with the cerium ion (CE1) present in the first molecule of the GI. Occupancy of the CE1 is 0.42. Similar interaction between ARG 266 and cerium ion (CE2) was also observed in the second molecule of the same protein GI. Occupancy of CE2 is 0.46. Nitrogen atom present in the amine group of *Arginine residue* (ARG 387) interacts with another cerium ion (CE3) present in the first molecule of the protein GI. Occupancy of CE3 is 0.45. Interactions or co-ordinations are also found between water molecules and cerium ions in both the molecules. Out of the eight cerium sites, three cerium ions did not exhibit any coordination with any of the amino acid residues or water molecules. Cerium sites having hydrogen bonding contacts with hydrogen-donor groups of protein or water molecules are shown in **Figures 3(a)-(e).**

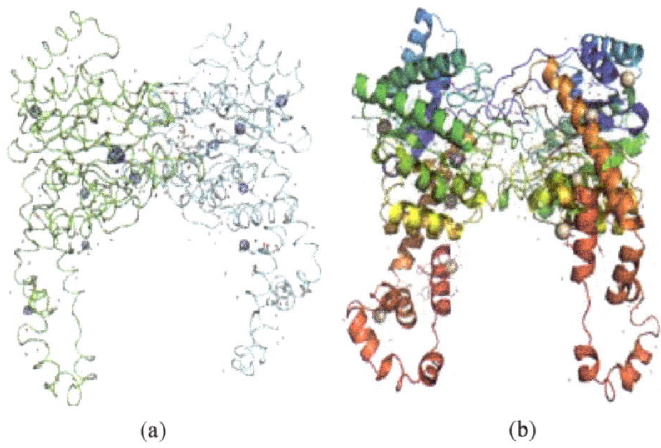

(a) (b)

Figure 1. (a) Anomalous difference fourier map and (b) Cartoon representation of homo-dimer GI with eight cerium ions, one manganese ion and one magnesium ion at 5σ with water molecules.

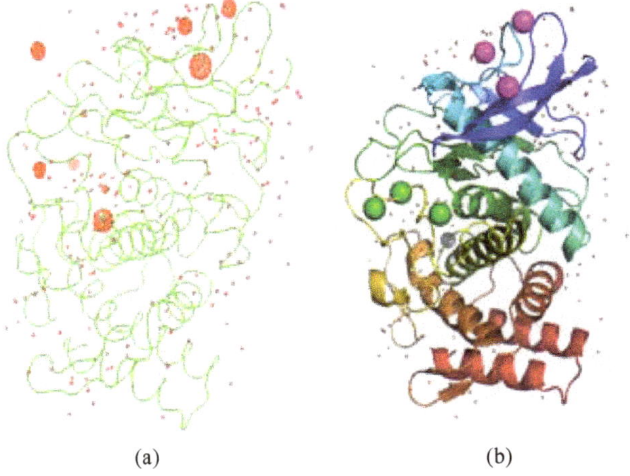

(a) (b)

Figure 2. (a) Anomalous difference Fourier map and (b) Cartoon representations of monomeric TL with three cerium ions, three calcium ions, one zinc ion and one sulfur atom at 5σ with water molecules.

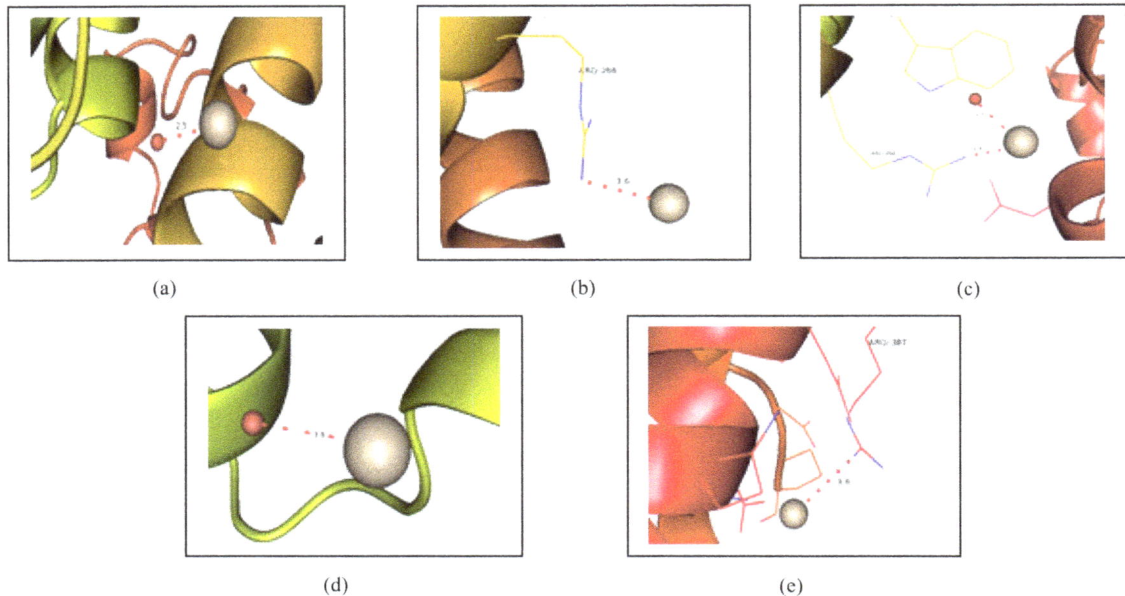

Figure 3. Picture showing (a) The coordination of First cerium ion with the water molecule in first molecule of GI, (b) The coordination of another cerium ion with the Arg 266 present in the first molecule of GI, (c) The coordination of another cerium ion with the Arg 263 and the water molecule present in the second molecule of GI, (d) The coordination of Fifth cerium ion with the water molecule present in the second molecule of GI, and (e) The coordination of another cerium ion with the Arg 387 present in the first molecule of GI.

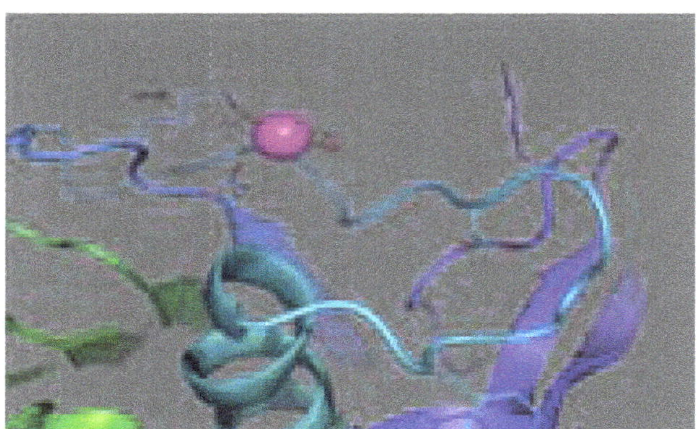

Figure 4. Picture showing the coordination binding of cerium ion with the Lys 45 and water molecule present in the TL.

For Ce-soaked TL data set, substructure solution determined anomalous scatterers for three cerium ions, three calcium ions, one zinc ion and two sulfur atoms. The cerium peaks showed spherical density with occupancies of 0.52, 0.58 and 0.63. The sulfur atom of methionine also showed spherical density. The positions of the three calcium ion peaks and one zinc ion peak were identified or validated by comparing their respective positions in the earlier reported TL structures [32]. The molecular structure remains the same after cerium ion binds to the protein. Coordination distances for cerium ions with respect to protein atoms and water molecules varied from 3.0Å to 3.2Å. Inspection of cerium ions in the structure of the protein reveals that they have occupied sites replacing the water molecules in the crystal lattice. Cerium ion (CE1) has coordination with *Lysine residue* (LYS 46) present in the protein TL. Occupancy of CE1 is 0.58. Cerium ion also shows a water mediated interaction. Cerium soaked data of Hen Egg White Lysozyme (HEWL) have already been solved using lab source method [33].

In the data, three bound cerium ions in the HEWL provided sufficient anomalous signal for phasing of 129 amino acid residues with the redundancy of 5.03. Cerium sites having hydrogen bonding contacts with hydrogen-donor groups of protein or water molecules are shown in **Figure 4**. The heavy metal ions (cerium) diffuse into protein crystals very quickly. A high salt concentration of the same increases the number of sites and their occupancy.

4. Conclusion

Cerium ions were incorporated separately using quick-soaking method in order to understand the binding of heavy atoms to the protein crystals and their interactions with the functional groups of amino acids. The concentration of salt in the soaking solution plays a more vital role than a prolonged soaking time. Owing to high anomalous scattering coefficient of the heavy metal ions used with respect to copper wavelength (1.54Å), ($f'' = 9.74$ ē), they prove to be excellent heavy atom markers for automated phasing. The well bound cerium ions in both the datasets provide sufficient anomalous signal for phasing of 772 (GI) and 316 (TL) amino acid residues, with the redundancy values of 14.7 and 40.40, respectively. The high solubility of cerium chloride in water and binding of the Ce (III) ions on the enzymes surface provide an added advantage for phasing. Present study also reveals that the dataset with prolonged soaking time has higher anomalous signal even at high resolution range. Metal ions present in the soaking solution in molar concentrations are well enough to rapidly diffuse or enter into protein crystals. This method fulfills the requirements of being first choice for high-throughput structural genomics projects and can be proposed as an alternative to prepare derivatives when a protein does not bind heavy metal atoms or ions. Data collected using copper wavelength with the incorporation of anomalously scattering heavy metal atoms may hence serve as a powerful tool for structural biologists to solve novel protein structures as well as in places where synchrotron beam line is not available.

5. Acknowledgements

SN and DV thank UGC (SAP), Government of India for the financial support for this research. DV thanks DST-FIST and UGC-SAP for funding facilities to the Centre for Advanced Study in Crystallography and Biophysics. SN and DV thank Department of Biotechnology, Govt. of India for the financial support for the in-house macromolecular data collection facility.

REFERENCES

[1] C. Lee and H. M. Berman *et al*., "The Protein Data Bank and the Challenge of Structural Genomics," *Nature Structure Biology*, Vol. 7, No. 1, 2000, pp. 957-959.

[2] W. A. Hendrickson and C. M. Ogata, "Phase Determination from Multiwavelength Anomalous Diffraction Measurements," *Methods in Enzymology*, *Macromolecular Crystallography*, *Part A*, Vol. 276, 1997, pp. 494-523.

[3] D. Blow, "Outline of Crystallography for Biologists," Oxford University Press, Oxford, 2002

[4] W. Yang, W. A. Hendrickson, R. J. Crouch and Y. Satow, "Structure of Ribonuclease H Phased at 2 A Resolution by MAD Analysis of the Selenomethionyl Protein," *Science*, Vol. 249, No. 4975, 1990, pp. 1398-1405.

[5] J. E. Debreczeni, G. Bunkóczi, B. Girmann and G. M. Sheldrick, "In-House Phase Determination of the Lima Bean Trypsin Inhibitor: A Low-Resolution Sulfur-SAD Case," *Acta Crystallographica*, Vol. D59, No. 2, 2003, pp. 393-395.

[6] Z. Dauter, "Estimation of Anomalous Signal in Diffraction Data," *Acta Crystallographica*, Vol. D62, No. 8, 2006, 867-876.

[7] W. A. Hendrickson, "Determination of Macromolecular Structures from Anomalous Diffraction of Synchrotron Radiation," *Science*, Vol. 254, No. 5028, 1991, pp. 51-58.

[8] J. L. Smith, "Determination of Three-Dimensional Structure by Multiwavelength Anomalous Diffraction," *Current Opinion in Structure Biology*, Vol. 1, No. 6, 1991, pp. 1002-1011.

[9] M. Yogavel, J. Gill and A. Sharma, "Iodide-SAD, SIR and SIRAS Phasing for Structure Solution of a Nucleosome Assembly Protein," *Acta Crystallographica*, Vol. D65, No. 6, 2009, pp. 618-622.

[10] D. K. Carugo, J. R. Helliwell, H. Stuhrmann and M. S. Weiss, "Softer and Soft X-Rays in Macromolecular Crystallography," *Journal of Synchrotron Radiation*, Vol. 12, No. 4, 2005, pp. 410-419.

[11] L. M. Rice, T. N. Earnest and A. T. Brunger, "Single-Wavelength Anomalous Diffraction Phasing Revisited," *Acta Crystallographica*, Vol. D56, No. 11, 2000, pp. 1413-1420.

[12] M. Yogavel, J. Gill, P. C. Mishra and A. Sharma, "SAD Phasing of a Structure Based on Cocrystallized Iodides Using an in-House Cu Kalpha X-Ray Source: Effects of Data Redundancy and Completeness on Structure Solution," *Acta Crystallographica*, D63, No. 8, 2007, pp. 931-934.

[13] T. L. Blundell, H. Jhoti, and C. Abell, "High-Throughput Crystallography for Lead Discovery in Drug Design," Nature Reviews Drug Discovery, Vol. 1, No. 1, 2002, pp. 45-54.

[14] B.-C. Wang, "Resolution of Phase Ambiguity in Macromolecular Crystallography," *Methods in Enzymology*, Vol. 115, Diffraction Methods for Biological Macromolecules, Part B, Academic Press, New York, 1985, pp. 90-111.

[15] J. E. Debreczeni, G. Bunkoczi, Q. Ma, H. Blaser and G. M. Sheldrick, "In-House Measurement of the Sulfur Anomalous Signal and Its Use for Phasing," *Acta Crystallographica*, D59, No. 4, 2003, pp. 688-696.

[16] G. Guncar, C.-I. A. Wang, J. K. Forwood, T. Teh, A.-M. Catanzariti, J.G. Ellis, P.N. Dodds and B. Kobe, "The use of Co^{2+} for Crystallization and Structure Determination, Using a Conventional Monochromatic X-Ray Source, of Flax Rust Avirulence Protein," *Acta Crystallographica*, Vol. F63, No. 3, 2007, pp. 209-213.

[17] M. Yogavel, N. Nithya, A. Suzuki, Y. Sugiyama, T. Yamane, D. Velmurugan and A. Sharma, "Structural Analysis of Actinidin and a Comparison of Cadmium and Sulfur Anomalous Signals from Actinidin Crystals Measured Using In-House Copper- and Chromium-Anode X-ray Sources," *Acta Crystallographica*, Vol. D66, No. 12, 2010, pp. 1323-1333.

[18] C. P. Hill, Z. Dauter, E. J. Dodson, G. G. Dodson and M. F. Dunn, "X-Ray Structure of an Unusual Calcium Site and the Roles of Zinc and calciUm in the Assembly, Sta-

bility, and Storage of the Insulin Hexamer," *Biochemistry*, Vol. 30, No. 4, 1991, pp. 917-924.

[19] L.-J. Baker, J. A. Dorocke, R. A. Harris and D. E. Timm, "The Crystal Structure of Yeast Thiamin Pyrophosphokinase," *Structure*, Vol. 9, No. 6, 2001, pp. 539-546.

[20] Sigma-Aldrich Corp., St. Louis, MO, USA.

[21] Bruker AXS GmbH, Ostliche Rheinbruckenstr, 49, 76187, Karlsruhe, Germany.

[22] Marresearch, GmbH, Hans-Böckler-Ring 17, D-22851, Norderstedt, Germany.

[23] P. D. Adams, *et al.*, "PHENIX: A Comprehensive Python-Based System for Macromolecular Structure Solution," *Acta Crystallographica*, Vol. D66, No. 2, 2010, pp. 213-221.

[24] G. M. Sheldrick, H. A. Hauptman, C. M. Weeks, M. Miller and I. Uson, "Ab Initio Phasing," In: E. Arnold and M. Rossman, Eds., *International Tables for Crystallography*, Vol. F, Chapter 16.1, IUCr and Kluwer Academic Publishers, Dordrecht, 2001, pp. 333-345.

[25] A. Perrakis, R. J. Morris and V. S. Lamzin, "Automated Protein Model Building Combined with Iterative Structure Refinement," *Nature Structural & Molecular Biology*, Vol. 6, No. 1, 1999, pp. 458-463.

[26] G. N. Murshudov, A. A. Vagin and E. J. Dodson, "Refinement of Macromolecular Structures by the Maximum-Likelihood Method," *Acta Crystallographica*, Vol. D53, No. 3, 1997, pp. 240-255.

[27] CCP4 (Collaborative Computational Project, Number 4), "The CCP4 Suite: Programs for Protein Crystallography," *Acta Crystallographica*, Vol. D50, No. 5, 1994, pp. 760-763.

[28] T. C. Don, "Calculation of Anomalous Scattering Factors at Arbitrary Wavelengths," *Journal of Applied Crystallography*, Vol. 16, No. 4, 1983, pp. 437-438.

[29] G. M. Sheldrick, "Experimental phasing with SHELXC/D/E: Combining Chain Tracing with Density Modification," *Acta Crystallographica*, Vol. D66, No. 4, 2010, pp. 479-485.

[30] P. Emsley and K. Cowtan, "Coot: Model-Building Tools for Molecular Graphics," *Acta Crystallographica*, Vol. D60, No. 12, 2004, pp. 2126-2132.

[31] H. L. Carrell, H. Hoier and J. P. Glusker, "Modes of Binding Substrates and Their Analogues to the Enzyme D-Xylose Isomerase," *Acta Crystallographica*, Vol. D50, No. 2, 1994, 113-123.

[32] C. Mueller-Dieckmann et al., "On the routine use of soft X-rays in Macromolecular Crystallography. Part IV. Efficient Determination of Anomalous Substructures in Biomacromolecules Using Longer X-ray Wavelengths," *Acta Crystallographica*, Vol. D63, 2007, 366-380.

[33] K. N. Vennila and D. Velmurugan, "In-House SAD Phasing with Surface-Bound Cerium Ions," *Acta Crystallographica*, Vol. F67, No. 12, 2011, pp. 1662-1665.

Crystal Structure of 3-Amino-5,6-Dimethyl-2-Nitrobiphenyl-4-Carbonitrile

Wanqiang Zhang
School of Chemistry and Chemical Engineering, Xuchang University, Xuchang, China

ABSTRACT

A kind of polysubstituted benzene, 3-amino-5,6-dimethyl-2-nitrobiphenyl-4-carbonitrile($C_{15}H_{13}N_3O_2$), was synthesized and its crystal structure was determined. The molecule is composed of two phenyl moieties, two methyl groups, a cyano group, an amino group and a nitro group. The methyl groups, cyano group and amino group are nearly coplanar with the connected benzene ring. Because of the large volume, the nitro group and the connected benzene ring are twisted. The dihedral angle between the two benzene rings is 83.51°. In the crystal, molecules are linked by N—H···N and C—H···O hydrogen bonds.

Keywords: Polysubstituted Benzene; Crystal Structure; Hydrogen Bonds

1. Introduction

Polysubstituted benzenes and aromatic compounds are widely used in the chemical industry as well in the laboratory [1]. Many liquid crystals belong to the group of polysubstituted benzenes and their derivatives [2-6]. Such compounds can be used for Organic Light Emitting Diode [7-10]. Heteroaromatic components and their derivatives play important parts in pharmaceutical and agrochemical chemicals [11-13]. Hence, more and more researchers have concerned about these compounds. The determination of crystal structure is helpful for its application and realization of property. In this paper, we report about the synthesis and structure of a kind of polysubstituted benzene, 3-amino-5,6-dimethyl-2-nitrobiphenyl-4-carbonitrile.

2. Experimental

2.1. Synthesis of the Compound

A 25 mL round bottom flask was charged with 120 mg 2-sec-Butylidene-malononitrile, 149 mg (2-nitrovinyl) benzene and 68 mg EtONa. The mixture was refluxed with vigorous stirring for 50 hours. The solvent was removed under reduced pressure. The yellow needle product was purified by column chromatography (EtOAc/ hexanes, 1:8). The product was recrystallized from n-hexane and single crystals suitable for X-ray diffraction were obtained.

2.2. Determination of the Crystal Structure

The crystal Structure was characterized by X-ray diffraction (XRD) using Bruker Smart-1000 X-ray single crystal Diffractometer equipped with graphite monochromatized MoKa radiation (λ = 0.71073 Å). The diffraction data were obtained at the temperature of 296 K and analyzed by the software of SHELXTL-97 [14]. The details were shown in **Table 1.**

3. Results and Discussions

3.1. Refinement Details

All reflections were defined based on F^2. The weighted R-factor wR and goodness of fit S are based on F^2, conventional R-factors R are based on F, with F set to zero for negative F^2. The threshold expression of $F^2 > \sigma(F^2)$ is used only for calculating R-factors (gt) etc., and is not relevant to the choice of reflections for refinement. R-factors based on F^2 are statistically about twice as large as those based on F, and R-factors based on all data will be even larger.

3.2. Geometry Details

All esds (except the esd in the dihedral angle between two l.s. planes) are estimated using the full covariance matrix. The cell esds are taken into account individually in the estimation of esds in distances, angles and torsion

Table 1. Data collection and handling.

Empirical formula	$C_{15}H_{13}N_3O_2$
Formula weight	267.28
Crystal external appearance	Yellow needle
Crystal size	0.28 × 0.20 × 0.07 mm
Diffractometer	Bruker Smart-1000
Scan mode	psi-scan
Theta range for data collection	2.14 to 25.10 deg
Wavelength	Mo Kα radiation (0.71073 Å)
Programs	SHELXS-97 , SHELXL-97
Refinement method	Full-matrix least-squares on F^2
Absorption correction	None

angles; correlations between esds in cell parameters are only used when they are defined by crystal symmetry. An approximate (isotropic) treatment of cell esds is used for estimating esds involving l.s. planes.

3.3. Structure Details

The structure was solved by direct methods [14], then refined by full-matrix least-squares technique, the position and anisotropic parameters of all the non-hydrogen atoms were obtained. **Table 2** showed some crystal and refinement parameters. The crystal system of the title compound belongs to Triclinic and the space group is P-1, with cell parameters a = 8.607(3)Å, b = 8.616(3)Å, c = 10.540(3)Å. Atomic coordinates and displacement parameters of the compound were shown in **Table 3**, from these data the structure was gained, which was shown in **Figure 1**. This compound is composed of two phenyl moieties, two methyl groups, a cyano group, an amino group and a nitro group. In the crystal, the methyl groups, cyano group and amino group are nearly coplanar with the connected benzene ring (C7-C12; A). Because of the large volume, the nitro group and the benzene ring A are twisted, the torsion angles are −80.2(6)° (O1-N3-C8-C7) and 99.3(5)° (O2-N3-C8-C7). To avoid steric conflicts, the two benzene rings are not coplanar, the dihedral angle between them is 83.51 (0.13)°. **Figure 2** showed packing diagram of the title compound. Hydrogen bonds are shown as dashed line.There are two kinds of hydrogen bonds in the crystal, N—H···N and C—H···O. In the crystal, Crosslinks among the molecules are provided by the hydrogen bonds **(Figure 2)**, leading to one-dimensional supermolecular structure.

4. Conclusion

A kind of polysubstituted benzene, $C_{15}H_{13}N_3O_2$, was synthesized and its crystal structure was determined. In the crystal, methyl groups, cyano group and amino group are nearly coplanar with the connected benzene ring.

Table 2. Crystal and refinement parameters.

Crystal system,	Triclinic,
Unit cell dimensions	a = 8.607(3)Å alpha = 69.067(5) deg b = 8.616(3)Å beta = 72.279(6) deg c = 10.540(3)Å gamma = 8.507(6) deg
Volume	691.8(4) Å3
Z, Calculated density	2, 1.283 mg/m^3
Absorption coefficient	0.088 mm^{-1}
Space group	P-1
Limiting indices	$-10 \le h \le 8$, $-10 \le k \le 9$, $-11 \le l \le 12$
Reflections collected/unique	3569/ 2429 [R(int) = 0.0217]
Completeness to theta= 25.10	98.4 %
Max. and min. transmission	0.9943 and 0.9755
Goodness-of-fit on F^2	1.061
Extinction coefficient	0.027(10)
Largest diff. peak and hole	0.240 and −0.204 e. Å^{-3}

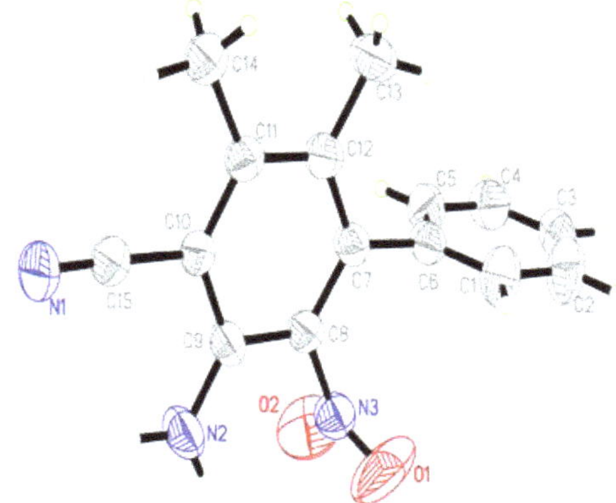

Figure 1. The molecular structure of the compound.

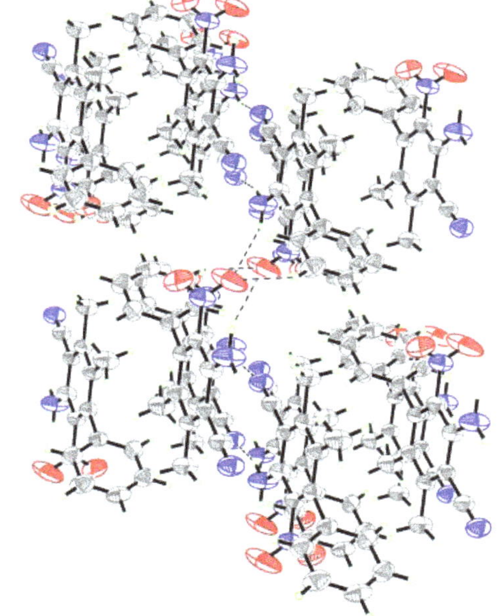

Figure 2. The packing diagram of the compound. Intermolecular hydrogen bonds are shown as dashed line.

Table 3. Atomic coordinates and displacement parameters (Å2).

	x	y	z	U11	U22	U33	U23	U13	U12
N1	0.4390(4)	0.0847(4)	1.1272(3)	0.129(3)	0.071(2)	0.077(2)	−0.0123(18)	−0.033(2)	0.0161(19)
N2	0.4277(5)	0.2894(4)	0.7831(4)	0.204(4)	0.067(2)	0.067(2)	−0.0254(17)	−0.040(2)	0.039(2)
O1	0.4262(7)	0.6743(7)	0.5386(4)	0.260(5)	0.265(5)	0.065(2)	−0.046(3)	0.020(3)	−0.146(5)
O2	0.2215(5)	0.5587(6)	0.5847(4)	0.163(4)	0.219(4)	0.099(3)	−0.049(3)	−0.065(2)	−0.007(3)
C1	0.2786(6)	0.9818(5)	0.6631(4)	0.105(3)	0.080(3)	0.090(3)	−0.003(2)	−0.038(2)	−0.010(2)
C2	0.2218(7)	1.1367(5)	0.5849(5)	0.141(5)	0.078(3)	0.102(4)	0.008(3)	−0.055(3)	−0.021(3)
C3	0.0619(7)	1.1742(5)	0.5825(4)	0.144(4)	0.063(2)	0.073(3)	−0.013(2)	−0.048(3)	0.009(3)
C4	−0.0445(6)	1.0569(5)	0.6577(4)	0.097(3)	0.086(3)	0.089(3)	−0.022(3)	−0.034(2)	0.019(3)
C5	0.0120(5)	0.9006(4)	0.7363(4)	0.087(3)	0.069(2)	0.093(3)	−0.019(2)	−0.025(2)	0.008(2)
C6	0.1724(5)	0.8618(4)	0.7409(4)	0.089(3)	0.059(2)	0.061(2)	−0.0178(18)	−0.0251(18)	0.006(2)
C7	0.2297(4)	0.6943(4)	0.8297(3)	0.078(2)	0.058(2)	0.061(2)	−0.0193(18)	−0.0188(17)	0.0050(17)
C8	0.2989(4)	0.5694(4)	0.7700(3)	0.090(3)	0.060(2)	0.053(2)	−0.0158(17)	−0.0145(17)	0.0047(18)
C9	0.3563(4)	0.4079(4)	0.8452(4)	0.104(3)	0.053(2)	0.062(2)	−0.0205(18)	−0.0186(19)	0.0078(18)
C10	0.3347(4)	0.3791(4)	0.9888(3)	0.074(2)	0.058(2)	0.056(2)	−0.0123(17)	−0.0125(16)	0.0011(17)
C11	0.2631(4)	0.5014(4)	1.0531(3)	0.085(3)	0.068(2)	0.053(2)	−0.0183(17)	−0.0140(16)	0.0020(18)
C12	0.2095(4)	0.6614(4)	0.9733(4)	0.093(3)	0.064(2)	0.061(2)	−0.0187(18)	−0.0204(18)	0.0079(18)
C13	0.1332(6)	0.7938(5)	1.0420(4)	0.160(4)	0.084(3)	0.078(3)	−0.039(2)	−0.029(3)	0.026(3)
C14	0.2446(5)	0.4591(5)	1.2074(4)	0.130(4)	0.077(2)	0.065(2)	−0.019(2)	−0.026(2)	0.005(2)
C15	0.3930(4)	0.2143(5)	1.0687(4)	0.095(3)	0.068(2)	0.058(2)	−0.0142(19)	−0.0206(19)	0.003(2)
N3	0.3166(6)	0.6046(4)	0.6211(4)	0.133(3)	0.077(2)	0.057(2)	−0.0226(18)	−0.022(2)	0.004(2)

Because of the large volume, the nitro group and the connected benzene ring are twisted. The dihedral angle between the two benzene rings is 83.51°. In the crystal, molecules are linked by N—H···N and C—H···O hydrogen bonds.

REFERENCES

[1] D. Astruc, "Modern Arene Chemistry," Wiley-VCH., Weinheim, 2002.

[2] G. Yeap, S. Balamurugan and S. Rakesh, "Synthesis and Mesomorphic Properties of Chiral Liquid Crystal Dimers Derived from Azobenzene and Substituted Naphthol," *Liquid Crystals*, Vol. 40, No. 4, 2013, pp. 555-563.

[3] G. Mao, J. Wang, R. S. Clingman and C. K. Ober, "Molecular Design, Synthesis, and Characterization of Liquid Crystal-Coil Diblock Copolymers with Azobenzene Side Groups," *Macromolecules*, Vol. 30, No. 9, 1997, pp. 2556-2567.

[4] S. Luo, J. Xiong and Z. Wang, "Design and Synthesis of 2(5H)-Furanone Liquid-Crystal Compounds Based on Natural Molecules and Biphenyl Derivatives," *Research on Chemical Intermediates*, Vol. 39, No. 4, 2013, pp. 1865-1876.

[5] D. Pauluth and K. Tarumi, "Advanced Liquid Crystals for Television," *Journal of Materials Chemistry*, Vol. 14, No. 8, 2004, pp. 1219-1227.

[6] H. S. Bhatt, P. D. Patel and J. S. Dave, "Synthesis of Liquid Crystals with Lateral Substituents in Central and Terminal Benzene Cores and Their Mesomorphic Characterization," *Molecular Crystals and Liquid Crystals*, Vol. 575, 2013, pp. 104-11.

[7] O. Lavastre, I. Illitchev, G. Jegou and P. H. Dixneuf, "Discovery of New Fluorescent Materials from Fast Synthesis and Screening of Conjugated Polymers," *Journal of American Chemistry Society*, Vol. 124, No. 19, 2002, pp. 5278-5279.

[8] C. W. Lee, K. S. Yook and J. Y. Lee, "Synthesis and Device Application of Hybrid Host Materials of Carbazole and Benzofuran for High Efficiency Solution Processed Blue Phosphorescent Organic Light-Emitting Diodes," *Organic Electronics*, Vol. 14, No. 3, 2013, pp. 1009-1014.

[9] X. Q. Lin, B. J. Chen, X. H. Zhang, C. S. Lee, H. L. Kwong, *et al.*, "A Novel Yellow Fluorescent Dopant for High-Performance Organic Electroluminescent Devices," *Chemistry of Materials*, Vol. 13, No. 2, 2001, pp. 456-458.

[10] J. Sun, H. Jiang and J. Zhang, Y. Tao and R. Chen, "Synthesis and Characterization of Heteroatom Substituted Carbazole Derivatives: Potential Host Materials for Phosphorescent Organic Light-Emitting Diodes," *New Journal of Chemistry*, Vol. 37, No. 4, 2013, pp. 977-985.

[11] T. Verdelet, G. Mercey, N. Correa, L. Jean and P. Y. Renard, "Straightforward and Efficient Synthesis of 3-Benzyloxy-4-bromopicolinate ester and 3-Benzyloxy-5-bromopicolinate Ester, Common Building Blocks for Pharmaceuticals and Agrochemicals," *Tetrahedron*, Vol. 67, No. 45, 2011, pp. 8757-8762.

[12] C. J. C. Connolly, J. M. Hamby, M. C. Schroeder, M. Barviana, G. H. Lub, *et al*., "Discovery and Structure-activity Studies of a Novel Series of Pyrido [2,3-d] Pyrimidine Tyrosine Kinase Inhibitors," *Bioorganic & Medicinal Chemistry Letters*, Vol. 7, No. 18, 1997, pp. 2415-2420.

[13] X. Zhang, G. C. G Paisa, E. S. Svarovskaiab, C. Marchandc, A. A. Johnson, *et al*., "Azido-Containing Aryl β-Diketo Acid HIV-1 Integrase Inhibitors," *Bioorganic & Medicinal Chemistry Letters*, Vol. 13, No. 6, 2003, pp. 1215-1219.

[14] G. M. Sheldrick, "SHELXL97. Program for the Refinement of Crystal Structures," University of Göttingen, Germany, 1997.

18

Crystal Structure Study on Non-Coplanarly Organized Accumulating Aromatic Rings Molecules: Spatial Organization of *C*,*C*,*N*-Triaryl Substituted Imines

Akiko Okamoto, Atsushi Nagasawa, Siqingaowa, Noriyuki Yonezawa
Department of Organic and Polymer Materials Chemistry, Tokyo University of Agriculture and Technology, Tokyo, Japan

ABSTRACT

The X-ray crystal structures of *C*,*C*,*N*-triaryl-substituted imine compounds, which have methoxy or hydroxy group adjacent to the imino moiety, are reported and discussed in comparison with those of the precursor ketone compounds, 1-(4-chlorobenzoyl)-2,7-dimethoxynaphthalene and 1-(4-chlorobenzoyl)-2-hydroxy-7-methoxynaphthalene. In crystals, three aromatic rings in a molecule of the methyl ether-retained imine compound are positioned almost perpendicularly to each other by giving non-coplanar spatial organization of the single molecular structure [dihedral angles: 85.32(18)° for *C*-linked phenyl ring and naphthalene ring; 79.27(17)° for *N*-linked phenyl ring and naphthalene ring; 84.78(17)° for *C*-linked phenyl ring and *N*-linked phenyl ring]. Spatial organization of the analogous methyl ether-cleaved imine compound has essentially same topology [dihedral angles 80.39(6)° for the *C*-linked phenyl ring and naphthalene ring; 82.35(6)° for the *N*-linked phenyl ring and naphthalene ring; 87.09(7)° for *C*- and *N*-linked phenyl rings]. These structural features of triarylimines apparently differ from those of the precursor ketones. Two aromatic rings in the methyl ether-cleaved ketone compound make smaller dihedral angle [58.10(6)°] by intramolecular hydrogen bond between ketonic carbonyl group and hydroxy group [2.5573(16) Å] than that of the methyl ether-retained ketone [72.06(7)°]. In molecular packing, the methyl ether-retained imine forms tubular molecular alignments composed of *R*—*S* dimeric molecular pairs, whereas the methyl ether-retained ketone affords consecutively stacks of one configurated molecules.

Keywords: Non-Coplanarly Accumulated Aromatic Rings; Spatial Organization; Triarylimine

1. Introduction

Non-coplanarly accumulated aromatic-rings compounds, e.g., biphenyls and binaphthyls, have been demonstrated as unique building blocks in construction for many functional materials such as molecular catalysis and functional polymers [1-12]. Thus, minute spatial structural characterization of these compounds [13-16] has attracted attention of the chemists in the wide-range of organic molecular science and polymer materials fields. However, intra- and inter-molecular interactions that afford various functions to such molecular units still remain ambiguous. As one of the protocols to estimate such interactions, the authors have been investigating synthesis and X-ray crystal structure analysis of congested spatial organization of aromatic rings accumulating molecules.

Recently, the authors have reported specific and characteristic electrophilic aromatic aroylation of naphthalene derivatives, *i.e.*, *two* aroyl groups are regioselectively and effectively introduced at the 1,8-positions of the naphthalene ring accompanying with simultaneously proceeding retroaroylation behavior [17,18]. The 1-aroylated naphthalenes, which correspond to the intermediates in the diaroylation, are also obtained by choice of acidic mediator.

X-ray crystal structure study has revealed that the aroyl groups in these *peri*-aroylated naphthalene molecules are non-coplanarly attached to the naphthalene rings by giving crowded molecular organization [19-22]. In a natural consequence, the authors have planned to introduce additional aromatic ring planes to the core part of the aroylnaphthalene molecules for realization of more crowded inner spatial situation in accumulated aro- matic-rings molecule. As one of the molecular transfor- mation approaches to obtain such spatial organization, the authors designed conversion of ketonic carbonyl group on 1-aroylnaphthalene to imino moiety by the re- action with aniline derivative. Imination of 1-aroylated 2,7-dimethoxynaphthalene with aromatic amines scarcely proceeded with conventional additives except for $TiCl_4$ and 1,4-diazabicyclo[2.2.2]octane (DABCO) mixture. In $TiCl_4$—DABCO mediated imination, triaryl-substituted imine compounds were formed in moderate conversion with/without preceding methyl ether cleavage reaction of the starting compound (**Scheme 1**) [23]. The neighboring ketonic carbonyl group of *peri*-aroylated 2,7-dimethoxynaphthalene derivatives plausibly accelerates $TiCl_4$-mediated scission of rather stable ether bonding.

In this article, the authors report and discuss the single molecular spatial organizations and the molecular packing characteristics of *C,C,N*-triarylated imine compounds by comparing with those of original ketone compounds: 1-aroyl-2,7-dimethoxynaphthalene and 1-aroyl-2-hydroxy-7-methoxynaphthalene.

2. Experimental

All reagents were of commercial quality and were used as received. Solvents were dried and purified using standard techniques.

2.1. Measurements

^{1}H NMR spectra were recorded on a JEOL JNM-AL300 spectrometer (300 MHz) and a JEOL ECX400 spectrometer (400 MHz). Chemical shifts are expressed in ppm relative to internal standard of Me_4Si (δ 0.00). ^{13}C NMR spectra were recorded on a JEOL JNM-AL300 spectrometer (75 MHz). Chemical shifts are expressed in ppm relative to internal standard of $CDCl_3$ (δ 77.0). IR spectra were recorded on a JASCO FT/IR-4100 spectrometer. Elemental analyses were performed on a Yanaco CHN CORDER MT-5 analyzer. High-resolution FAB mass spectra were recorded on a JEOL MStation (MS700) ion trap mass spectrometer in positive ion mode.

2.2. Synthetic Procedure

Starting material **1** and triarylimines were prepared as follows.

2.2.1. Electrophilic Aromatic Substitution Aroylation of 2,7-Dimethoxynaphthalene by $AlCl_3$

To a solution of 2,7-dimethoxynaphthalene (0.200 mmol, 68.2 mg) and 4-chlorobenzoyl chloride (0.220 mmol, 38.5 mg) in dichloromethane (0.5 mL), $AlCl_3$ (0.220 mmol, 29.3 mg) was added by portions at 0°C under nitrogen atmosphere. After the reaction mixture was stirred at r. t. for 3 h, it was poured into iced water (20 mL) and the mixture was extracted with $CHCl_3$ (15 mL × 3). The combined extracts were washed with 2 M NaOH aq., sat. NaCl aq. and dried over anhydrous magnesium sulfate. The solvent was removed under reduced pressure to give powdery product. The crude product of 1-momoaroylnaphthalene **1** was purified by recrystallization (hexane, isolated yield 78%).

1-(4-Chlorobenzoyl)-2,7-dimethoxynaphthalene (1): Colourless needle (hexane), Mp 121.5°C - 122°C; IR (KBr): 1667, 1628, 1586, 1512 cm^{-1}; ^{1}H NMR δ (300 MHz, $CDCl_3$): 7.87 (1H, d, J = 9.0 Hz), 7.78 (2H, d, J = 8.4 Hz), 7.72 (1H, d, J = 9.0 Hz), 7.39 (2H, d, J = 8.4 Hz), 7.16 (1H, d, J = 9.0 Hz), 7.02 (1H, dd, J = 2.4, 9.0 Hz), 6.78 (1H, d, J = 2.4 Hz), 3.79 (3H, s), 3.73 (3H, s) ppm; ^{13}C NMR δ (75 MHz, $CDCl_3$): 196.81, 158.96, 155.02, 139.71, 136.45, 132.94, 131.28, 130.87, 129.72, 128.86, 124.34, 121.06, 117.15, 110.05, 101.88, 56.239, 55.168 ppm; Calcd for $C_{19}H_{15}O_3Cl$: C, 69.83%; H, 4.63%; Found: C, 69.61%; H, 4.74%.

2.2.2. $TiCl_4$—DABCO Mediated Imination of 1-(4-Chlorobenzoyl)-2,7-dimethoxynaphthalene (1)

To a solution of 1-(4-chlorobenzoyl)-2,7-dimethoxynaphthalene (**1**, 0.200 mmol, 65.4 mg) in monochlorobenzene (1 mL), mixtures of aniline (0.220 mmol, 20.5 mg), $TiCl_4$ (0.330 mmol, 62.4 mg), DABCO (1.320 mmol, 148 mg) and monochlorobenzene (1 mL) were added by portions at 90°C under nitrogen atmosphere. After the reaction mixture was stirred at 125°C for 1.5 h, the resulting solution was filtrated to remove the precipitate. The solvent was removed under reduced pressure to give crude material. The crude product was purified by silicagel column chromatography (Chloroform; isolated yield: imine **3**, 10%; imine **4**, 10%, 2-hydroxy compound **5**, 8%).

Imine 3: Colourless block ($CHCl_3$/hexane) Mp 174°C - 175°C, IR (KBr) 1625, 1502, 1238, 1029, 830 cm^{-1}; ^{1}H NMR δ (300 MHz, $CDCl_3$): 7.72 (1H, d, J = 9.0 Hz), 7.66 (2H, d, J = 8.4 Hz), 7.60 (1H, d, J = 9.0 Hz), 7.29 (2H, d, J = 8.4 Hz), 7.25 (1H, d, J = 9.0 Hz), 7.02 (1H, d, J = 9.0 Hz), 6.92 (1H, dd, J = 9.0, 2.4 Hz), 6.74 (2H, d, J = 8.8 Hz), 6.68 (1H, d, J = 2.4 Hz), 6.53 (2H, d, J = 8.8

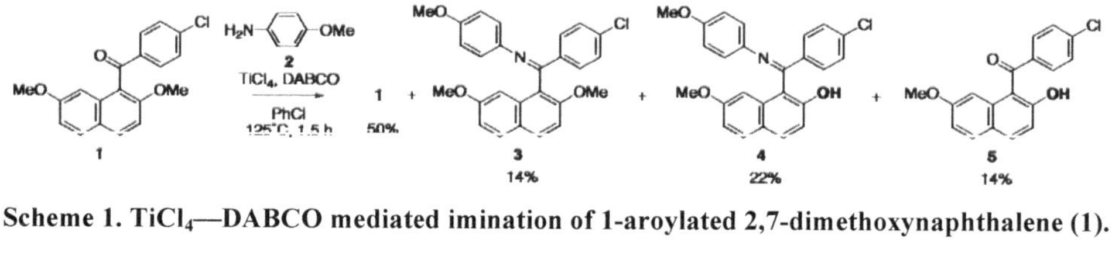

Scheme 1. $TiCl_4$—DABCO mediated imination of 1-aroylated 2,7-dimethoxynaphthalene (1).

Hz), 3.72 (3H, s), 3.70 (3H, s), 3.60 (3H, s) ppm; ^{13}C NMR δ (75 MHz, $CDCl_3$): 163.86, 158.73, 156.27, 154.96, 144.33, 138.11, 136.46, 132.80, 130.46, 129.80, 129.51, 128.64, 124.06, 121.15, 118.58, 116.85, 113.40, 109.87, 102.72, 56.11, 55.32, 55.23 ppm; HRMS (FAB; *m*-nitrobenzyl alcohol [*m*-NBA]) m/z: $[M + H]^+$; Calcd for $C_{26}H_{23}O_3NCl$, 432.1371; Found 432.1366; Anal. Calcd for $C_{26}H_{23}O_3NCl$: C 72.15%, H 5.11%. Found: C 72.30%, H 5.13%.

Imine 4: Colourless block ($CHCl_3$/hexane), Mp 184°C - 185°C; IR (KBr) 3407, 1626, 1502, 1225, 1207, 826 cm^{-1}; ^{1}H NMR δ (300 MHz, $CDCl_3$): 7.71 (d, J = 8.8 Hz, 1H), 7.69 (d, J = 8.8 Hz, 2H), 7.64 (d, J = 8.8 Hz, 1H), 7.62 (d, J = 9.2 Hz, 1H), 7.53 (d, J = 9.2 Hz, 1H), 7.31 (d, J = 8.8 Hz, 2H), 7.27 (d, J = 8.8 Hz, 2H), 7.09 (d, J = 8.4 Hz, 2H), 7.03 (d, J = 8.8 Hz, 1H), 6.94 (dd, J = 2.4, 9.2 Hz, 1H), 6.87 (d, J = 7.6 Hz, 1H), 6.84 (d, J = 10.0 Hz, 2H), 6.77 (dd, J = 2.4, 8.6 Hz, 1H), 6.72 (m, 4H), 6.66 (d, J = 2.4 Hz, 1H), 6.57 (d, J = 9.2 Hz, 2H), 6.21 (d, J = 2.4 Hz, 1H), 3.76 (s, 3H), 3.69 (s, 3H), 3.64 (s, 3H), 3.20 (s, 3H) ppm; ^{1}H NMR δ (300 MHz, DMSO-d_6): 10.01 (s, 1H), 7.67 - 7.56 (m, 5H), 7.41 (d, J = 8.4 Hz, 2H), 6.97 (d, J = 8.7 Hz, 1H), 6.84 - 6.75 (m, 3H), 6.57 (d, J = 8.7 Hz, 2H), 6.46 (d, J = 2.1 Hz, 1H), 3.59 (s, 3H), 3.52 (s, 3H) ppm; ^{13}C NMR δ (75 MHz, $CDCl_3$): 169.19, 166.95, 162.91, 158.94, 157.82, 157.11, 157.01, 150.87, 143.50, 137.98, 137.64, 137.03, 135.45, 135.35, 135.20, 134.36, 133.38, 131.05, 130.68, 130.33, 129.88, 129.69, 129.24, 128.85, 124.55, 124.14, 123.72, 121.73, 118.59, 116.33, 116.15, 114.96, 114.52, 114.14, 113.84, 111.37, 106.53, 103.58, 55.45, 55.26, 55.24, 54.44 ppm; HRMS (FAB; *m*-NBA) m/z: $[M + H]^+$; Calcd for $C_{25}H_{21}O_3NCl$, 418.1162; Found 418.2110; Anal. Calcd for $C_{26}H_{23}O_3NCl$: C 71.97%, H 4.87%. Found: C 71.85%, H 4.82%.

1-(4-Chlorobenzoyl)-2-hydroxy-7-methoxynaphthalene (5): Yellow platelet (hexane), Mp 118°C - 118.5°C; IR (KBr): 3434, 1623, 1583, 1513, 1214, 843 cm^{-1}; ^{1}H NMR δ (300 MHz, $CDCl_3$): 11.35 (s, 1H), 7.85 (d, 1H, J = 9.0 Hz), 7.63 (d, 1H, J = 9.0 Hz), 7.58 (d, 2H, J = 8.7 Hz), 7.40 (d, 2H, J = 8.7 Hz), 7.07 (d, 1H, J = 9.0 Hz), 6.91 (dd, 1H, J = 2.4, 9.0 Hz), 6.58 (d, 1H, J = 2.4 Hz), 3.37 (s, 3H) ppm; ^{13}C NMR δ (75 MHz, $CDCl_3$): 199.1, 162.6, 158.2, 138.8, 138.7, 136.5, 133.8, 130.7, 130.2, 128.9, 123.7, 116.4, 115.8, 113.4, 106.5, 54.5 ppm; Anal. Calcd for $C_{18}H_{13}ClO_3$: C 69.13, H 4.19. Found: C 69.11, H 4.09.

Imine 6: Colourless block ($CHCl_3$/hexane), Mp 172°C - 173°C; IR (KBr): 3407, 2937, 2592, 1625, 1585, 1509, 1227 cm^{-1}; ^{1}H NMR δ (300 MHz, DMSO-d_6): 10.13, (s, 1H), 7.66 - 7.60 (m, 4H), 7.44 (d, 2H), 7.00 (t, 2H), 6.95 (d, H), 6.86 - 6.76 (m, 4H), 6.52 (d, 1H), 3.64 (s, 3H), 3.29 (s, 6H) ppm; ^{13}C NMR δ (75 MHz, DMSO-d_6): 164.4, 158.2, 153.7, 151.0, 137.6, 135.7, 132.2, 130.3, 130.0, 129.7, 128.7, 128.2, 123.8, 122.9, 119.2, 115.1, 115.0, 114.9, 102.6, 55.1, 47.3 ppm; HRMS (FAB; *m*-NBA) m/z: $[M + H]^+$; Calcd for $C_{24}H_{19}ClNO_2$, 388.1110; Found, 388.1104.

2.3. X-Ray Crystallography

For the crystal structure determination, the single-crystal of the title compounds were used for data collection on a four-circle Rigaku R-AXIS RAPID diffractometer (equipped with a two-dimensional area IP detector). The graphite-mono-chromated Cu Kα radiation (λ = 1.54187 Å) was used for data collection. The lattice parameters were determined by the least-squares methods on the basis of all reflections with F2 > 2σ(F2). The data collection and cell refinement were performed using *PROCESS-AUTO* software [24]. The data reduction was performed using *CrystalStructure* [25]. The structures were solved by direct methods using *SIR*2004 [26] and refined by a full-matrix least-squares procedure using the program *SHELXL*97 [27]. All H atoms were found in a difference map and were subsequently refined as riding atoms, with the aromatic C-H = 0.95 Å and methyl C-H = 0.98 Å, and with U_{iso}(H) = 1.2 U_{eq}(C). Molecular structures of the title compounds showing the atomic numbering schemes are shown in **Figures 1** and **2**. The crystallography details for the structure determination of the compounds are displayed in **Tables 1** and **2**.

3. Results and Discussion

Tables 1 and **2** show the crystallographic data of triarylimines and the precursor compounds. *C*,*C*,*N*-triarylimines were prepared by $TiCl_4$—1,4-diazabicyclo [2.2.2] octane (DABCO)-mediated imination (see Experimental section).

Figure 1 shows single molecular structures of 1-aroyl-2,7-dimethoxynaphthalene **1** [28], imine **3**, and 1-aroyl-2-hydroxy-7-methoxynaphthalene **5** [29] in crystal.

The aroyl group of 1-aroylnaphthalene **1** is non- coplanarly attached to the naphthalene ring. The dihedral angle between the benzene ring and the naphthalene ring is 72.06(7)°. On the other hand, 1-aroylated 2-hydroxy-7-methoxynaphthalene (**5**) has an intramolecular O-H···O=C hydrogen bond between the carbonyl group and the hydroxy substituent on the naphthalene ring system [O2-H2···O1; 2.5573(16) Å]. The angle between the C=O bond plane and the naphthalene ring system is relatively small [20.96(8)°]. Naturally, the angle between the benzene ring and the carbonyl group is rather large [35.65(9)°] compared to that in the original ketone compound [**1**, 3.43(11)°], which has 2-methoxy group instead of 2-hydroxy substituent. Consequence of this, two aromatic rings in the methyl ether-*cleaved* ketone **5** make smaller dihedral angle [58.10(6)°] than the precursory ketone compound [**1**, 72.06(7)°] in crystal.

Table 1. Crystallographic data and structure refinement parameters of molecule 1 and 3.

	1	imine 3
Empirical formula	$C_{19}H_{15}ClO_3$	$C_{26}H_{22}ClNO_3$
Formula weight (g·mol^{-1})	326.76	431.90
Crystal shape, colour	Platelet, colorless	Platelet, yellow
Melting point (K)	296	447.2 - 447.7
Temperature (K)	193(2)	193(2)
Radiation type	Cu *Kα*	Cu *Kα*
Wavelength (Å)	1.54187	1.54187
Crystal sytem Space group	Orthorhombic Pbca	Monoclinic $P2_1/a$
a, *b*, *c* (Å)	6.6033 (3), 16.0751 (7), 130.2216 (12)	10.8534(6), 20.6421(12), 11.1449(8)
°	90.00, 90.00, 90.00	90.00, 118.335(3), 90.00
Volume (Å^3)	3208.0(2)	2197.7(2)
Z, Calculated density (Mg·m^{-3})	8, 1.353	4, 1.305
Absorption coefficient (mm^{-1})	2.213	1.761
F(000)	1360	904
Crystal size (mm)	0.40 × 0.15 × 0.10	0.40 × 0.40 × 0.10
Theta range for data collection	5.5° to 68.1°	4.3° to 68.2°
Limiting indices	$-7 \leq h \leq 7$ $-19 \leq k \leq 19$ $-36 \leq l \leq 36$	$-13 \leq h \leq 13$ $-24 \leq k \leq 24$ $-13 \leq l \leq 13$
Reflections collected/unique	54984/2919 [R_{int} = 0.032]	15973/4023 [R_{int} = 0.127]
Completeness to theta (%)	100 [68.13°]	99.9 [68.25°]
Max. and min. transmission	0.802 and 0.617	0.844 and 0.539
Refinement method	Full-matrix least-squares on F^2	Full-matrix least-squares on F^2
Data/restraints/parameters	2910/0/210	4023/0/281
Goodness-of-fit on F^2	1.11	1.01
Final *R* indices [I > 2 sigma (I)]	R1 = 0.040, wR2 = 0.113	R1 = 0.074, wR2 = 0.187
R indices (all data)	R1 = 0.046, wR2 = 0.118	R1 = 0.10, wR2 = 0.213
Largest diff. peak and hole	0.13 e Å^{-3} and −0.33 e Å^{-3}	0.44 e Å^{-3} and −0.36 e Å^{-3}

Table 2. Crystallographic data and structure refinement parameters of molecule 5 and 6.

	5	imine 6
Empirical formula	$C_{18}H_{13}ClO_3$	$C_{24}H_{18}ClNO_2 \cdot 0.5C_6H_{12}N_2$
Formula weight (g·mol^{-1})	312.73	443.93
Crystal shape, colour	Platelet, yellow	Block, colorless
Melting point (K)	391.0 - 391.5	445.6 - 446.0
Temperature (K)	123	193(2)
Radiation type	Cu *Kα*	Cu *Kα*
Wavelength (Å)	1.54187	1.54187
Crystal sytem Space group	Orthorhombic Pbca	Monoclinic *C*2/*c*
a, *b*, *c* (Å)	17.8030 (3), 8.68121 (10), 18.8683 (3)	25.0027(5), 9.92298(18), 20.0052(4)
°	90.00, 90.00, 90.00	90.00, 114.6210(10), 90.00
Volume (Å^3)	2916.14(8)	4512.07(16)
Z, Calculated density (Mg·m^{-3})	8, 1.425	8, 1.307
Absorption coefficient (mm^{-1})	2.41	1.71
F(000)	1296	1864
Crystal size (mm)	0.60 × 0.15 × 0.05	0.60 × 0.50 × 0.40
Theta range for data collection	4.7° to 68.2°	3.6° to 68.2°
Limiting indices	$-21 \leq h \leq 21$ $-10 \leq k \leq 10$ $-22 \leq l \leq 22$	$-30 \leq h \leq 30$ $-11 \leq k \leq 11$ $-24 \leq l \leq 24$
Reflections collected/unique	49864/2669 [R_{int} = 0.033]	39753/4125 [R_{int} = 0.026]
Completeness to theta (%)	100 [68.23°]	100 [68.23°]
Max. and min. transmission	0.886 and 0.485	0.548 and 0.381
Refinement method	Full-matrix least-squares on F^2	Full-matrix least-squares on F^2
Data/restraints/parameters	2669/0/205	40125/0/495
Goodness-of-fit on F^2	1.08	1.04
Final *R* indices [I > 2 sigma (I)]	R1 = 0.0328 wR2 = 0.0932	R1 = 0.0341, wR2 = 0.0955
R indices (all data)	R1 = 0.0370, wR2 = 0.0963	R1 = 0.0359, wR2 = 0.0969
Largest diff. peak and hole	0.17 e Å^{-3} and −0.25 e Å^{-3}	0.44 e Å^{-3} and −0.32 e Å^{-3}

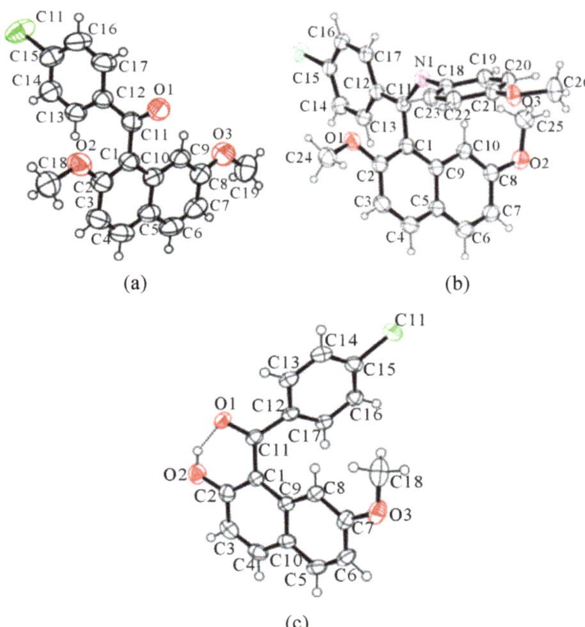

Figure 1. Molecular structures of 1-aroyl-2,7-dimetho- xy-naphthalene 1 (a), imine 3 (b), and 1-aroyl-2-hydroxy-7-methoxynaphthalene 5 (c) with the atom-labeling scheme and displacement ellipsoids drawn at the 50% probability level.

Figure 2. Molecular structures of analogous imine 6 with the atom-labeling scheme and displacement ellipsoids drawn at the 50% probability level.

About methyl ether-*retained* imine **3**, each of the aromatic rings is connected almost perpendicularly against both of other aromatic rings. The dihedral angles between two of the aromatic rings are close to 90° [85.32 (18)° for *C*-linked phenyl ring and naphthalene ring; 79.27(17)° for *N*-linked phenyl ring and naphthalene ring; 84.78(17)° for *C*-linked phenyl ring and *N*-linked phenyl ring, in imine **3**] compared to that of precursor 1-aroylnaphthalene **1**. These angle values indicate that three aromatic rings in methyl ether-*retained* imine **3** are situated to avoid steric hindrance to each other. However, the spatial organization of triarylimine **3** based on perpendicular arrangement of aromatic rings is essentially same to that of precursor ketone **1**.

Figure 2 shows the single molecular structures of methyl ether-*cleaved* 1-aroylnaphthalene (**5**) originated triarylimine (**6**) [30] in crystal. Though preparation of satisfactorily qualified crystal for X-ray crystal analysis of *N*-(4-methoxyphenoxy)imine compound **4** was unsuccessful, the crystal structure of analogous imine compound (**6**) was determined. In the crystal of analogous methyl ether-*cleaved* imine **6**, two molecules of imine **6** form a 2:1 set with a DABCO molecule. However, the spatial organization of the aromatic rings in methyl ether-*cleaved* imine **6** has essentially same topology to methyl ether-*retained* imine **3**. The dihedral angles of the *C*-linked 4-chlorophenyl ring and the *N*-linked phenyl ring with the naphthalene ring are 80.39(6)° and 82.35 (6)°, respectively. The dihedral angle between *C*- and *N*-linked benzene rings is 87.09(7)°. The structural similarities between imines **3** and **6** strongly suggest that the single molecular spatial organization of three aromatic rings in perpendicular fashion is satisfactorily stable, regardless of whether triarylimine has a methoxy group at the 2-position of the naphthalene or a hydroxy one. Although both ketone **5** and imine **6** have 2-hydroxynaphthalene unit, imine **6** has molecular organization of perpendicular-based aromatic rings arrangement, which is clearly distinguishable against rather planar structure of ketone **5**.

The molecular packing of the methyl ether-*retained* imine compound **3** is compared with those of methyl ether-*retained* ketone **1** and methyl ether-*cleaved* ketone **5**. In molecular packing, 1-aroylnaphthalene **1** is mainly stabilized by van der Waals interactions. The molecules of methyl ether-*retained* ketone **1** are aligned consecutively in stacks along the *a* axis (**Figure 3**). Adjacent 4-chlorophenyl groups are exactly parallel, and the perpendicular distance between these planes is 3.660 (1) Å (**Figure 4**). **Figure 5** shows the herringbone packing of

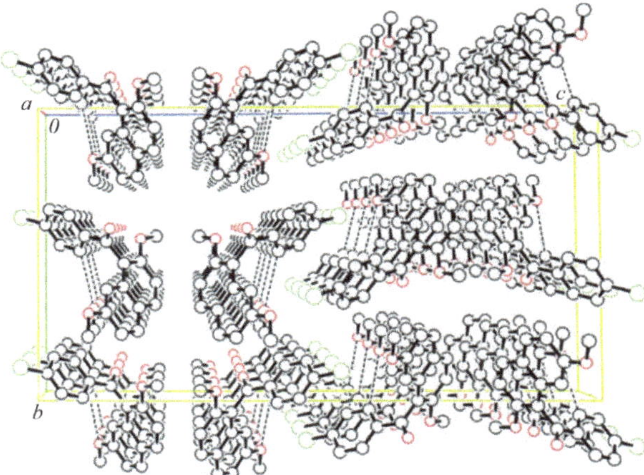

Figure 3. The molecular alignment of 1-monoaroylnaphhthalene 1, viewed along the *a* axis. H atoms are omitted.

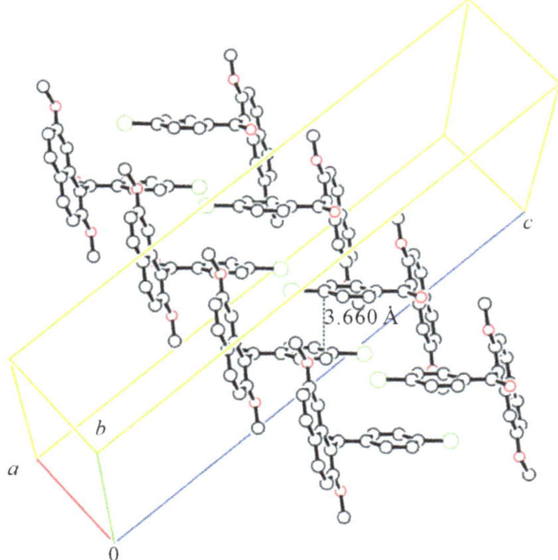

Figure 4. The alignment of the molecule 1, viewed in an oblique direction. H atoms are omitted.

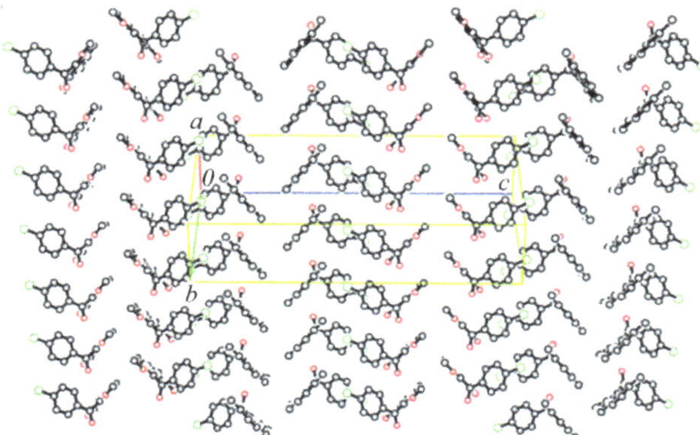

Figure 5. The alignment of the molecule 1, showing the herringbone packing. H atoms are omitted.

the naphthalene ring in the crystal of ketone **1**. One pile is composed of one configuration of molecule **1**. The adjacent pile is composed of the other configuration of molecule **1**. The piles of molecules are aligned alternately to vanish Cl···Cl electrostatic repulsion. The crystal packing is additionally stabilized by intermolecular (benzene)C-H···O(methoxy) hydrogen bonding between the hydrogen atom of the neighboring 4-chlorophenyl group and a methoxy oxygen of the adjacent molecule (C13-H13···O3; **Figure 3**). In the molecular packing of methyl ether-*cleaved* ketone **5**, a three-dimensional molecular network in which the alternate arrangement of *R*- and *S*-configurated compounds is formed by loose van der Waals interactions. The naphthalene rings interact with the phenyl rings [C5···C13 = 3.363 (2) Å] and the carbonyl groups [H6···O1 = 2.70 Å] along the *a*-axis. They also interact with the methyl groups [H3···C18 = 2.79 Å] and aroyl groups [H6···Cl1 = 2.88 Å] along the *c*-axis (**Figure 6**). On the other hand, the naphthalene rings also interact with the methyl groups [C6···H18B = 2.81 Å, C7···H18B = 2.70 Å] and the phenyl rings [C6···H17 = 2.88 Å, C7···H17 = 2.79 Å] along the *b*-axis. The naphthalene rings are almost perpendicular to the phenyl rings of the adjacent molecules along the *b*-axis. In addition, the hydroxy groups interact with the phenyl rings [O2···H14 = 2.71 Å] along the *b*-axis (**Figure 7**).

In the molecular packing of methyl ether-*retained* imine **3**, one *R*-configurated molecule of imine **3** and an *S*-counterpart make a pair by the aid of (*N*-phenyl) C-H···π (*C*-phenyl) interactions (C20-H20···Cg3; Cg3 is *C*-linked benzene ring of the adjacent molecule) and C-H···N (C25-H25B···N1) ones, and then the dimeric units stack along a *ac* diagonal through (*C*-phenyl) C-Cl···O (naphthalene) (C15-Cl1···O2) interactions (**Figures 8** and **9**). The tubular molecular alignments are connected by two types of C-H···O interaction (C7-H7···O3 and C14-H14···O1, **Figure 10**). According to **Table 3**, the molecular packing structures of methyl

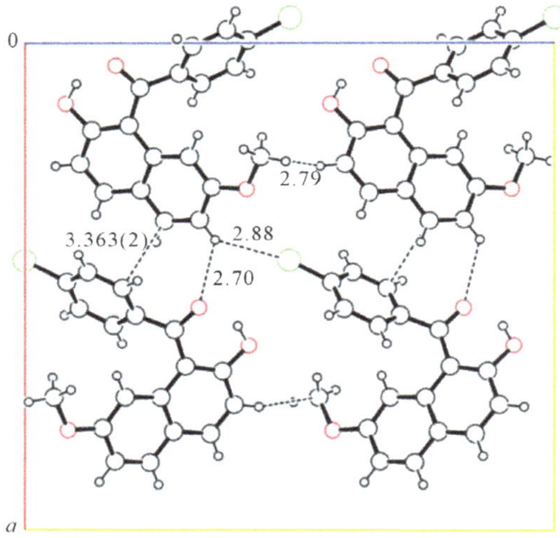

Figure 6. A partial crystal packing diagram of compound 4, viewed down the *b*-axis (the intermolecular C-H···O and C-H···π interactions are shown as dashed lines).

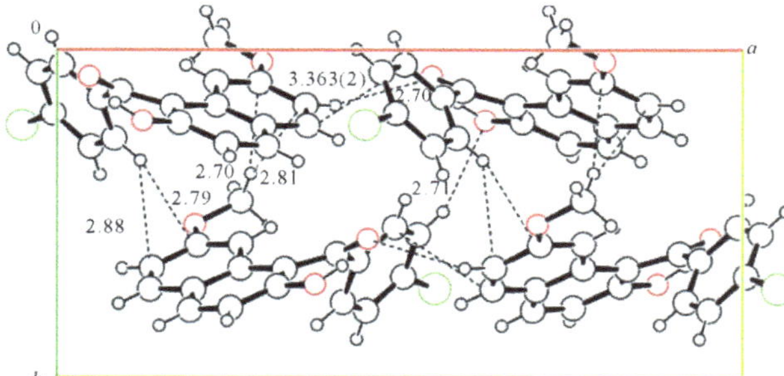

Figure 7. A partial crystal packing diagram of compound 4, viewed down the *c*-axis (the intermolecular C-H···O and C-H···π interactions are shown as dashed lines).

methyl ether-*retained* imine **3** are apparently directed by various kinds of effective interactions compared to the precursor ketone **1**. Especially, C-H···π and C-H···N interactions affording molecular pairs play a key role to govern the whole molecular packing. These interactions presumably maintain the spatial organization of an *R-S* pair of imine **3** molecules with minimized inner steric repulsions.

Methyl ether-*retained* ketone **1** has enough flexible molecular skeleton to perturb the spatial organization so that the suitable stabilized molecular stack is achieved leading the optimal molecular packing. On the other hand, the rigid conformation of methyl ether-*retained* imine **3** molecule should have little space for perturbation of configuration. As a result, predominant two interactions function within the same pair of imine **3** instead of sequential interactions resulting in formation of dimeric pairs. Although the semi-rigid conformation of the methyl ether-*cleaved* ketone **5** is similar to methyl ether-*retained* imine **3**, loose van der Waals interactions might restrict roughly the perturbation of configuration.

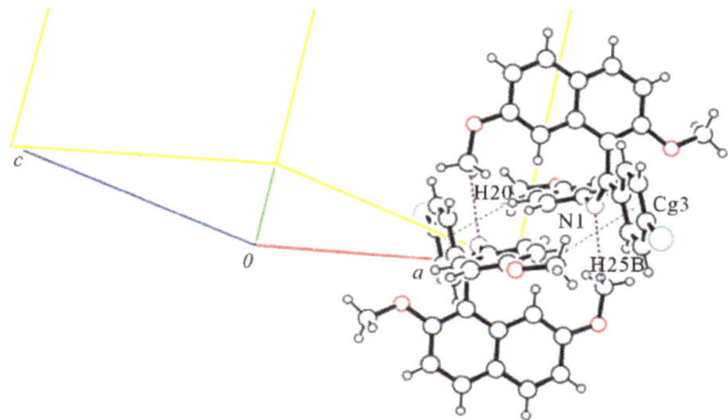

Figure 8. Molecular pair of imine 3 by forming C-H···N and C-H···π interactions.

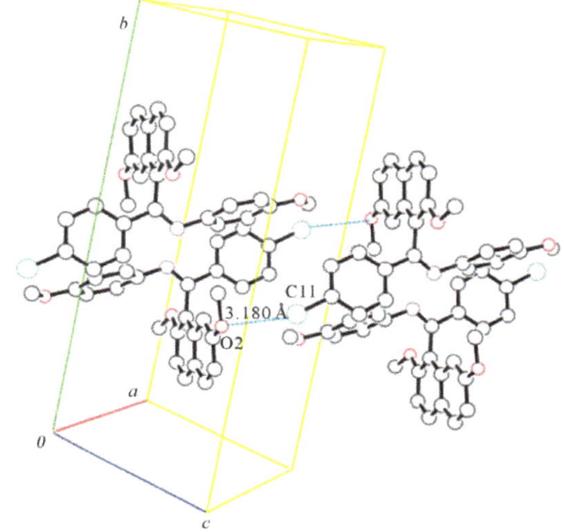

Figure 9. Molecular alignment of imine 3 via C-Cl···O interactions, viewed in *ac* diagonal. H atoms are omitted.

4. Conclusion

Conclusively, the single molecular organization of the two types of *C,C,N*-triarylimine compounds with 2- methoxy or 2-hydroxy group in crystal is displayed topologically same. The three aromatic rings are situated almost perpendicularly to each other, regardless whether triarylimine has a methoxy group at the 2-position of the naphthalene or a hydroxy one. On the other hand, the crystal structure of methyl ether-*retained* ketone clearly differs from the methyl ether-*cleaved* counterpart. There-

Table 3. Crystallographic data and structure refinement parameters of molecule 1, 3, 5, and 6.

D-X···A	D-X	X···A	D···A	D-X···A
1-aroyl-2-OCH_3 **1**				
C13-H13···O3^{i}	0.93	2.58	3.401(2)	148
Imine **3**				
C20-H20···Cg3^{ii}	0.95	2.90	3.719(4)	145
C7-H7···O3^{iii}	0.95	2.51	3.244(5)	134
C14-H14···O1^{iv}	0.95	2.67	3.525	150
C25-H25B···N1^{ii}	0.98	2.70	3.756	122
C15-Cl1···O2^{v}	1.739(4)	3.180(3)	4.795	153.00(15)
1-aroyl-2-OH **5**				
O2-H2···O1	0.94(2)	1.71(2)	2.5573(16)	148(2)
Imine **6**				
O1-H1···N2^{vi}	0.89 (2)	1.86 (2)	2.7401(18)	167.2 (18)
C20-H20···Cl1^{vii}	0.95	2.78	3.6071(17)	146

Symmetry code: (i) $x + 1, y, z$, (ii) $1 - x, 2 - y, 1 - z$, (iii) $-1/2 + x$, 1.5 − y, $-1 + z$, (iv) $-x, 2 - y, 1 - z$, (v) $-x, 2 - y, -z$, (vi) $-x + 1, -y, -z+1$, (vii) $x + 1/2, -y + 1/2$, z + 1/2.

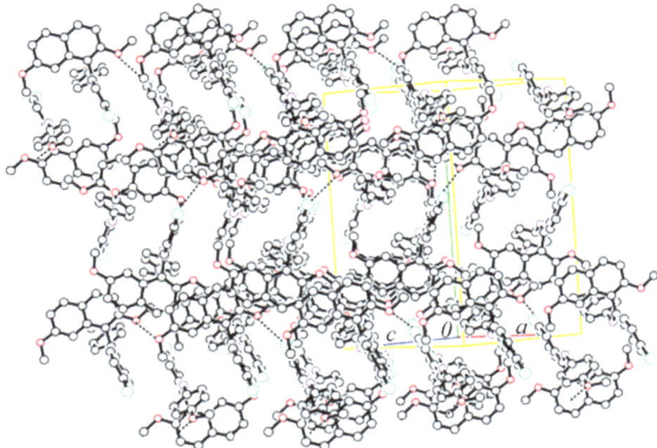

Figure 10. Tubluar molecular alignments of imine 3, viewed down *ac* diagonal. H atoms are omitted.

fore, triarylimine compounds have proved enough stable by the aid of the adapting of steric hindrance releasing molecular organization, where three aromatic rings situate perpendicularly to each other. The molecular packings of 1-aroyl-2,7-dimethoxynaphthalene and methyl ether-*retained* triarylimine clearly differ to each other, *i.e.*, the piles composed of one configurated molecules for the ketone and the tubular molecular alignments composed of *R-S* dimeric molecules for the imine. The difference is interpreted on the basis of flexibility of the molecular skeletons governing the number and the strength of effective intermolecular interactions.

5. Acknowledgements

This work was partially supported by the Iron and Steel Institute of Japan (ISIJ) Research Promotion Grant, Tokyo, Japan.

REFERENCES

[1] A. J. Neel, J. P. Hehn, P. F. Tripet and F. D. Toste, "Asymmetric Cross-Dehydrogenative Coupling Enabled by the Design and Application of Chiral Triazole-Containing Phosphoric Acids," *Journal of American Chemical Society*, Vol. 135, No. 38, 2013, pp. 14044-14047.

[2] R. Sun, C. Xue, X. Ma, M. Gao, H. Tian and Q. Li, "Light-Driven Linear Helical Supramolecular Polymer Formed by Molecular-Recognition-Directed Self-Assembly of Bis(*p*-sulfonatocalix[4]arene) and Pseudorotaxane," *Journal of American Chemical Society*, Vol. 135, No. 16, 2013, pp. 5990-5993.

[3] B. Jose, S. Matsushita and K. Akagi, "Lyotropic Chiral Nematic Liquid Crystalline Aliphatic Conjugated Polymers Based on Disubstituted Polyacetylene Derivatives That Exhibit High Dissymmetry Factors in Circularly Polarized Luminescence," *Journal of American Chemical Society*, Vol. 134, No. 48, 2012, pp. 19795-19807.

[4] Z. Furen, S. Haibin and Z. Guofu, "Synthesis and Catalytic Activity of Group 5 Metal Amides with Chiral Biaryldiamine-Based Ligands," *Dalton Transactions*, Vol. 40, No. 7, 2011, pp. 1547-1566.

[5] C. Chan, N.-W. Tseng, J. Ram, J. Liu, R. Kwok and B. Tang, "Construction of Functional Macromolecules with Well-Defined Structures by Indium-Catalyzed Three-Component Polycoupling of Alkynes, Aldehydes, and Amines," *Macromolecules*, Vol. 46, No. 9, 2013, pp. 3246-3256.

[6] O. Lucchi, "High Symmetry Chiral Auxiliaries Containing Heteroatoms," *Pure and Applied Chemistry*, Vol. 68, No. 4, 1996, pp. 945-950.

[7] K. Maruoka, "Asymmetric Phase Transfer Catalysis," Wiley-VCH, Weinheim, 2008.

[8] M. T. Scerba, C. M. Leavitt, M. E. Diener, A. F. DeBlase and T. Lectka, "NH^+-F Hydrogen Bonding in a Fluorinated 'Proton Sponge' Derivative: Integration of Solution, Solid-State, Gas-Phase, and Computational Studies," *Journal of Organic Chemistry*, Vol. 76, No. 19, 2011, pp. 7975-7984.

[9] S. Shinamura, E. Miyazaki and K. Takiyama, "Synthesis, Properties, Crystal Structures, and Semiconductor Characteristics of Naphtho[1,2-b:5,6-b']dithiophene and -Diselenophene Derivatives," *Journal of Organic Chemistry*, Vol. 75, No. 4, 2010, pp. 1228-1234.

[10] Z. Y. Wang and A. L. Guen, "Synthesis and Properties of Poly(arylene ether)s Containing 1,8-Dibenzoylnaphthalene Units," *Macromolecules*, Vol. 28, No. 10, 1995, pp. 3728-3732.

[11] Y. L. Jiang, X. Gao, G. Guannan, A. Patel and A. Javer, "Selective Recognition of Uracil and Its Derivatives Using a DNA Repair Enzyme Structual Mimic," *Journal of Organic Chemistry*, Vol. 75, No. 2, 2010, pp. 324-333.

[12] X. Mei, R. M. Martin and C. Wolf, "Synthesis of Sterically Crowded Atropisomeric 1,8-Diacridylnaphthalene for Dual-Mode Enantioselective Fluorosensing," *Journal of Organic Chemistry*, Vol.71, No.7, 2006, pp. 2854-2861.

[13] S. Cohen, M. Thirumalaikumar, S. Pogodin and I. Agranat, "Peri Interactions in Naphthalene Diketones: A Preference for (Z,Z) Conformations," *Structure Chemistry*, Vol. 154, No. 4, 2004, pp. 339-346.

[14] L.-H. Jing, D.-B. Qin, L. He, S.-J. Gu, H.-X. Zhang and G. Lei, "Dimethyl Naphthalene-1,8-dicarboxylate," *Acta Crystallographica Section E*, Vol. 61, 2005, pp. o3595-o3596.

[15] A. Okamoto, S. Watanabe, K. Nakaema and N. Yonezawa, "Crystal Structure and Solution Structural Dynamic Feature of 1,8-Dibenzoyl-2,7-dimethoxynaphthalene," *Crystal Structure Theory and Applications*, Vol. 1, No. 3, 2012, pp. 121-127.

[16] P. H. Gore and K. Henrick, "1,8-Dibenzoyl-2,7-dimethylnaphthalene," *Acta Crystallographica Section B*, Vol. B36, 1980, pp. 2462-2465.

[17] A. Okamoto and N. Yonezawa, "Reversible ArS_E Aroylation of Naphthalene Derivatives," *Chemistry Letters*, Vol. 38, No. 9, 2009, pp. 914-915.

[18] A. Okamoto, R. Mitsui, H. Oike and N. Yonezawa, "Lewis Acid-Mediated ArS_E Aroylation of Naphthalene Derivative: Distinct Second Aroylation Behavior of Naphthyl Ketone," *Chemistry Letters*, Vol. 40, No. 11, 2011, pp. 1283-1284.

[19] A. Okamoto, A. Nagasawa and N. Yonezawa, "Preparation and Structure of *C,C,N*-Triaryl Substituted Imine: $TiCl_4$-DABCO-Mediated Imination of 1-Aroyl-2,7-dimethoxynaphthalene and Spatial Organization of the Produced Imine Molecule in Crystal," 2013, in press.

[20] K. Sasagawa, R. Takeuchi, T. Kusakabe, N. Yonezawa and A. Okamoto, "{2,7-Dimethoxy-8-[4-(propan-2-yloxy)-benzoyl]naphthalen-1-yl}[4-(propan-2-yloxy)phenyl]methanone," *Acta Crystallographica Section E*, Vol. 69, 2013, pp. o444-445.

[21] S. Mouri, D. Hijikata, K. Isozaki, N. Yonezawa and A. Okamoto, "[2,7-Diethoxy-8-(4-fluorobenzoyl)naphthalen-1-yl](4-fluorophenyl)methanone," *Acta Crystallographica Section E*, Vol. 69, 2013, p. o637.

[22] S. Yoshiwaka, D. Hijikata, K. Sasagawa, N. Yonezawa and A. Okamoto, "[8-(4-Phenoxybenzoyl)-2,7-bis(propan-2-yloxy)naphthalen-1-yl](4-phenoxyphenyl)methanone," *Acta Crystallographica Section E*, Vol. 69, 2013, p. o242.

[23] A. Okamoto, A. Nagasawa and N. Yonezawa, "Preparation and Structure of *C,C,N*-Triaryl Substituted Imines: $TiCl_4$-1,4-Diazabicyclo[2.2.2]octane-mediated Imination of 1-Aroyl-2,7-dimethoxynaphthalene and Spatial Organization of the Produced Imine Molecule in Crystal," *European Chemical Bulletin*, Vol. 3, 2014, pp. 13-17.

[24] Rigaku, "PROCESS-AUTO," Rigaku Corporation, Tokyo, 1998.

[25] Rigaku/MSC, "CrystalStructure," Rigaku/MSC, The Woodlands, 2004.

[26] M. C. Burla, R. Caliandro, M. Camalli, B. Carrozzini, G. L. Cascarano, L. De Caro, C. Giacovazzo, G. Polidori and R. Spagna, "*SIR*2004: An Improved Tool for Crystal Structure Determination and Refinement," *Journal of Applied Crystallography.*, Vol. 38, 2005, pp. 381-388.

[27] G. M. Sheldrick, "A Short History of SHELX," *Acta Crystallographica Section A*, Vol. A64, 2008, pp. 112-122.

[28] R. Mitsui, K. Nakaema, K. Noguchi, A. Okamoto and N. Yonezawa, "1-(4-Chlorobenzoyl)-2,7-dimethoxynaphthalene," *Acta Crystallographica Section E*, Vol. E64, 2008, p. o1278.

[29] R. Mitsui, K. Nakaema, K. Noguchi and N. Yonezawa, "(4-Chlorophenyl)(2-hydroxy-7-methoxynaphthalen-1-yl) methanone," *Acta Crystallographica Section E*, Vol. E64, 2008, p. o2497.

[30] A. Nagasawa, R. Mitsui, Y. Kato, A. Okamoto and N. Yonezawa, "1-[(4-Chlorophenyl)(phenylimino)methyl]-7-methoxy-2-naphthol-1,4-diazabicyclo[2.2.2]octane (2/1)," *Acta Crystallographica Section E*, Vol. E66, 2010, p. o2497.

A Hirshfeld Surface Analysis and Crystal Structure of 2'-[1-(2-Fluoro-Phenyl)-1H-tetrazol-5-Yl]-4-Methoxy-Biphenyl-2-Carbaldehyde

S. Madan Kumar[1], B. C. Manjunath[1], G. S. Lingaraju[2], M. M. M. Abdoh[3], M. P. Sadashiva[2], N. K. Lokanath[1*]

[1]Department of Studies in Physics, University of Mysore, Mysore, India
[2]Department of Studies in Chemistry, University of Mysore, Mysore, India
[3]Department of Physics, Faculty of Science, An Najah National University, Nabtus West Bank, Palestinian Territories

ABSTRACT

The title compound, $C_{21}H_{15}FN_4O_2$ is synthesized and characterized by ^1H NMR, LC-MS and finally confirmed by single crystal X-ray diffraction method. This molecule crystallizes in the monoclinic crystal system and space group *P*21/*c*, with crystal parameters a = 9.4386(5) Å, b = 20.8082(1) Å, c = 9.4338(6) Å, β = 99.566(2)^0, Z = 4 and V = 1826.98(19) Å^3. The mean planes of fluro-phenyl moiety makes a dihedral angle of 21.51 (7)^0 with biphenyl moiety. The molecules are connected by hydrogen bonds of the type C---H...O and C---H...F. In addition, crystal structure is stabilized with $\pi \ldots \pi$ (exhibits intramolecular interaction) and C---O... π interactions. The intercontacts in the crystal structure are analyzed using Hirshfeld surfaces computational method.

Keywords: Crystal Structure; Intermolecular Interactions; Hirshfeld Surfaces

1. Introduction

Tetrazoles and its derivatives are the most important in the field of medicinal chemistry and found wide spectrum of applications in coordination chemistry because of their multiple coordination status, acting as ligands to metal ions and for the construction of novel metal-organic frameworks [1-3]. And they exhibit biological activities like antibacterial [4,5], antifungal and anticonvulsant [6], analgesic [7], antitubercular activity [8] and anti-cancer activity [9]. Also, 1,5-disubstituted tetrazoles used as anti-inflammatory and anti-hypertensive agents [10,11], such as Losartan [12,13]. Biphenyl tetrazoles have also demonstrated activities as stimulators of growth hormone release [14], metallo-protease inhibitors [15,16] and chloride channel blockers [17]. And, the 5-substituted 1H-tetrazole moiety has been used in the drug discovery as a bioisotere for the corboxylic acid group [18]. In addition, tetrazole compounds are used as new energetic materials because of their good thermal stability due to the presence of aromatic ring system (5-Azido-1*H*-tetrazole) [19]. Synthesizing the organic compounds in the Suzuki-Miyaura cross-coupling is one of the powerful methods for aromatic C-C bond formation [20]. We report here, the synthesis, spectroscopic studies, structural studies by X-ray diffraction method and analysis of intercontacts by Hirshfeld surfaces computational method of 2'-[1-(2-Fluoro-phenyl)-1H-tetrazol-5-yl]-4-methoxy- biphenyl-2-carbaldehyde.

*Corresponding author.

2. Experimental

All reagents were purchased as reagent grade and used without further purification. The reaction was monitored and determination of product was accomplished by TLC technique. The melting point was determined on SELACO-650 hot stage apparatus. Elemental analysis (C, H, N) were determined with Vario-EL instrument. ^1H NMR spectra were recorded on a bruker DRX 300 MHz spectrometer using DMSO-d6 as solvent and TMS as internal standard. Chemical shifts are given in δ (ppm).

3. Synthesis and Crystallization of the Title Compound

The title compound is obtained using the Suzuki-Miyaura

coupling (**Figure 1**) of the compound 1-(2-Flurophenyl)-1*H*-tetrazole (1mmol), 1), with 2-formyl-4-methoxy phenyl boronic acid (1 mmol) in presence of sodium carbonate (15 mmol) and palladium catalyst in a mixture of dimethyl ether (DME) and water in the ratio 3:1. Then the mixture was degasified by bubbling with nitrogen for 15 minutes. After degasify $PdCl_2$ $(PPh_3)_2$ [dikis] (0.05 mmol) was added. The resultant mixture was heated at 80˚C under nitrogen atmosphere for 5 hours. After completion of reaction (monitored by TLC), the reaction mixture was concentrated under reduced pressure to remove DME. Then residue was dissolved with ethyl acetate (25 mL), washed with 0.1 N hydrochloric acid (2 * 25 mL), followed by brine solution (2 * 25 mL). Then, the organic layer was dried over anhydrous sodium sulfate, filtered and concentrated under reduced pressure to afford crude 2'-[1-(2-Fluorophenyl)-1H-tetrazol-5-yl]-4-methoxy-biphenyl-2-carbaldehyde. 2), which was purified by column chromatography over silica gel (60 - 120 mesh) using Hexane: Ethyl acetate mixture in 8:2 ratios as eluent. The pure compound 2 was crystallized in ethyl acetate and hexane to obtain colorless single crystals.

4. Spectral Analysis

^{1}H NMR ($CDCl_3$, 400 MHz): δ 9.90 (s, 1H), 7.57 - 7.52 (m, 2H), 7.35 - 7.32 (m, 2H), 7.20 (s, 1H), 7.14 (t, J = 6.0 Hz, 2H), 6.98 (t, J = 7.8 Hz, 2H), 6.69 - 6.60 (m, 2H), 3.80 (s, 3H) (**Figure 2**). Mass, calculated: 374.36 found: 375 (M^+ + 1) Elemental analyses, calculated: C, 67.37; H, 4.04; F, 5.07; N, 14.97; O, 8.55. Found: C, 67.57; H, 4.28; F, 5.02; N, 14.773; O, 8.76 (**Figure 3**). Melting point (˚C): 103 - 105 (Uncorrected).

5. Crystal Structure Determination

A good single crystal of the title compound with dimension 0.30 × 0.35 × 0.35 mm was chosen for X-ray diffraction study. Data collection and cell refinement were carried out using Bruker Kappa ApexII CCD diffractometer [21] with Mo*Kα* radiation. The absorption correction was applied using multi-scan technique for data collection. The lattice parameters were determined by the least-squares methods on the basis of all reflections with $F^2 > 2\sigma (F^2)$. The structure was solved by the direct methods using SHELXS-97 [22,23]. All the non-hydrogen atoms were revealed in the Fourier map itself. Full-matrix least squares refinement using SHELXL-97 [22,23] with isotropic temperature factors for all the atoms was done. Refinement of non-hydrogen atoms with anisotropic parameters was started at this stage. The hydrogen atoms were placed at chemically acceptable positions and were allowed to ride on their parent atoms. About 165 parameters were refined with 3213 unique reflections which saturated the residuals to R1 = 0.0375 and wR2 = 0.1071. The details of the crystal data and refinement are given in **Table 1**. **Table 2** lists the hydrogen bonds. All the figures (*ORTEP*, packing and hydrogen bonding) were plotted using *MERCURY* [24]. Hirshfeld surface analyses were carried out and finger print plots were plotted using *CRYSTALEXPLORER* [25]. Electrostatic potentials were calculated using *TONTO* [26,27].

6. Results and Discussion

The dihedral angle between mean planes of fluoro-phenyl moiety and benzeze ring (C2/C3/C4/C5/C6/C7 attached with methoxy and carbaldehyde species) is $8.03(8)^0$. And, the mean planes of rings making dihedral angle with each other is as follows; the tetrazole ring (N1/N2/N3/N4/C15) makes $49.16(11)^0$ with benzene ring (C14/C13/C12/C11/C10/C9), $57.38(10)^0$ with fluro-phenyl moiety and $54.86(10)^0$, with phenyl moiety [(C2/C3/C4/C5/C6/C7) attached with methoxy and carbaldehyde species]. Similarly, the benzene ring (C14/C13/C12/C11/C10/C9) makes a dihedral angle of $64.67(10)^0$ with fluoro-phenyl moiety. Also, it makes an angle of $56.86(10)^0$ with phenyl moiety (C2/C3/C4/C5/C6/C7) attached with methoxy and carbaldehyde species. The overall geometry of the title compound is similar to 1-(4-nitrophenyl)-1*H*-tetrazol-5-amine and {(*E*)-[1-(4-ethoxyphenyl)-1*H*-tetrazol-5-yl] iminomethyl} dimethylamine [28].

Figure 4 represents the *ORTEP* diagram of the title molecule. The molecules in the unit cell are connected by hydrogen bonds C10-H10...F1, C12-H12...O2 and C20-H20...O2 (**Table 2**). And, **Figure 5** shows the packing of the molecules are arranged in the fishing net pattern. The observed weak interactions π...π and C---O...π helps in crystal structure stabilization. The intramolecular π...π interactions exists between centroid (*Cg*4: C16/C17/C18/C19/C20/C21) of fluoro-phenyl moiety and benzene ring (*Cg*2: C2/C3/C4/C5/C6/C7) with a distance 3.7806(10) Å [*x*, *y*, *z*] (**Figure 6**). And, inter molecular π...π exists between face to face (*Cg*4 and *Cg*4) interactions with a distance of 3.6875(11) Å [$2 - x, -y, 1 - z$]. In addition to this C---O...π (*Cg*4) interaction exists between carbaldehyde moiety (C8-O2) and *Cg*4 with a distance of 3.9551(15) Å [*x*, *y*, *z*].

7. Hirshfeld Surface Analysis

The intermolecular interactions of the title compound are quantified using Hirshfeld surface analysis. This approach is a graphical tool for visualization and understanding of intermolecular interactions [25]. Here, we estimate the intermolecular contacts, which are shown in **Figure 7**. The chart indicates that the contribution of inter-contacts to the Hirshfeld surfaces, H...H (36%), N...H (19%), C...H (16%), O...H (14%), F...H (7%) and others (C...C, N...N, C...O, N...F; 8%). These inter-contacts

Figure 1. Schematic diagram and synthesis pathway of the title compound.

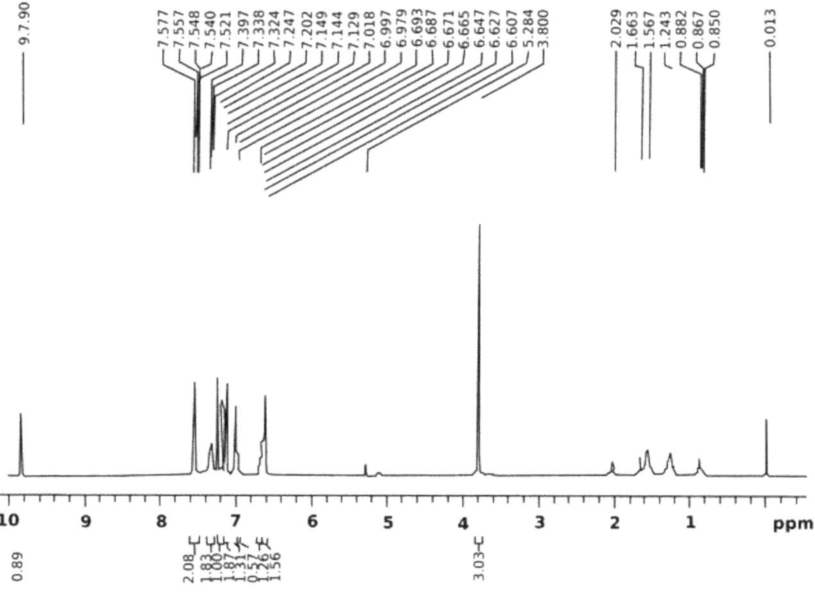

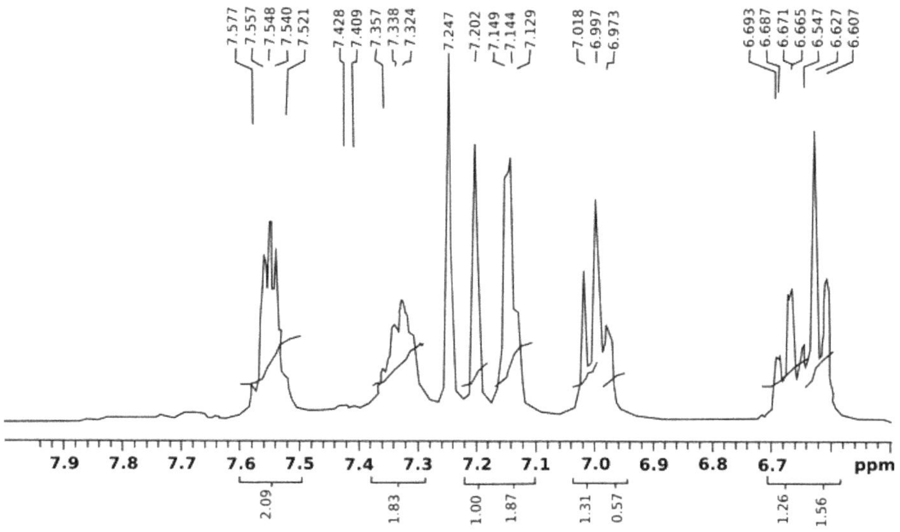

Figure 2. ^{1}H NMR spectra of the title compound.

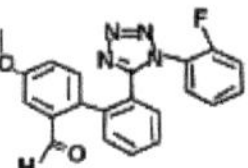

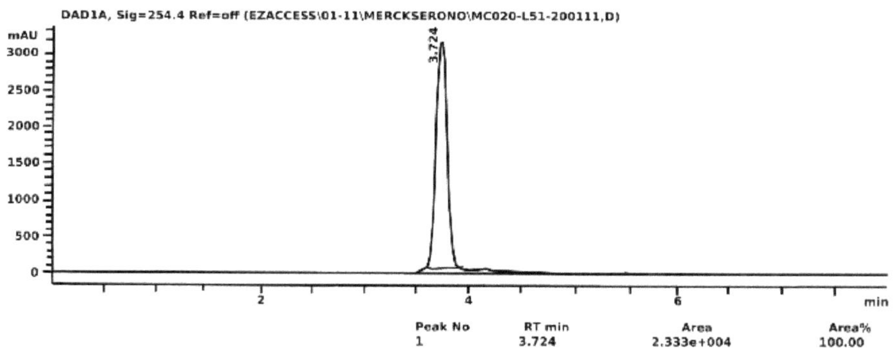

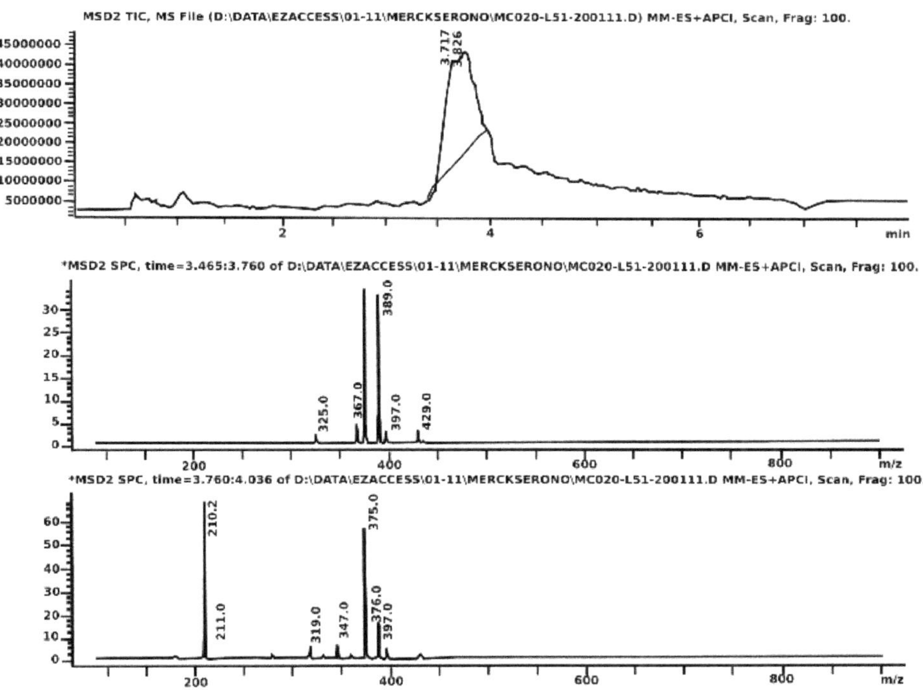

Figure 3. LC-MS spectra of the title compound.

are highlighted by conventional mapping of d_{norm} on molecular Hirshfeld surfaces are shown in **Figure 8**. The red spots over the surface indicate the intercontacts involved in the hydrogen bonds. Further, intercontacts were plotted with fingerprint plots (**Figure. 9**). H...H intercontacts, (**Figure 9(a)**) shows large surfaces, whereas the O...H plot (**Figure 9(d)**) shows the presence of O...H contact with the two characteristic wings. The N...H contact plot is shown in **Figure 9(b)**. The intercontacts F...H **(Figure 9(e))** showing two narrow pointed wings provide evidence for C-H...F non-classical hydrogen bonds. And, C...H plot reveals the information of inter molecular hydrogen bonds.

The electrostatic potential is mapped on Hirshfeld surface using STO-3G basis set at the Hartree-Fock theory over the range of ± 0.025 au (**Figure 10**). The positive electrostatic potential (blue region) over the surface indicates hydrogen donor potential, whereas the hydrogen bond acceptors are represented by negative electrostatic potential (red region) [27]. The crystal geometries were

Table 1. Crystal data, data collection and structure refinement.

CCDC	933058
Empirical formula	$C_{21}H_{15}N_4O_2$
Formula weight	374.37
Temperature	293 K
Wavelength	0.71073Å
Crystal system	monoclinic
Space group	*P*21/*c*
Unit cell dimensions	*a* = 9.4348(5) Å b = 20.8082(1) Å c = 9.4338(6) Å β = 99.566(2) °
Volume	1826.98(19) $Å^3$
Z	4
Calculated density	1.361 Mg/m^3
Absorption coefficient	0.098 mm^{-1}
$F_{(000)}$	776
Crystal size	0.30 × 0.35 × 0.35 mm
Theta range for data collection	2.4° to 25.0°
Limiting indices	−11 <= h <= 11, −24 <= k <= 24, −11 <= l <= 11
Reflections collected / unique	16483/3213 [$R_{(int)}$ = 0.029]
Refinement method	Full-matrix least-squares on F^2
Data/restraints/parameters	3213/0/255
Goodness-of-fit on F^2	1.04
Final *R* indices	$[I > 2\sigma(I)] R_1 = 0.0375, wR_2 = 0.1071$
Largest diff. peak and hole	0.16 and −0.18 e.$Å^{-3}$

Table 2. Hydrogen bonds [Å, 0].

D-H…A	D-H*	H-A*	D-A	D-H…A	Symmetry codes
C10-H10...F1	0.93	2.50	3.388(2)	160	$x, 1/2 - y, -1/2 + z$
C12-H12...O2	0.93	2.57	3.352(2)	143	$-1 + x, y, z$
C20-H20...O2	0.93	2.52	3.430(2)	167	$x, 1/2 - y, 1/2 + z$

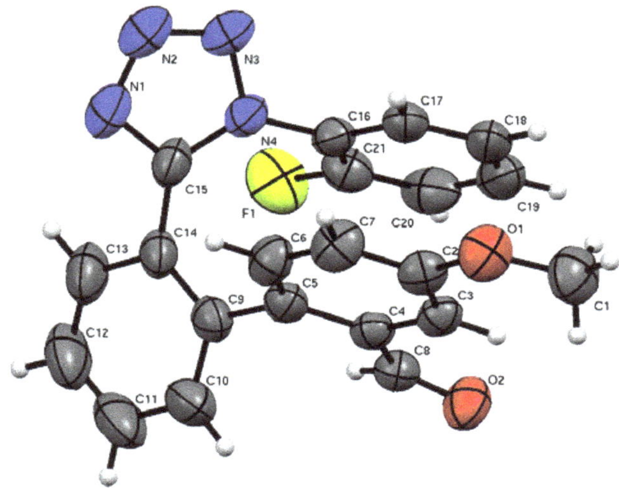

Figure 4. Molecular structure of the title compound showing the atomic numbering system. Displacement ellipsoids are drawn at the 50% probability.

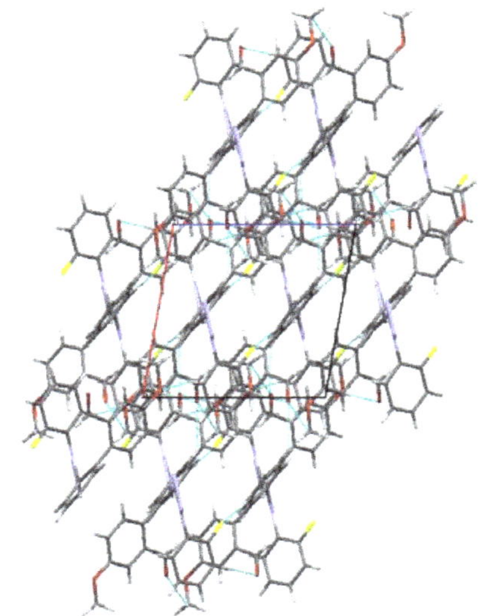

Figure 5. Packing diagram of the title molecule along b-axis. Dotted lines represents hydrogen bonds.

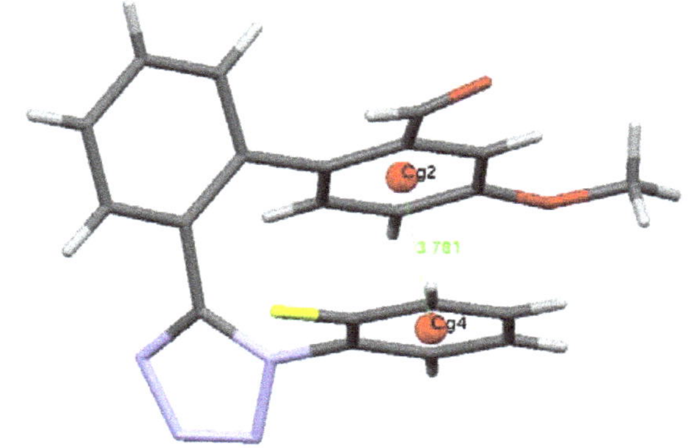

Figure 6. Intramolecular interaction between *Cg*2 and *Cg*4. Dotted lines (green) represents interaction.

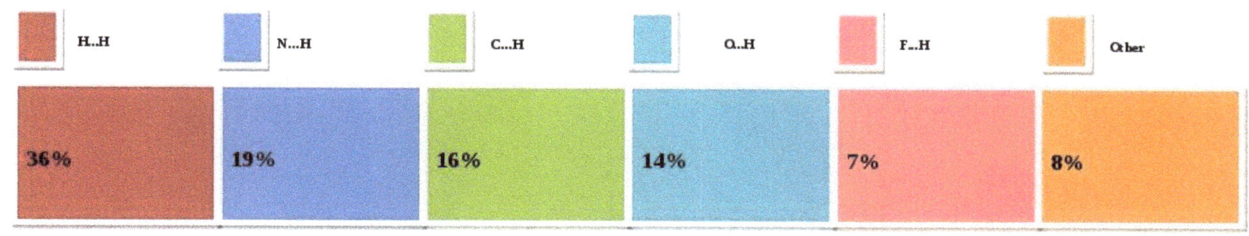

Figure 7. Hirshfeld surface: Percentage of various intermolecular contacts contributed to the Hirshfeld surface.

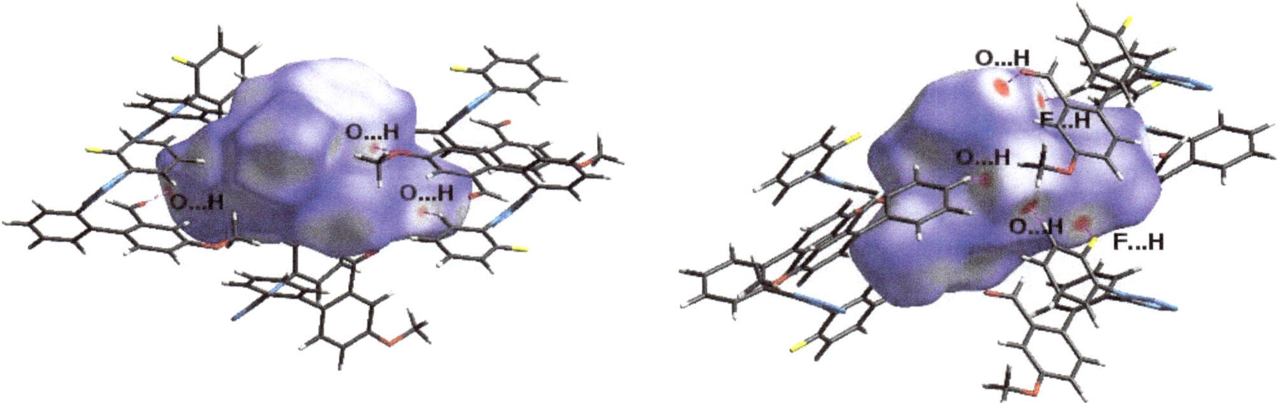

Figure 8. d_{norm} mapped on Hirshfeld surface for visualizing the intercontacts of the title compound. Color scale in between −0.18 au (blue) to 1.4 au (red). Dotted lines (magenta) represent hydrogen bonds.

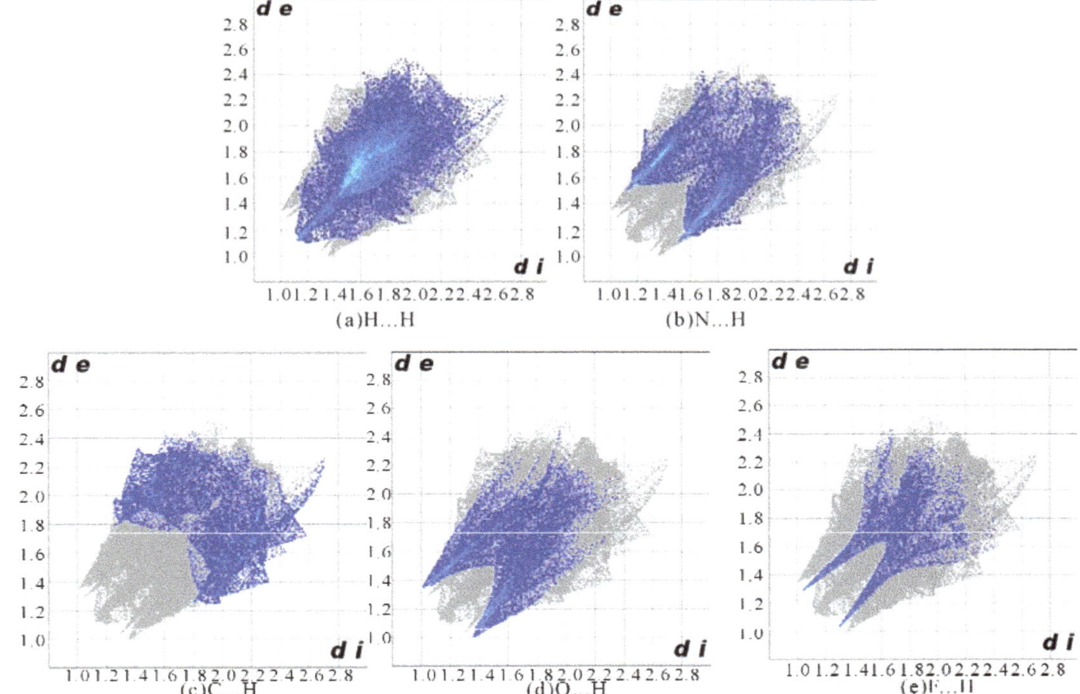

Figure 9. Fingerprint of the title compound, (a) H...H, (b) C...H, (c) C...H, (d) O...H and (e) F...H. The outline of the full fingerprint is shown in gray. d_i is the closest internal distance from a given point on the Hirshfeld surface and de is the closest external contacts.

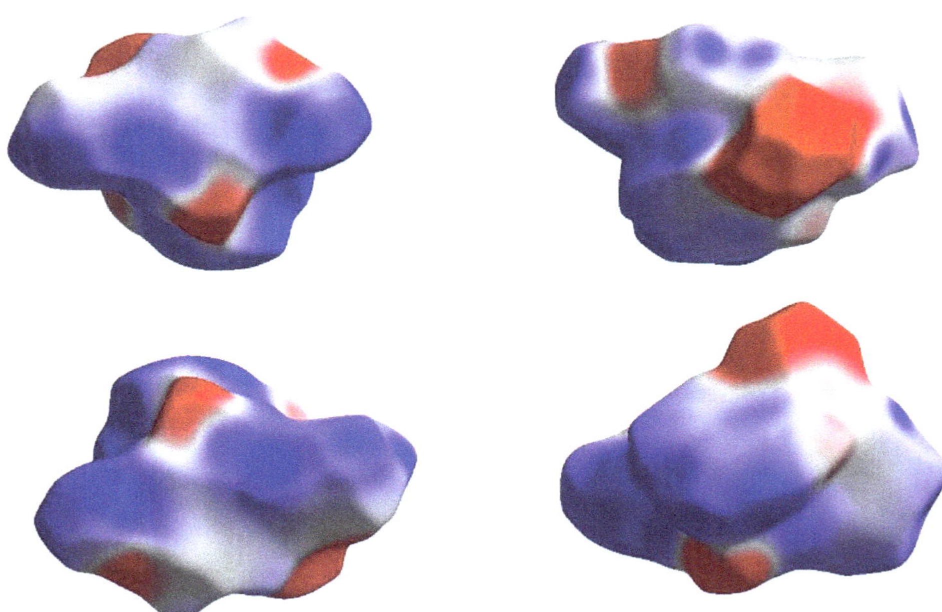

Figure 10. Electrostatic potential mapped on Hirshfeld surface (different orientation) with ± 0.25 au. Blue region corresponds to positive electrostatic potential and red region to negative electrostatic potential.

used as input to the *TONTO* [26] integrated with *Crystal explorer* [25].

8. Conclusion

The dihedral angle between fluro-phenyl moiety and bi-phenyl moiety is 21.51 (7)^0. The molecules are connected by C-H...O and C-H...F hydrogen bonds. In addition, the short contacts of the type π...π and C---O...π help in crystal stabilization. The Hirshfeld surface analysis with finger plots and electrostatic potential map reveals the percentage of intermolecular contacts and distribution of electrostatic potential of the title compound.

9. Acknowledgments

Madan Kumas S. thanks UGC-BRS and University of Mysore for providing fellowship. MPS gratefully acknowledge the financial support from University Grants Commission, New Delhi, India.

REFERENCES

[1] A. R. Katritzky, C. Cai and N. K. Meher, "Efficient Synthesis of 1,5-Disubstituted Tetrazoles," *Synthesis*, Vol. 8, 2007, pp. 1204-1208.

[2] X.-S. Wang, Y. Z. Tang, X-F, Huang, Z.-R. Qu, C.-M. Che, C. W. H. Chan and R.-G. Xiong, "Syntheses, Crystal Structures, and Luminescent Properties of Three Novel Zinc Coordination Polymers with Tetrazolyl Ligands," Inorganic Chemistry, Vol. 44, No. 15, 2005, pp. 5278-5285.

[3] R.-G. Xiong, X. Xue, H. Zhao, X.-Z. You, B. F. Abrahams and Z.-L. Xue, "Novel, Acentric Metal-Organic Coordination Polymers from Hydrothermal Reactions Involving *in Situ* Ligand Synthesis," *Angewandte Chemie International Edition*, Vol. 41, No. 20, 2005, pp. 3800-3803.

[4] K. Kemajl, V. Idriz, H. Arben, G. Sevdije, I. Muharrem and M. Sefkija, "Fabrication of 3D Hepatic Tissues by Additive Photopatterning of Cellular Hydrogels," *The FASEB Journal*, Vol. 21, No. 3, 2007, pp. 790-795.

[5] V. V. Mulwad, R. B. Pawar and A. C. Chaskar, "Synthesis and Antibacterial Activity of New Tetrazole Derivatives," *Journal of Korean Chemical Society*, Vol. 52, No. 3, 2008, pp. 249-256.

[6] R. S. Upadhayaya, S. Jain, N. Sinha, N. Kishore, R. Chandra and S. K. Arora, "Synthesis of Novel Substituted Tetrazoles Having Antifungal Activity," *European Journal of Medicinal Chemistry*, Vol. 39, No. 7, 2004, pp. 579- 592.

[7] A. Rajasekaran, N. Sankar, A. Murugesh, Kalasalingam and A. Rajagopal, "Antibacterial, Antifungal and Anticonvulsant Evaluation of Novel Newly Synthesized 1[2--(1H-Tetrazol-5-yl)ethyl]-1H-benzo[d][1,2,3]triazoles," *Archives of Pharmacal Research*, Vol. 29, No. 7, 2006, pp. 535-540.

[8] J. Adamec, K. Waisser, J. Kunes and J. Kaustova, "A Note on the Antitubercular Activities of 1-Aryl-5-benzylsulfanyltetrazoles," *Pharmaceutical & Medicinal Chemistry* (Weinheim), Vol. 338, No. 8, 2005, pp. 385-389.

[9] A. O. De, Souza, M. T. Pedrosa, J. B. Alderete, A. F. Cruz, M. A. Prado, R. B. Alves and C. L. Silva, "Cyto-

toxicity, Antitumoral and Antimycobacterial Activity of Tetrazole and Oxadiazole Derivatives," *Pharmazie*, Vol. 60, No. 5, 2005, pp. 396-397.

[10] R. J. Herr, "5-Substituted-1*H*-tetrazoles as Carboxylic Acid Isosteres: Medicinal Chemistry and Synthetic Methods," *Bioorganic and Medicinal Chemistry*, Vol. 10, No. 11, 2002, pp. 3379.

[11] L. V. Myznikov and G. I. Koldobskii, "Drugs in the Tetrazole Series (Review)," *Chemistry of Heterocyclic Compounds*, Vol. 43, No. 1, 2007, pp. 1-9.

[12] R. D. Smith, C. S. Sweet, A. Goldberg and P. B. W. M. Timmermans, "Losartan Potassium Cozaar (TM) a Nonpeptide Antagonist of Angiotensin-II," *Drugs Today*, Vol. 32, 1996, pp. 1-12.

[13] K. Dickstein, P. B. M. W. M. Timmermans and R. Segal, "Losartan: A Selective Angiotensin II Type 1 (AT1) Receptor Antagonist for the Treatment of Heart Failure," *Expert Opinion on Investigational Drugs*, Vol. 7, No. 11, 1998, pp. 1897-1914.

[14] R. G. Smith, K. Cheng, W. R. Schoen, S. S. Pong, G. Hickey, T, Jacks, B. Butler, W. W. S. Chan, L. Y.-P, Chaung, F. Judith, J, Taylor, M. J. Wyvratt and M. H. Fisher, "A Nonpeptidyl Growth-Hormone Secretagogue," *Science*, Vol. 260, No. 5114, 1993, pp. 1640-1643.

[15] B. G. Green, J. H. Toney, J. W. Kozarich and S. K. Grant, "Inhibition of Bacterial Peptide Deformylase by Biaryl Acid Analogs," *Archives of Biochemistry and. Biophysics*, Vol. 375, No. 2, 2000, pp. 355-358.

[16] J. H. Toney, P. M. D. Fitzgerald, N. Grover-Sharma S. H. Olson, W. J. May, J. G. Sundelof, D. E. Vanderwall, K. A. Cleary, S. K. Grant, J. K. Wu, J. W. Kozarich, D. L. Pompliano and G. G. Hammond, "Antibiotic Sensitization Using Biphenyl Tetrazoles as Potent Inhibitors of Bacteroides Fragilis Metallo-Beta-Lactamase," *Chemistry & Biology*, Vol. 5, No. 4, 1998, pp. 185-196.

[17] P. Christophersen and B. H. Dahl, WO Patent, 0024707, 2000.

[18] J. Stierstorfer, M. T. Klapotke, A. Hammerl and R. D. Chapman, "5-Azido-1H-Tetrazole—Improved Synthesis, Crystal Structure and Sensitivity Data," *Journal of Inorganic and General Chemistry*, Vol. 634 No. 6-7, 2008, pp. 1051-1057.

[19] J. Spencer, H. Patel, J. J. Deadman, R. A. Palmer, L. Male, J. S. Coles, G. O. Uzoh and S. L. Price, "The Unexpected But Predictable Tetrazole Packing in Flexible 1-Benzyl-1H-Tetrazole," *Crystal Engineering Communications*, Vol. 14, No. 20, 2012, pp. 6441-6446.

[20] N. Miyaura, "Organobom Compounds," *Topics in Current Chemistry*, Vol. 219, 2002, pp. 11-59.

[21] Bruker, "APEX2 and SAINT-Plus," Bruker AXS Inc., Madison, 2004.

[22] G. M. Sheldrick, "SHELXS-97 and SHELXL-97, Program for Crystal Structure Solution and Refinement," University of Gottingen, Gottingen, 1997.

[23] G. M. Sheldrick, "A Short History of SHELX," *Acta Crystallographica A*, Vol. 64, Part 1, 2008, pp. 112- 122.

[24] C. F. Macrae, I. J. Bruno, J. A. Chisholm, P. R. Edgington, P. McCabe, E. Pidcock, L. Rodriguez-Monge, R. Taylor, J. van de Streek and P. A. Wood, "Mercury CSD 2.0—New Features for the Visualization and Investigation of Crystal Structures," *Journal of Applied Crystallography*, Vol. 41, Part 2, 2008, pp. 466-470.

[25] S. K. Wolff, D. J. Grimwood, J. J. McKinnon, D. Jayatilaka and M. A. Spackamn, "Crystal Explorer 3.0," University of Westren Australia, Perth, 2007.

[26] D. Jayatilaka, D. J. Grimwood, A. Lee, A. Lemay, A. J. Russel, C. Taylo, S. K. Wolff, Cassam-Chenai and A. Whitton, "TONTO—A System for Computational Chemistry," 2005.

[27] M. A. Spackmann, J. J. McKinnon and D. Jayatilaka, "Electrostatic Potentials Mapped on Hirshfeld Surfaces Provide Direct Insight into Intermolecular Interactions in Crystals," *Crystal Engineering Communications*, Vol. 10, No. 4, 2008, pp. 377-388.

[28] A. S. Lyakhov, A. N. Vorobiov, L. S. Inashkevich and P. N. Gapoink, "Two Derivatives of 1,5-Disubstituted Tetrazoles: 1-(4-Nitrophenyl)-1*H*-tetrazol-5-amine and {(*E*)-[1-(4-ethoxyphenyl)-1*H*-tetrazol-5-yl]iminomethyl}dimet hylamine," *Acta Crystallographica*, Vol. C64, Part 8, 2008, pp. o414-o416.

Purification, Crystallization and Preliminary X-Ray Diffraction Analysis of Exodeoxyribonuclease III from Crenarchaeon *Sulfolobus tokodaii* Strain 7

Shuichi Miyamoto[1*], Chieko Naoe[2], Masaru Tsunoda[3], Kazuo T. Nakamura[2]
[1]Faculty of Pharmaceutical Sciences, Sojo University, Kumamoto, Japan
[2]School of Pharmacy, Showa University, Tokyo, Japan
[3]Faculty of Pharmacy, Iwaki Meisei University, Iwaki, Japan

ABSTRACT

Exodeoxyribonuclease III (EXOIII) acts as a 3'→5' exonuclease and is homologous to purinic/apyrimidinic (AP) endonuclease (APE), which plays an important role in the base excision repair pathway. To structurally investigate the reaction and substrate recognition mechanisms of EXOIII, a crystallographic study of EXOIII from *Sulfolobus tokodaii* strain 7 was carried out. The purified enzyme was crystallized by using the hanging-drop vapor-diffusion method. The crystals belonged to space group *C*2, with unit-cell parameters a = 154.2, b = 47.7, c = 92.4 Å, β = 125.8° and diffracted to 1.5 Å resolution.

Keywords: Crenarchaeon; Crystallization; Exodeoxyribonuclease; *Sulfolobus tokodaii*; X-Ray Diffraction

1. Introduction

A variety of mechanisms exist to repair damaged DNA and maintain a high degree of genomic stability within cells. Endonucleases specific for abasic or apurinic/ apyrimidinic (AP) sites are a major component of the cellular DNA repair machinery. AP sites are generated by spontaneous base loss [1], direct action of reactive oxygen species [2] and damage-specific DNA glycosylases as repair intermediates in the DNA base excision repair (BER) pathway [3,4]. It is estimated that as many as 10,000 purine bases alone are lost in a mammalian cell each day [5]. The generated AP sites are potentially mutagenic and cytotoxic, because of the inhibition of DNA replication and transcription [6,7]. AP sites are specifically corrected by the BER pathway: AP endonucleases (APE) first recognize the AP site and then cleave the phosphodiester backbone immediately 5' of the AP site, leaving a free 3'-hydroxyl nucleotide end and a deoxyribose 5'-phospate as termini [8,9].

Exodeoxyribonuclease III (EXOIII; EC 3.1.11.2) is homologous to APE and acts as a 3'→5' exonuclease. It is considered that the AP site undergoes a unique conformational change upon protein binding that permits complex formation and activation of attacking water, leading to incision, in the presence of Mg^{2+} [10,11]. However, the detailed reaction and substrate recognition mechanisms of AP endonucleases are not fully understood yet. Although EXOIII and APE belong to the same family, they have different substrate specificities. The three dimensional structures of EXOIII from *Escherichia coli* (*E. coli*) [12] and human APE [13] have been elucidated, but structures of these two enzymes from the same species have yet to be determined. To reveal the structural basis for the reaction and substrate recognition mechanisms of these enzymes from the same organism, we crystallized EXOIII and APE from the crenarchaeon, *Sulfolobus tokodaii* strain 7 [14]. *Sulfolobus tokodaii* is an aerobic thermoacidophilic crenarchaeon that optimally grows at 353 K and at low pH. Therefore, we also hope to obtain structural insights into the thermostability of these enzymes by comparing their structures with those of corresponding enzymes from other mesophilic organisms.

Here, we report the expression, purification, crystallization and preliminary X-ray diffraction analysis of EXOIII from *Sulfolobus tokodaii* strain 7, that is *sto*- EX-

*Corresponding author.

OIII. The enzyme consists of 241 amino acids (NCBI reference code NP_377894; UniProt ID F9VNP2), and has a molecular weight of 28.08 kDa and an isoelectric point of 8.72.

2. Materials and Methods

All chemicals were special grade reagents purchased from Wako Pure Chemical Industries, Japan unless stated otherwise.

2.1. Expression and Purification

The *sto*EXOIII protein was expressed in *E. coli* Rosetta-gami (DE3) using the pET-11a expression system (Merck KGaA, Germany). Since neither affinity tag nor extra residue was included, the sequence of the expressed protein was the same as that of the database entry. Bacterial clones were cultivated at 310 K in Luria-Bertani medium (Sigma-Aldrich, USA) with 100 $\mu g \cdot ml^{-1}$ ampicillin and 30 $\mu g \cdot ml^{-1}$ chloramphenicol. The cells were harvested by centrifugation at 8000 g, resuspended in 20 mM Tris-HCl pH 8.0, 50 mM NaCl and sonicated on ice. The mixture was heated at 343 K for 30 min and the soluble fraction was collected by centrifugation at 8000 g for 15 min. Ammonium sulfate was added to the supernatant to 50% saturation and the mixture was centrifuged at 10,000 g for 15 min. Ammonium sulfate was again added to the supernatant, to a final concentration of 60% saturation. After centrifugation at 10,000 g for 15 min, the pellet was dissolved in 50 mM sodium phosphate buffer (pH = 7.0) and dialyzed against the same buffer. The desalted mixture was loaded onto a SP-Sepharose column (1 ml; GE Healthcare Life Sciences) pre-equilibrated with 50 mM sodium phosphate buffer (pH = 7.0). The protein was eluted with a linear gradient of 0 - 1 M NaCl in the same buffer and protein purity was confirmed by SDS-PAGE. Simultaneous removal of NaCl and concentration of the protein was achieved by several rounds of diafiltration using an Amicon Ultra-10 Centrifugal Filter device (Millipore) with 20 mM sodium phosphate buffer (pH = 7.0) to a final protein concentration of 10 $mg \cdot ml^{-1}$. The final yield of pure protein was around 0.9 mg per liter of culture.

2.2. Crystallization

Initial crystallization conditions were screened by the hanging-drop vapour-diffusion technique at 293 K using Index, Crystal Screen and Crystal Screen 2 crystal screening kits (Hampton Research, USA). The crystallization droplets were prepared on siliconized cover slides by mixing 1 μl protein solution with 1 μl reservoir solution; the drops were then equilibrated against 800 μl of reservoir solution. Several single crystals appeared within a few days in several solutions containing PEG 4000 - 8000 (1st grade, Wako, Japan) as precipitant. Crystallization conditions were further optimized manually by refinement of buffer choice and PEG concentration. The quality of the crystals were checked from diffraction images obtained using an R-AXIS IV^{++} X-ray diffractometer (Rigaku, Japan). Finally, the most suitable precipitant solution for crystallization was 15% (w/v) PEG 6000, 0.1 M Bis-Tris pH 5.5 or 0.1 M HEPES pH 7.0, and 0.1 M L(-)-proline. *sto*EXOIII protein crystals with approximate dimensions of 1.0 × 0.1 × 0.1 mm were obtained in about one week at 293 K (**Figure 1**).

2.3. Data Collection

A *sto*EXOIII crystal was picked from the crystallization droplet using a mounted nylon loop and placed directly into a cold nitrogen-gas stream at 100 K. X-ray diffraction data were collected with a CCD detector (Quantum 210, Area Detector Systems Co., USA) on beamline BL-NW12 of the Photon Factory-Advanced Ring (PF-AR), Tsukuba, Japan, using a wavelength of 1.00 Å. The distance between the crystal and detector was 110 mm. A total of 360 images with 0.5 degrees oscillation were collected with an exposure time of 10 seconds.

3. Results and Discussion

Diffraction data were obtained in the resolution range 100 - 1.5 Å (**Figure 2**) and were then indexed and scaled using *DENZO* and *SCALEPACK* from the *HKL*-2000 program suite [15]. The crystals belong to the monoclinic space group *C*2, with unit-cell parameters a = 154.2, b = 47.7, c = 92.4 Å, β = 125.8°. A total of 87532 independent reflections obtained with a multiplicity of 5.1 at 1.5 Å resolution reached 99.3% completeness with an R_{merge} of 6.7%. Assuming that two monomers are present in the asymmetric unit, the Matthews coefficient was calculated to be 2.4 $Å^3 \cdot Da^{-1}$ [16]. This corresponds to an estimated

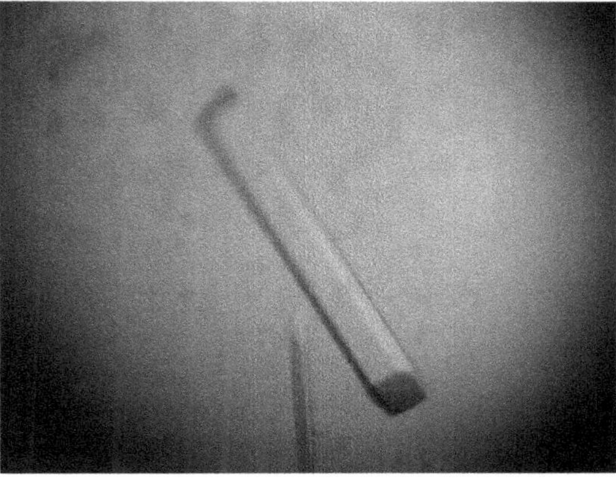

Figure 1. Crystals of *sto*EXOIII grown in 15% (w/v) PEG 6000, 0.1 M HEPES pH 7.0 and 0.1 M L(-)-proline using the hanging-drop vapor-diffusion method at 293 K. The size of the main crystal is approximately 1.0 × 0.1 × 0.1 mm.

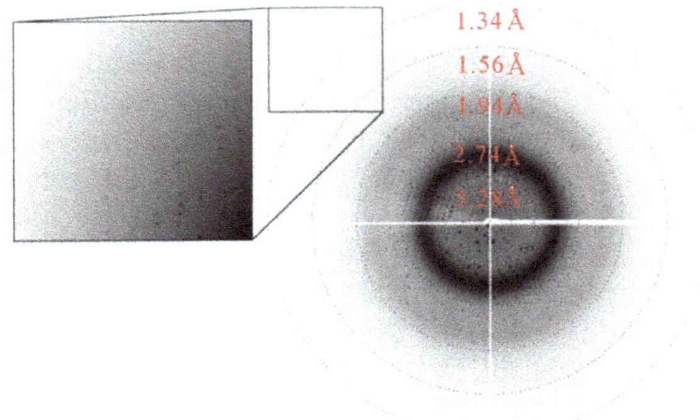

Figure 2. X-ray diffraction image. A corner section at the upper left is enlarged and shown in the inset.

Table 1. Data-collection statistics. Values in parentheses are for the highest resolution shell.

X-ray source	PF-AR BL-NW12
Wavelength (Å)	1.00
Space group	*C*2
Unit-cell parameters	a = 154.2, b = 47.7, c = 92.4 Å, β = 125.8°
Resolution range (Å)	100 - 1.5 (1.55 - 1.50)
No. of observed reflections	446783
No. of unique reflections	87532 (8513)
Multiplicity	5.1 (3.9)
Completeness (%)	99.3 (97.2)
R_{merge}[a] (%)	6.7 (20.5)
$\langle I/\sigma(I)\rangle$	40.5 (8.1)

[a] $R_{merge} = \sum_{hkl}\sum_i \left|I_i(hkl) - \langle I(hkl)\rangle\right| / \sum_{hkl}\sum_i I_i(hkl)$, where $I_i(hkl)$ is the intensity of the ith observation of reflection *hkl* and $\langle I(hkl)\rangle$ is the average intensity of reflection *hkl*.

solvent content of 49%, which is in the range normally observed for protein crystals. The relevant data-collection statistics are summarized in **Table 1**. The crystal structure determination of *sto*EXOIII is under way by the molecular replacement method using the MOLREP program [17] in the CCP4 suite [18] at 1.5 Å resolution.

4. Conclusion

Exodeoxyribonuclease III from the crenarchaeon, *Sulfolobus tokodaii* strain 7, was purified and then crystallized by using the hanging-drop vapor-diffusion method. The crystals belonged to space group C2, with unit-cell parameters a = 154.2, b = 47.7, c = 92.4 Å, β = 125.8° and diffracted to 1.5 Å resolution.

5. Acknowledgements

We are grateful to the scientists and staff at the Photon Factory for use of the synchrotron. This work was partly supported by the National Project on Protein Structural and Functional Analysis by the Ministry of Education, Culture, Sports, Science and Technology of Japan (Protein 3000 project). We thank Professor Seiki Kuramitsu for organizing the research group in the program.

REFERENCES

[1] T. Lindahl, "Instability and Decay of the Primary Structure of DNA," *Nature*, Vol. 362, No. 6422, 1993, pp. 709-715.

[2] B. Demple and L. Harrison, "Repair of Oxidative Damage to DNA: Enzymology and Biology," *Annual Review of Biochemistry*, 1994, Vol. 63, 1994, pp. 915-948.

[3] H. E. Krokan, R. Standal and G. Slupphang, "DNA Glycosylases in the Base Excision Repair of DNA," *Biochemical Journal*, 1997, Vol. 325, pp. 1-16.

[4] J. Baute and A. Depicker, "Base Excision Repair and Its Role in Maintaining Genome Stability," *Critical Reviews in Biochemistry and Molecular Biology*, Vol. 43, No. 4, 2008, pp. 239-276.

[5] T. Lindahl and B. Nyberg, "Rate of Depurination of Native Deoxyribonucleic Acid," *Biochemistry*, Vol. 11, No. 19, 1972, pp. 3610-3618.

[6] L. A. Loeb and B. D. Preston, "Mutagenesis by Apurinic/Apyrimidinic Sites," *Annual Review of Genetics*, Vol. 20, 1986, pp. 201-230.

[7] S. L. Yu, S. K. Lee, R. E. Johnson, L. Prakash and S. Prakash, "The Stalling of Transcription at Abasic Sites Is Highly Mutagenic," *Molecular and Cellular Biology*, Vol. 23, No. 1, 2003, pp. 382-388.

[8] P. W. Doetsch and R. P. Cunningham, "The Enzymology of Apurinic/Apyrimidinic Endonucleases," *Mutation Research*, Vol. 236, No. 2-3, 1990, pp. 173-201.

[9] G. Barzilay and I. D. Hickson, "Structure and Function of Apurinic/Apyrimidinic Endonucleases," *Bioessays*, Vol. 17, No. 8, 1995, pp. 713-719.

[10] J. P. Erzberger, D. Barsky, O. D. Schärer, M. E. Colvin and D. M. Wilson III, "Elements in Abasic Site Recognition by the Major Human and *Escherichia coli* Apurinic/Apyrimidinic Endonucleases," *Nucleic Acids Research*, Vol. 26, No. 11, 1998, pp. 2771-2778.

[11] D. Lu, J. Silhan, J. T. MacDonald, E. P. Carpenter, K. Jensen, C. M. Tang, G. S. Baldwin and P. S. Freemont, "Structural Basis for the Recognition and Cleavage of Abasic DNA in *Neisseria meningitidis*," *Proceedings of the National Academy of Sciences of the United States of America*, Vol. 109, No. 42, 2012, pp. 16852-16857.

[12] C. D. Mol, C.-F. Kuo, M. M. Thayer, R. P. Cunningham

and J. A. Tainer, "Structure and Function of the Multifunctional DNA-Repair Enzyme Exonuclease III," *Nature*, Vol. 374, No. 6520, 1995, pp. 381-386.

[13] M. A. Gorman, S. Morera, D. G. Rothwell, E. de La Fortelle, C. D. Mol, J. A. Tainer, I. D. Hickson and P. S. Freemont, "The Crystal Structure of the Human DNA Repair Endonuclease HAP1 Suggests the Recognition of Extra-Helical Deoxyribose at DNA Abasic Sites," *EMBO Journal*, Vol. 16, No. 21, 1997, pp. 6548-6558.

[14] Y. Kawarabayasi, Y. Hino, H. Horikawa, K. Jin-no, M. Takahashi, M. Sekine, S. Baba, A. Ankai, H. Kosugi, A. Hosoyama, S. Fukui, Y. Nagai, K. Nishijima, R. Otsuka, H. Nakazawa, M. Takamiya, Y. Kato, T. Yoshizawa, T. Tanaka, Y. Kudoh, J. Yamazaki, N. Kushida, A. Oguchi, K. Aoki, S. Masuda, M. Yanagii, M. Nishimura, A. Yamagishi, T. Oshima and H. Kikuchi, "Complete Genome Sequence of an Aerobic Thermoacidophilic Crenarchaeon, *Sulfolobus tokodaii* Strain 7," *DNA Research*, Vol. 8, No. 4, 2001, pp. 123-140.

[15] Z. Otwinowski and W. Minor, "Processing of X-Ray Diffraction Data Collected in Oscillation Mode," *Methods in Enzymology*, Vol. 276, 1997, pp. 307-326.

[16] B. W. Matthews, "Solvent Content of Protein Crystals," *Journal of Molecular Biology*, Vol. 33, No. 2, 1968, pp. 491-497.

[17] A. Vagin, and A. Teplyakov, "*MOLREP*: An Automated Program for Molecular Replacement," *Journal of Applied Crystallography*, Vol. 30, Part 10, 1997, pp. 1022-1025.

[18] Collaborative Computational Project, Number 4, "The *CCP4* Suite: Programs for Protein Crystallography," *Acta Crystallographica Section D*, Vol. 50, Part 5, 1994, pp. 760-763.

Synthesis, Characterization, Crystal Studies of (E)-3-(3-(4-Fluorophenyl)-1-isopropyl-1H-indol-2-yl) Acrylaldehyde

Veerendra Kumar A. Kalalbandi, J. Seetharamappa*

Department of Chemistry, Karnatak University, Dharwad, India

ABSTRACT

We have synthesized and developed single crystals of the title compound (E)-3-(3-(4-fluorophenyl)-1-isopropyl-1H-indol-2-yl) acrylaldehyde, which is a key intermediate of anti-cholesterol fluvastatin drug. It crystallized under orthorhombic system with a space group $Pna2_1$. The dihedral angle between the indole mean plane and 4-F-phenyl ring was observed to be 111.5 (3)° in the title molecule. Further, strong hydrogen bonds were not found in the crystal structure.

Keywords: Single Crystal Structure; X-Ray Diffraction; Intermediate of Fluvastatin; Indoles

1. Introduction

Indole and its derivatives have been a topic of research interest and continue to be one of the dynamic areas of heterocyclic chemistry, particularly due to their natural occurrence and pharmacological activities [1]. A large number of indole derivatives are at the fore as pharmacologically active lead compounds for drug development [2]. Many drugs contain indole moiety, either as a basic prototype or as an attached group to invoke particular properties [3]. The incorporation of indole nucleus, a biologically accepted pharmacophore in medicinal compounds, has made it versatile heterocyclic possessing wide spectrum of biological activities [4,5] like anticancer, antimicrobial, anti-HIV, antitubercular, antiviral, antidepressant, cardiovascular activity, antihypertensive, etc., [6-9]. Furthermore, fluvastatin, which is a synthetic member of the statin class, contains indole moiety in its molecular makeup and the title molecule is its key intermediate.

In the course of synthesizing some fluvastatin derivatives, we have developed single crystals of key intermediate, (E)-3-(3-(4-fluorophenyl)-1-isopropyl-1H-indol-2-yl) acrylaldehyde and its molecular structure were determined. In the present paper, we report the synthesis, characterization by FTIR, ^1H NMR and ^{13}C NMR and its crystal properties.

*Corresponding author.

2. Experimental

2.1. Synthesis of the Title Compound

The title compound (E)-3-(3-(4-fluorophenyl)-1-isopropyl-1H-indol-2-yl) acrylaldehyde was synthesized by following the reported method [10,11]. Friedel-Crafts condensation of fluorobenzene with chloroacetyl chloride in the presence of $AlCl_3$ gave 4-Fluoro phenacyl chloride. Phenacyl chloride was then condensed with N-isopropylaniline in the presence of DMF to yield 1-(4-fluorophenyl)-2-(isopropyl(phenyl)amino) ethanone. Further, this was cyclized by means of $ZnCl_2$ to yield 3-(4- fluorophenyl)-1-isopropyl-1H-indole. The condensation of 3-(4-fluorophenyl)-1-isopropyl-1H-indole with 3-(N-methyl-N-phenylamino) acrolein in the presence of $POCl_3$ in acetonitrile yielded the title compound. The synthetic route for the synthesis of title compound is portrayed in **Scheme 1**. The title compound was characterized by FT-IR, ^1H NMR and ^{13}C NMR spectra. Further, pale yellow colored crystals suitable for X-ray diffraction analysis were grown by slow evaporation of a dilute solution of title compound in chloroform (m. p.136°C - 138°C) at room temperature.

2.2. Physical and Spectral Measurements

Melting point was determined in open capillary tubes on

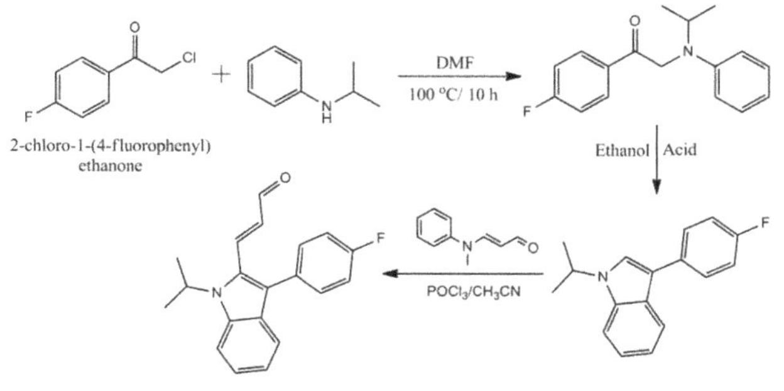

Scheme 1. Synthetic route for the synthesis of title compound.

a melting point apparatus of Concord Instruments (P) Ltd., Bangalore and is uncorrected. The IR absorption spectrum was recorded on a Shimadzu FT-IR-8400S Spectrophotometer using KBr pellets and is reported as wave numbers (ν cm^{-1}). The ^1H NMR spectrum was determined on a Bruker AV400 II at 400 MHz. The ^{13}C NMR spectrum was recorded on Bruker DSX-300(S) AV-III 400(L) model at 100 MHz. All materials were purchased from commercial companies and are used directly. The solvents were dried by refluxing with appropriate drying agents and distilled before use.

2.3. Single-Crystal X-Ray Crystallography

Pale yellow colored crystals of the size of 0.30 × 0.20 × 0.20 mm was selected for data collection under a polarizing microscope and it was mounted on glass fiber for X-ray diffraction data collection. The high resolution X-ray diffraction data sets were collected on a Bruker SMART APEX2 CCD Diffractometer using Mo K_α radiation (λ = 0.71073 Å) at ambient temperature. The crystal-to-detector distance was fixed at 40 mm. The diffraction data have been scaled for absorption effect by the multi-scanning method. The total exposure time was 0.98 h. The frames were integrated with the Bruker SAINT Software package using a narrow-frame algorithm. The structure was solved and refined using the Bruker SHELXTL [12] Software Package, using the space group $Pna2_1$, with Z = 4 for the formula unit, $C_{20}H_{18}FNO$. The ORTEP, packing and planes diagrams are generated using the Mercury 3.1.

3. Results and Discussion

3.1. Crystal Structure

The integration of the crystal data of the title compound (E)-3-(3-(4-fluorophenyl)-1-isopropyl-1H-indol-2-yl) acrylaldehyde using an orthorhombic unit cell yielded a total of 5558 reflections to a maximum θ angle of 23.92° (0.88 Å resolution), of which 2408 were independent (average redundancy 2.308, completeness = 99.5%, R_{int} = 2.25%, R_{sig} = 2.93%) and 2108 (87.54%) were greater than $2\sigma(F^2)$. The final cell constants of a = 12.4637(4) Å, b = 9.9386(3) Å, c = 13.0272(3) Å, volume = 1613.70(8) $Å^3$, are based upon the refinement of the XYZ-centroids of 1826 reflections above 20 σ(I) with 6.10° < 2θ < 47.84°. Data were corrected for absorption effects using the multi-scan method (SADABS). The ratio of minimum to maximum apparent transmission was 0.978. The crystal data and structure refinement parameters are given in **Table 1**.

The final anisotropic full-matrix least-squares refinement on F^2 with 210 variables converged at R1 = 4.26%, for the observed data and wR2 = 8.48% for all data. The goodness-of-fit was 1.028. The largest peak in the final

Table 1. Crystal data and structure refinement.

Empirical formula	$C_{20}H_{18}FNO$
Formula weight	307.35
Temperature	296 K
Crystal size	0.30 × 0.20 × 0.20 mm
Crystal color	Pale yellow
Crystal system	Orthorhombic
Space group	$Pna2_1$
a	12.4637 (4) (Å)
b	9.9386 (3) (Å)
c	13.0272 (3) (Å)
α, β and γ	90°, 90° and 90°
Limiting indices	$-10 \le h \le 14$, $-11 \le k \le 11$, $-14 \le l \le 14$
Volume	1613.70 (8) ($Å^3$)
Z, Calculated density	4, 1.265 mg/m^3
Reflections collected/unique	5558/2408 [R(int) = 0.0225]
θ range for data collection and completeness	θ_{max} = 23.92°, θ_{min} = 3.05° and 99.5%
Maximum and minimum transmission	0.972 and 0.951
Absorption correction	Semi-empirical from equivalents
Data/Restraints/Paramaeters	2408/2/211
Final R indices [I > 2σ(I)]	R1 = 0.0347, wR2 = 0.0804
R indices (all data)	R1 = 0.0426, wR2 = 0.0848
Absolute structure parameter	−2.3 (12)
Extinction coefficient	0.0071 (14)
Largest diff. peak and hole	0.151 and −0.105 e. $Å^{-3}$
Measurements	Bruker SMART APEX2 CCD Diffractometer
Refinement	Full-matrix leastsquares on F^2
Goodness-of-fit on F^2	1.028

difference electron density synthesis was 0.151 e. $Å^{-3}$ and the largest hole was −0.105 e. $Å^{-3}$ with an RMS deviation of 0.026 e. $Å^{-3}$. On the basis of the final model, the calculated density was 1.265 mg/m^3 and F (000), 648 e^-.

3.2. Molecular Skeleton

The structure of the compound consisted of butenal and fluorobenzene fragments that connected to N-isopropyl-indole moiety. The ORTEP diagram of the title compound is given in **Figure 1**. The 4-F-phenyl ring C4-C3 form dihedral angle with the mean plane of the nine membered indole ring system was 111.5 (3)° (**Figure 2**). The sum of the bond angles around N1 [359.9 (3)°] indicated sp^2 hybridization [13]. The packing diagram (**Figure 3**) of the title molecule showed 4 molecules pack in a unit cell and no strong hydrogen bonds in the crystal structure were observed. However, weak inter molecular hydrogen bonds observed between C11H11···O1 by a distance of 2.692 Å are shown in **Figure 4**. All bond lengths and bond angles are in normal ranges and are given in **Table 2**. Atomic coordinates and equivalent isotropic displacement parameters are shown in **Table 3**. Anisotropic dis-placement parameters are shown in **Table 4**. Hydrogen coordinates and isotropic displacement parameters are shown in **Table 5**. Torsion angles are shown in **Table 6**.

3.3. Spectral Studies

3.3.1. FTIR Spectrum

The FT-IR spectrum of the title compound shows the strong characteristic absorption band at 1666.38 cm^{-1} due to aldehyde carbonyl stretching frequency. The shift in the band to lower wavenumber was due to the presence of α, β unsaturation. The band at 1610.45 cm^{-1} was observed due to C=C stretching.

3.3.2. ^{1}H NMR Spectrum

The ^{1}H NMR spectrum of the title compound showed two singlets at δ 1.64 and 1.66 ppm due to two methyl protons. The doublet at δ 9.61 ppm was attributed to aldehyde proton; it coupled with adjacent C8-CH protons with coupling constant of 7.6 Hz and the doublet at δ 7.95 ppm was accounted for vinylic proton with J value of 16 Hz. Further, C7 proton splits C8 proton signal into doublet (J = 16 Hz) which is split by C9 aldehyde proton into new doublets (J = 7.6 Hz) producing four-line spectrum (doublet of doublet) at δ 6.115 ppm. The signals appeared around δ 7.70 - 7.78 ppm were attributed to aromatic protons. ^{1}H NMR (400 MHz, DMSO-d6) δ ppm: 1.649 (s, 3H, C19-CH_3), 1.666 (s, 3H, C20-CH_3), 5.107 (m, 1H, C18-CH), 6.115 (dd, 1H, J = 16, 7.6 Hz, C8-CH), 7.959 (d, 1H, J = 16Hz, C7-CH), 9.619 (d, 1H, J = 7.6Hz, C9-CHO), 7.709 - 7.782 (m, 8H, Ar-H,).

3.3.3. ^{13}C NMR Spectrum

The ^{13}C NMR of the title compound exhibited a signal at δ 194.02 ppm for carbonyl carbon of the aldehyde group. The low intensity signal at δ 162.62 ppm was assigned to C1 carbon attached fluorine atom and the methyl carbons were observed at δ 21.37 ppm. The signal at δ 47.39 ppm was attributed to CH carbon of the isopropyl group. The peaks appeared between δ 112.60 - 160.20 ppm were assigned to aromatic and vinylic carbon atoms. ^{13}C NMR ($CDCl_3$, 100 MHz, δ ppm): 21.37 CH_3, 47.39 C18-CH, 162.62 C1-CF, 194.02 C9-CHO, 112, 115, 119, 120, 124, 127, 130, 132, 136, 141, 160 Ar-C.

4. Conclusion

The title compound, a key starting material for the syn-

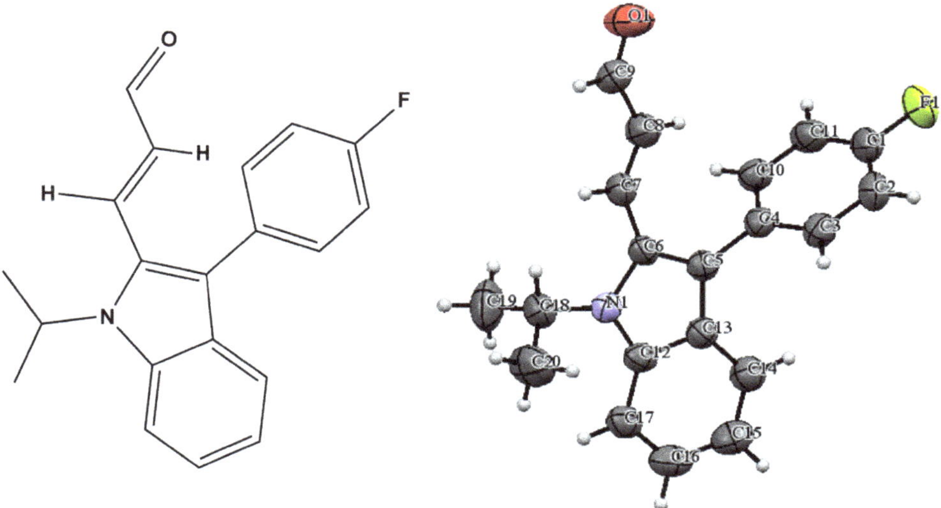

Figure 1. ORTEP diagram of the title molecule.

Table 2. Bond lengths [Å] and angles [°].

Bond Lengths		Bond Angles	
F(1)-C(1)	1.363(3)	C(12)-N(1)-C(6)	108.34(18)
O(1)-C(9)	1.196(3)	C(12)-N(1)-C(18)	126.95(19)
N(1)-C(12)	1.381(3)	C(6)-N(1)-C(18)	124.66(19)
N(1)-C(6)	1.401(3)	C(11)-C(1)-C(2)	123.2(2)
N(1)-C(18)	1.477(3)	C(11)-C(1)-F(1)	118.8(3)
C(1)-C(11)	1.350(4)	C(2)-C(1)-F(1)	118.0(3)
C(1)-C(2)	1.363(4)	C(1)-C(2)-C(3)	117.7(3)
C(2)-C(3)	1.381(3)	C(1)-C(2)-H(2)	121.1
C(2)-H(2)	0.9300	C(3)-C(2)-H(2)	121.1
C(3)-C(4)	1.388(3)	C(2)-C(3)-C(4)	121.9(2)
C(3)-H(3)	0.9300	C(2)-C(3)-H(3)	119.1
C(4)-C(10)	1.389(3)	C(4)-C(3)-H(3)	119.1
C(4)-C(5)	1.475(3)	C(3)-C(4)-C(10)	117.7(2)
C(5)-C(6)	1.382(3)	C(3)-C(4)-C(5)	120.49(19)
C(5)-C(13)	1.414(3)	C(10)-C(4)-C(5)	121.7(2)
C(6)-C(7)	1.447(3)	C(6)-C(5)-C(13)	107.7(2)
C(7)-C(8)	1.295(3)	C(6)-C(5)-C(4)	128.8(2)
C(7)-H(7)	0.9300	C(13)-C(5)-C(4)	123.4(2)
C(8)-C(9)	1.429(3)	C(5)-C(6)-N(1)	108.57(19)
C(8)-H(8)	0.9300	C(5)-C(6)-C(7)	129.9(2)
C(9)-H(9)	0.9300	N(1)-C(6)-C(7)	121.5(2)
C(10)-C(11)	1.390(3)	C(8)-C(7)-C(6)	131.8(2)
C(10)-H(10)	0.9300	C(8)-C(7)-H(7)	114.1
C(11)-H(11)	0.9300	C(6)-C(7)-H(7)	114.1
C(12)-C(17)	1.396(3)	C(7)-C(8)-C(9)	122.8(2)
C(12)-C(13)	1.412(3)	C(7)-C(8)-H(8)	118.6
C(13)-C(14)	1.398(3)	C(9)-C(8)-H(8)	118.6
C(14)-C(15)	1.371(4)	O(1)-C(9)-C(8)	128.4(3)
C(14)-H(14)	0.9300	O(1)-C(9)-H(9)	115.8
C(15)-C(16)	1.399(4)	C(8)-C(9)-H(9)	115.8
C(15)-H(15)	0.9300	C(4)-C(10)-C(11)	120.8(2)
C(16)-C(17)	1.370(4)	C(4)-C(10)-H(10)	119.6
C(16)-H(16)	0.9300	C(11)-C(10)-H(10)	119.6
C(17)-H(17)	0.9300	C(1)-C(11)-C(10)	118.7(2)
C(18)-C(20)	1.485(4)	C(1)-C(11)-H(11)	120.7
C(18)-C(19)	1.502(4)	C(10)-C(11)-H(11)	120.7
C(18)-H(18)	0.9800	N(1)-C(12)-C(17)	131.6(2)
C(19)-H(19A)	0.9600	N(1)-C(12)-C(13)	107.92(19)
C(19)-H(19B)	0.9600	C(17)-C(12)-C(13)	120.5(2)
C(19)-H(19C)	0.9600	C(14)-C(13)-C(12)	119.9(2)
C(20)-H(20A)	0.9600	C(14)-C(13)-C(5)	132.7(2)
C(20)-H(20B)	0.9600	C(12)-C(13)-C(5)	107.4(2)
C(20)-H(20C)	0.9600	C(15)-C(14)-C(13)	119.1(2)

Continued

Bond Angles			
		C(15)-C(14)-H(14)	120.4
C(13)-C(14)-H(14)	120.4	C(14)-C(15)-C(16)	120.3(3)
C(14)-C(15)-H(15)	119.9	C(16)-C(15)-H(15)	119.9
C(17)-C(16)-C(15)	122.1(2)	C(17)-C(16)-H(16)	118.9
C(15)-C(16)-H(16)	118.9	C(16)-C(17)-C(12)	118.0(2)
C(16)-C(17)-H(17)	121.0	C(12)-C(17)-H(17)	121.0
N(1)-C(18)-C(20)	113.0(2)	N(1)-C(18)-C(19)	110.1(2)
C(20)-C(18)-C(19)	115.4(2)	N(1)-C(18)-H(18)	105.8
C(20)-C(18)-H(18)	105.8	C(19)-C(18)-H(18)	105.8
C(18)-C(19)-H(19A)	109.5	C(18)-C(19)-H(19B)	109.5
H(19A)-C(19)-H(19B)	109.5	C(18)-C(19)-H(19C)	109.5
H(19A)-C(19)-H(19C)	109.5	H(19B)-C(19)-H(19C)	109.5
C(18)-C(20)-H(20A)	109.5	C(18)-C(20)-H(20B)	109.5
H(20A)-C(20)-H(20B)	109.5	C(18)-C(20)-H(20C)	109.5
H(20A)-C(20)-H(20C)	109.5	H(20B)-C(20)-H(20C)	109.5

Table 3. Atomic coordinates ($\times 10^4$) and equivalent isotropic displacement parameters ($Å^2 \times 10^3$). U(eq) is defined as one third of the trace of the orthogonalized U^{ij} tensor.

	X	Y	Z	U(eq)
F(1)	6616(2)	1344(2)	10599(2)	100(1)
O(1)	8210(2)	7666(2)	11792(2)	93(1)
N(1)	4331(2)	8390(2)	9129(1)	48(1)
C(1)	6155(2)	2534(3)	10322(2)	63(1)
C(2)	6182(2)	2889(3)	9311(2)	63(1)
C(3)	5712(2)	4094(2)	9035(2)	53(1)
C(4)	5236(2)	4936(2)	9754(2)	45(1)
C(5)	4724(2)	6205(2)	9430(2)	45(1)
C(6)	5045(2)	7513(2)	9621(2)	45(1)
C(7)	5979(2)	8008(2)	10162(2)	51(1)
C(8)	6651(2)	7421(2)	10773(2)	61(1)
C(9)	7554(2)	8099(3)	11210(2)	58(1)
C(10)	5217(2)	4509(2)	10769(2)	54(1)
C(11)	5684(2)	3295(3)	11053(2)	63(1)
C(12)	3543(2)	7631(2)	8660(2)	48(1)
C(13)	3789(2)	6260(2)	8817(2)	46(1)
C(14)	3131(2)	5264(3)	8396(2)	60(1)
C(15)	2250(2)	5638(3)	7836(2)	68(1)
C(16)	2005(2)	7001(3)	7703(2)	67(1)
C(17)	2628(2)	8003(3)	8111(2)	57(1)
C(18)	4405(2)	9873(2)	9159(2)	60(1)
C(19)	3702(3)	10420(3)	9997(3)	98(1)
C(20)	4263(3)	10505(3)	8135(3)	94(1)

Table 4. Anisotropic displacement parameters (Å × 10^3). The anisotropic displacement factor exponent takes the form: $-2\pi^2\left[h^2a*^2 U_{11} + \cdots + 2hka*b*U_{12}\right]$.

	U11	U22	U33	U23	U13	U12
F(1)	137(2)	45(1)	118(2)	6(1)	−40(1)	20(1)
O(1)	91(2)	97(2)	91(2)	14(1)	−42(1)	−12(1)
N(1)	58(1)	44(1)	44(1)	−3(1)	−6(1)	6(1)
C(1)	75(2)	38(2)	75(2)	2(1)	−20(2)	1(1)
C(2)	74(2)	47(2)	69(2)	−9(1)	−4(1)	6(1)
C(3)	60(1)	50(1)	48(1)	−2(1)	2(1)	−3(1)
C(4)	49(1)	42(1)	43(1)	−1(1)	−2(1)	−3(1)
C(5)	50(1)	47(1)	37(1)	2(1)	2(1)	1(1)
C(6)	52(1)	46(1)	36(1)	1(1)	1(1)	8(1)
C(7)	58(1)	43(1)	51(1)	−1(1)	−3(1)	2(1)
C(8)	71(2)	49(2)	63(2)	7(1)	−14(2)	−3(1)
C(9)	62(2)	62(2)	51(2)	−4(1)	−4(1)	−3(1)
C(10)	64(2)	53(2)	44(1)	0(1)	−3(1)	−1(1)
C(11)	83(2)	53(2)	54(1)	13(1)	−16(1)	−8(1)
C(12)	49(1)	59(2)	37(1)	0(1)	1(1)	6(1)
C(13)	50(1)	51(2)	39(1)	−1(1)	2(1)	2(1)
C(14)	63(2)	58(2)	59(2)	2(1)	−2(1)	−4(1)
C(15)	62(2)	80(2)	63(2)	−3(2)	−10(1)	−11(1)
C(16)	53(2)	91(2)	57(2)	4(2)	−9(1)	6(1)
C(17)	58(2)	66(2)	48(1)	0(1)	−1(1)	10(1)
C(18)	67(2)	48(2)	66(2)	−2(1)	−10(1)	9(1)
C(19)	103(2)	74(2)	118(3)	−35(2)	14(2)	16(2)
C(20)	113(3)	66(2)	101(3)	30(2)	−23(2)	−9(2)

Table 5. Hydrogen coordinates ($\times 10^4$) and isotropic displacement parameters ($Å^2 \times 10^3$).

	X	Y	Z	U(eq)
H(2)	6505	2338	8825	76
H(3)	5714	4349	8348	63
H(7)	6123	8914	10049	76
H(8)	6542	6520	10935	73
H(9)	7642	8993	11016	70
H(10)	4888	5041	11264	64
H(11)	5673	3012	11734	76
H(14)	3290	4358	8494	72
H(15)	1812	4984	7543	82
H(16)	1399	7234	7326	81
H(17)	2447	8904	8025	69
H(18)	5143	10077	9367	72
H(19A)	3888	10002	10637	147
H(19B)	2964	10232	9839	147
H(19C)	3803	11375	10050	147
H(20A)	3512	10544	7972	140
H(20B)	4630	9979	7626	140
H(20C)	4553	11399	8146	140

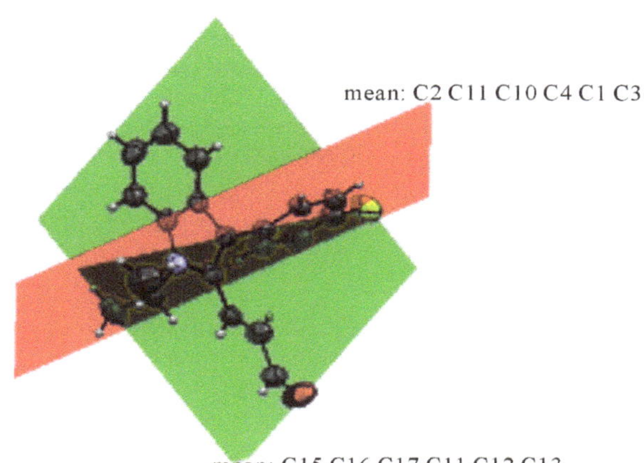

Figure 2. Dihedral angle formed between indole plane and benzene plane.

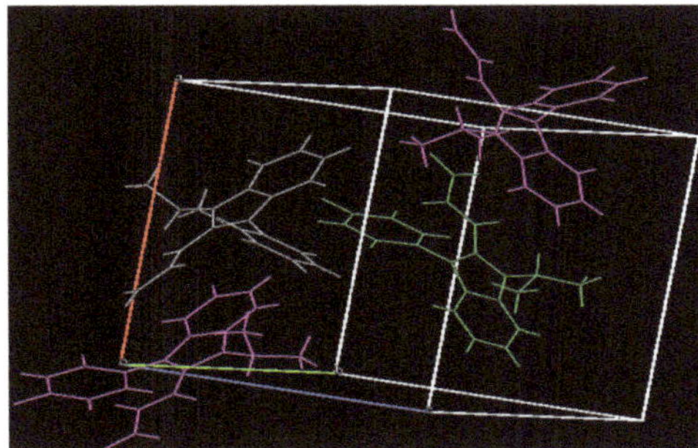

Figure 3. Packing diagram of title compound, ab Plane, colored by symmetry operation.

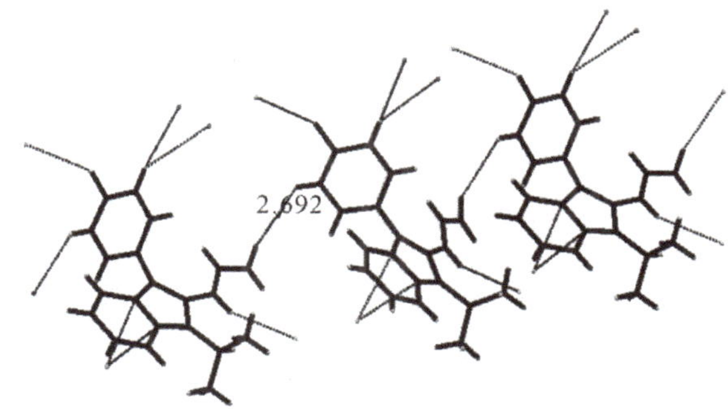

Figure 4. Diagram showing inter molecular weak Hydrogen bonding.

Table 6. Torsion angles [°].

C(11)-C(1)-C(2)-C(3)	0.8(4)
F(1)-C(1)-C(2)-C(3)	179.9(2)
C(1)-C(2)-C(3)-C(4)	0.7(4)
C(2)-C(3)-C(4)-C(10)	−1.9(4)
C(2)-C(3)-C(4)-C(5)	−178.9(2)
C(3)-C(4)-C(5)-C(6)	−111.5(3)
C(10)-C(4)-C(5)-C(6)	71.6(3)
C(3)-C(4)-C(5)-C(13)	66.7(3)
C(10)-C(4)-C(5)-C(13)	−110.2(3)
C(13)-C(5)-C(6)-N(1)	−0.6(2)
C(4)-C(5)-C(6)-N(1)	177.8(2)
C(13)-C(5)-C(6)-C(7)	−176.5(2)
C(4)-C(5)-C(6)-C(7)	1.9(4)
C(12)-N(1)-C(6)-C(5)	2.1(2)
C(18)-N(1)-C(6)-C(5)	179.6(2)
C(12)-N(1)-C(6)-C(7)	178.39(19)
C(18)-N(1)-C(6)-C(7)	−4.1(3)
C(5)-C(6)-C(7)-C(8)	−15.2(4)
N(1)-C(6)-C(7)-C(8)	169.3(3)
C(6)-C(7)-C(8)-C(9)	178.1(2)
C(7)-C(8)-C(9)-O(1)	178.6(3)
C(3)-C(4)-C(10)-C(11)	1.6(4)
C(5)-C(4)-C(10)-C(11)	178.6(2)
C(2)-C(1)-C(11)-C(10)	−1.0(4)
F(1)-C(1)-C(11)-C(10)	179.9(2)
C(4)-C(10)-C(11)-C(1)	−0.2(4)
C(6)-N(1)-C(12)-C(17)	177.1(2)
C(18)-N(1)-C(12)-C(17)	−0.4(4)
C(6)-N(1)-C(12)-C(13)	−2.7(2)
C(18)-N(1)-C(12)-C(13)	179.8(2)
N(1)-C(12)-C(13)-C(14)	−178.29(19)
C(17)-C(12)-C(13)-C(14)	1.9(3)
N(1)-C(12)-C(13)-C(5)	2.4(2)

Continued

C(17)-C(12)-C(13)-C(5)	−177.5(2)
C(6)-C(5)-C(13)-C(14)	179.7(2)
C(4)-C(5)-C(13)-C(14)	1.2(4)
C(6)-C(5)-C(13)-C(12)	−1.1(2)
C(4)-C(5)-C(13)-C(12)	−179.62(19)
C(12)-C(13)-C(14)-C(15)	−0.2(3)
C(5)-C(13)-C(14)-C(15)	178.9(2)
C(13)-C(14)-C(15)-C(16)	−1.0(4)
C(14)-C(15)-C(16)-C(17)	0.6(4)
C(15)-C(16)-C(17)-C(12)	1.1(4)
N(1)-C(12)-C(17)-C(16)	178.0(2)
C(13)-C(12)-C(17)-C(16)	−2.3(3)
C(12)-N(1)-C(18)-C(20)	−48.7(3)
C(6)-N(1)-C(18)-C(20)	134.3(2)
C(12)-N(1)-C(18)-C(19)	82.0(3)
C(6)-N(1)-C(18)-C(19)	−95.0(3)

thesis of fluvastatin was synthesized, its spectral data were discussed and single crystals were grown. The crystal parameters were evaluated. Bond lengths and bond angles are in normal ranges. No classical hydrogen bonding was observed. The 4-F-phenyl ring C4-C3 form dihedral angle with the mean plane of the nine membered indole ring system was 111.5 (3)°.

5. Acknowledgements

The authors acknowledge the University Scientific Instrumentation Centre (USIC), Karnatak University, Dharwad for SC-XRD data and NMR Research Centre, Indian Institute of Science (IISc), Bengaluru, India for carrying out the spectral analyses.

REFERENCES

[1] R. J. Sundberg, "Indoles," Academic Press, New York, 1996.

[2] F. R. de sa alves, E. Z. Barreiro and C. A. M. Fraga, "From Nature to Drug Discovery: The Indole Scaffold as a Privileged Structure," *Mini-Reviews in Medicinal Chemistry*, Vol. 9, No. 7, 2009, pp. 782-793.

[3] N. K. Kaushik, N. Kaushik, P. Attri, N. Kumar, C. H. Kim, A. K. Verma and E. H. Choi, "Biomedical Importance of Indoles," *Molecules*, Vol. 18, No. 6, 2013, pp. 6620-6662.

[4] D. A. Horton, G. T. Bourne and M. L. Smythe, "The Combinatorial Synthesis of Bicyclic Privileged Structures or Privileged Structures," *Chemical Reviews*, Vol. 103, No. 3, 2003, pp. 893-930.

[5] D. M. Wang, M. N. Sun and G. Liu, "Substituent Diversity-Directed Synthesis of Indole Derivatives," *Journal of Combonatorial Chemistry*, Vol. 11, No. 4, pp. 556-575.

[6] V. Sharma, P. Kumar and D. Pathak, "Biological Importance of the Indole Nucleus in Recent Years: A Comprehensive Review," *Journal of Heterocyclic Chemistry*, Vol. 47, No. 3, 2010, pp. 491-502.

[7] S. Rossiter, L. K. Folkes and P. Wardman, "Halogenated Indole-3-Acetic Acid as Oxidatively Activated Prodrugs with Potential for Targeted Cancer Therapy," *Bioorganic and Medicinal Chemistry Letters*, Vol. 12, No. 2002, pp. 2523-2526.

[8] E. Abele, R. Abele, O. Dzenitis and E. Lukevics, "Indole and Isatin Oximes: Synthesis, Reactions and Biological Activity," *Chemistry of Heterocyclic Compounds*, Vol. 39, No. 1, 2003, pp. 3-35.

[9] H. Panwar, R. S. Verma, V. K. Srivastava and A. Kumar, "Synthesis of Some Substituted Azetidinonyl and Thiazolidinonyl-1,3,4-thiadiazino[6,5-b]indoles as Prospective Antimicrobial Agents," *Indian Journal of Chemistry*, Vol. 45B, No. 9, 2006, pp. 2099-2104.

[10] R. E. Walkup and J. Linder, "2-Formylation of 3-Arylindoles," *Tetrahedron Letters*, Vol. 26, No. 18, 1985, pp. 2155-2158.

[11] G. T. Lee, J. C. Amedio Jr., R. Underwood, K. Prasad and O. Repic, "Vinylformylation Utilizing Propeniminium Salts," *Journal of Organic Chemistry*, Vol. 57, No. 11, 1992, pp. 3250-3252.

[12] G. M. Sheldrick, "A Short History of SHELX," *Acta Crystallographica Section A*, Vol. A64, 2008, pp. 112-122.

[13] R. Senthamizhselvi, G. Bhaskar, P. R. Sheshadri, P. T. Perumal and K. Illangovan, "3-(1,2-Di-p-tolylvinyl)-2-methyl-1H-indole," *Acta Crystallographica Section E*, Vol. E68, 2012, p. o1815.

Synthesis, Spectroscopic Investigations, Quantum Chemical Studies (*Ab-initio* & DFT) and Antimicrobial Activities of 3-(3-Chloro-4,5-dimethoxy-phenyl)-1-(4, 5-dimethoxy-2-methyl-Phenyl) prop-2-en-1-one

Urmila H. Patel[1], Sahaj A. Gandhi[1*], Vijay M. Barot[2], Mitesh C. Patel[2]
[1]Department of Physics, Sardar Patel University, Vallabh Vidyanagar, India
[2]P. G. Center in Chemistry, Smt. S. M. Panchal Science College, Talod, India

ABSTRACT

The chalcones (1,3-diaryl-2-propenones) and their derivatives are important intermediates in organic synthesis and have widespread applications in medicinal industry. The title choloro chalcone derivative, 3-(3-chloro-4,5-dimethoxy-phenyl)-1-(4,5-dimethoxy-2-methyl phenyl) prop-2-en-1-one, has been synthesized. It is characterized by FTIR, ^{1}H NMR, ^{13}C NMR and single crystal X-ray diffraction. Title compound crystallizes in monoclinic space group C2/c with a = 23.540(11) Å, b = 9.738(4) Å, c = 17.305(7) Å, β = 106.37 (3)°, V = 3806(3) Å^3 and Z = 8. The mean plane of the two substituted benzene rings is twisted by 66.29 (12)° with respect to each other. *Ab-initio* and density functional Theory (DFT) calculations have been carried out for the title molecule using RHF/6-311G and B3LYP/6-311G basis set respectively. The calculated results show that the predicted geometry can well reproduce structural parameters. In addition, frontier molecular orbitals and Mullikan charge distributions are carried out by using RHF and B3LYP methods. The calculated HOMO and LUMO energies show that charge transfer occurs in the molecule. Numbers of weak but significant interactions like C-H···O, C-H···π and π-π are involved in the stability of the structure. The weak π-π stacked interaction involves the centroids of the methyl phenyl rings with Cg-Cg separation distance of 3.857(2) Å. Synthesized compound has been screened for its antimicrobial activity against different panels of organisms.

Keywords: Chalcones; FTIR; NMR; Single Crystal X-Ray Diffraction; Hydrogen Bond Interactions; Quantum Chemical Calculations; Antimicrobial Activities

1. Introduction

In chalcones, two aromatic rings are linked by an aliphatic three-carbon chain. The synthesis of chalcone derivative has generated vast interest to organics as well as for medicinal chemists. Chalcones are also key precursors in the synthesis of many biologically important heterocycles such as benzothiazepine, pyrazolines, 1,4-diketones and flavones. Chalcones, belonging to flavonoid family, synthesized or the natural one, displayed many interesting properties including antimalarial [1,2], anticancer [3,4], antiviral [5], antibacterial [6], antifungal [7], antihyperglycemic [8] and photocytotoxicity [9] activities. In the chemical structure, three-carbon $\alpha-\beta$ unsaturated carbonyl system, the back bone of the open chain flavonoids, joins two aromatic rings. The *ab-initio* and density functional theory (DFT) are applied for the investigation of the optimized molecular structure and few significant spectroscopy properties. As part of our ongoing research on X-ray Crystallographic investigations of drug molecules and theoretical quantum computational studies of synthesized drugs [10,11], we have synthesized and investigated molecular structure of a novel methoxy-chloro substituted chalcone derivative by IR, NMR and X-ray diffraction techniques and optimized the structure by quantum chemistry.

2. Experimental

2.1. Materials and Instrumentations

General chemicals are purchased from Merck, SD Fine

*Corresponding author.

and commercial source. All non-aqueous reactions are performed in dry glass ware. Thin layer chromatography (TLC) is performed on pre-coated plates, silica gel 60-F254 (Merck 1.16834, layer thickness 0.25 mm) using toluene/methanol mixtures (8:2) as developing system. The detection of the products on TLC is carried out in iodine vapor. Melting points are determined on a Polmon Instrument Model: M.P.: 96, Range: 25˚C to 350˚C, Resolution: 0.1˚C fitted with a microscope.

2.2. Synthesis

A mixture of 1-(2-methyl-4,5-dimethoxyphenyl) ethanone (0.01 mole) and 3-chloro-4,5-dimethoxy benzaldehyde (0.01 mole) in ethanol (30 ml) are added to a solution of potassium hydroxide (40 ml, 40%) with constant shaking of the reaction flask. The reaction mixture is stirred for a 24 hours on a magnetic stirrer and poured in to crushed ice and acidified with diluted HCl (2N). The solid mass which separated out is filtered, washed with water, dried and crystallized from methanol to give light yellow needles. The reaction scheme of the title molecule is shown in **Figure 1**.

2.3. FTIR and NMR Analysis

The FTIR spectrum of the compound is recorded in the KBr phase in the frequency region of 400 - 4000 cm^{-1} using Bruker FTIR spectrometer. ^{1}H-NMR and ^{13}C-NMR spectra are recorded with a Fourier transform instrument at 400 MHz (Bruker AVANCE 400). The details of FTIR and NMR analysis are given below.

IR (KBr): ν 1447 (C-H def (asym) alkyl), 1375 (C-H def (sym) alkyl), (C-H str. arom.), 1565 (C=C str. arom.), 1113 (C-H i.p.def arom.), 820 (C-H o.o.p.def. arom.), 1217 (C-O-C (sym) ether), 1077 (C-O-C (asym) ether), 1641 (C=O str., chalcone), 1565 (C=C, chalcone), 3273 (OH, phenol), 978 (CH=CH def. chalcone), 3087 (CH=CH str. chalcone), 1641 (C=C str. chalcone), 694 (C-Clstr.) cm^{-1}.

^{1}H NMR ($CDCl_3$) δ ppm: 2.44 (s, 3H), 3.89 (s, 3H, OCH_3), 3.91 (s, 6H, OCH_3), 3.93 (s, 3H, OCH_3), 6.754 (s, 1H), 6.99 (d, 1H, J = 1.8 Hz) ,7.05 (m, 1H + 1H chalcone), 7.22 (d, 1H, J = 1.6 Hz), 7.38 (d, 1H, J = 16 Hz, chalcone).

^{13}C NMR ($CDCl_3$) δ ppm: 56.01 (C-1), 56.04 (C-2), 55.09 (C-3), 55.04 (C-4), 15.38 (C-5), 194.56 (C-6), 122.75 (C-7), 141.03 (C-8),150.75 (C-9), 146.25 (C-10), 155.27 (C-11),144.89 (C-12), 114.05 (C-13), 131.39 (C-14), 112.21 (C-15), 110.51 (C-16), 130.68 (C-17), 130.20 (C-18), 127.08 (C-19), 126.31 (C-20). (mp: 142˚C - 148˚C; Yield: 80%. Analysis: $C_{20}H_{21}ClO_5$, Found: C: 63.74%, H: 5.61%, O: 21.24%, Cl: 9.41%, Calculated: C: 63.75%, H: 5.62%, O: 21.23%, Cl: 9.42%).

2.4. X-Ray Diffraction Study

A suitable sample of single crystal of size (0.6 × 0.4 × 0.1) mm^3 is selected for the crystallographic study. All diffraction measurements are performed at room temperature (296 K) using graphite monochromated MoKα radiation of wavelength 0.71073 Å. The crystal structure is solved by direct methods and refined by full-matrix least square technique on F^2, using SHELX-97 set of program [12]. All non-hydrogen atoms are refined anisotropically. The structure is refined to R = 0.0612 for the observed reflections 4383 and Goodness of fit S = 1.048 by using the intensity (I) values of 3132 reflections satisfying the I > 2σ (I) criterion and 4383 reflections in refinement for 235 crystallographic parameters. Highest and lowest electron density peaks "$\Delta\rho$" are 0.758 and −0.546 eÅ^{-3} respectively. The crystallographic data and details of the data collection and structure refinements are listed in **Table 1**.

2.5. Computational Details

The quantum chemical study of the title compound has been performed within the framework of Hartree Fock [13] and the density functional theory with Becke's three-parameter hybrid exchange functional with Lee-Yang-Parr correlation functional (B3LYP) employing 6-311G basis set [14,15]. All quantum chemical calculations are performed using computer software Gaussian-09 [16] and Gauss-View molecular visualization program [17].

3. Results and Discussion

3.1. X-Ray Crystallography

The ORTEP diagram of the title compound, 3-(3-chloro-4,

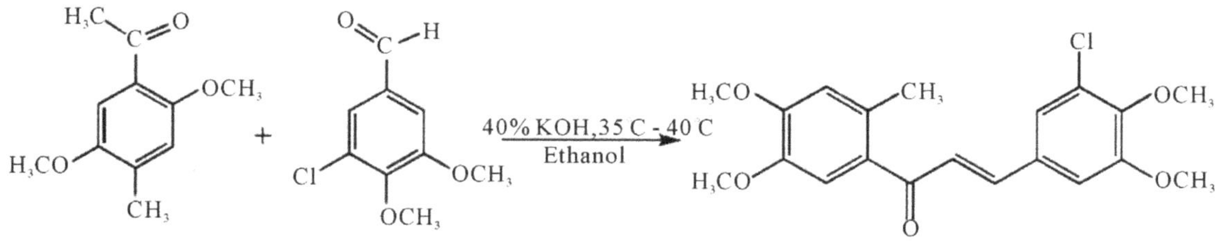

Figure 1. Chemical reaction scheme of the title compound.

Table 1. Crystal data and structure refinement parameters of the title compound.

Empirical formula	$C_{20}H_{21}ClO_5$
Formula weight	376.82
Temperature (K)	296(2)
Wavelength (Å)	0.71073
Crystal system	Monoclinic
Space group	C2/c
Crystal size (mm^3)	0.6 × 0.4 × 0.1
a (Å)	23.540(11)
b (Å)	9.738(4)
c (Å)	17.305(7)
β(°)	106.37(3)
Volume ($Å^3$)	3806(3)
Z	8
Calculated density (g/cm^3)	1.315
Absorption coefficient (mm^{-1})	0.228
F(000)	1584
θ range for data collection (°)	1.80 to 27.72
Limiting indices	$-30 \le h \le 30$ $-12 \le k \le 9$ $-22 \le l \le 22$
Reflections collected/unique [R(int)]	15634/4383 [0.0311]
Completeness to θ = 27.72 (%)	0.978
Absorption correction	N. A.
Refinement method	Full Matrix Least Square of $\|F\|^2$
Data/restrains/parameters	4383/0/235
Goodness-of-fit on F^2	1.048
Final R indices	$R_1 = 0.0612$, $wR_2 = 0.1670$
R indices (all data)	$R_1 = 0.0855$, $wR_2 = 0.1852$
Largest diff. peak and hole (e $Å^{-3}$)	0.758 and −0.546

5-dimethoxy-phenyl)-1-(4,5-dimethoxy-2-methyl phenyl) prop-2-en-1-one with thermal ellipsoids drawn at a 50% probability is shown in **Figure 2**.

Single crystal X-ray diffractions confirms the molecular structure of the title molecule, $C_{20}H_{21}ClO_5$, 2-methoxy, 1-methyl substituted phenyl ring is joined by a prop-2-en-1-one group to 3-chloro-4,5-dimethoxy substituted phenyl ring. Both the phenyl rings (C1-C6) and (C16-C21) are planar. The mean plane of two phenyl rings are twisted by 66.30(12)° with respect to each other. The

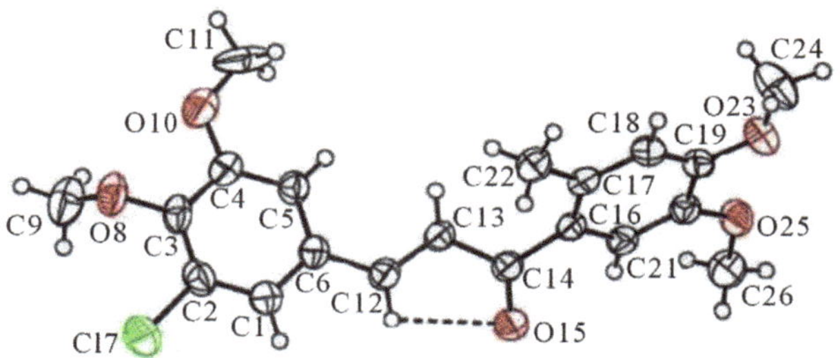

Figure 2. The ORTEP view of the title molecule shows atomic labelling scheme and 50% probability level displacement ellipsoids.

dihedral angle between the mean plane of prop-2-en1-one group (C12-C14\O15) with the mean plane of the chloro phenyl ring (C1-C6) and methyl phenyl ring (C16-C21) are 29.85 (11)° and 38.46 (9)° respectively. Molecular conformation about C6-C12, C12-C13, C13-C14 and C14-C16 bonds described by the torsional angles C5-C6-C12-C13, C6-C12-C13-C14, C12-C13-C14-C16 and C13-C14-C16-C21 are 18.1 (4)°, −176.6 (2)°, −166.0 (2)° and −142.3 (2)° respectively and the geometry of the molecule about these bonds are very well comparable with other reported structure [10]. Intra molecular interaction involving C12-H12 with carbonyl oxygen O15 in C12-H12···O15 generates a pseudo ring of S(5) graph set motif [18] (**Figure 2**). Molecular packing is due to C-H···O, C-H···π and π-π intramolecular hydrogen bond interactions in the structure (**Table 2**).

In the crystal structure, the Chloro phenyl ring of methoxy group carbon C11 via H113 acts as potential donor to the prop-2-en-1-one group of oxygen O15 of x, $-1-y$, $-\frac{1}{2}+z$ molecule and other phenyl ring of methoxy group carbon C26 via H261 link to the oxygen O15 ($\frac{1}{2}-x$, $-\frac{1}{2}+y$, $\frac{5}{2}-z$) forming a chain parallel to the *ac* plane (**Figure 3**). The C-H···π interaction of type-I [19] involves chloro phenyl ring carbon C1 via H1 to the centroid (*Cg*(2)) of the symmetry ($\frac{1}{2}-x$, $-\frac{1}{2}+y$, 2-*z*) related methoxy phenyl ring (C16-C21) where C1-H1···Cg(2) = 3.6613(3) Å.

Face to face π-π stacked interaction observed between the centroids of symmetry related [$\frac{1}{2}-x$, $-\frac{3}{2}-y$, 2-*z*] methyl phenyl ring (C16-C21), with Cg–Cg separation distance of 3.857(2) Å (α = 0°), further contributes to the molecular packing (**Figure 4**).

3.2. *Ab-Initio* and DFT Studies

The *ab-initio* and Density Functional Theory (DFT) with Gaussian-09 program package employing B3LYP (Becke

Table 2. Intra and intermolecular interactions (distances in Å, angles in °).

A. π···π interaction					
Cg(I)-Cg(J)	Cg(I)...Cg(J) Å	α	β	γ	Cg(I)...P Å
2-2(ii)	3.857(2)	0.0	12.3	12.3	3.768
B. C-H···π interactions					
C- H(I) ···Cg (J)	d(H-Cg) Å	d(C-Cg) Å	Y-X...Cg°	γ°	H...P Å
C1-H1···Cg(2)(iii)	2.80	3.661(3)	154	1.08	2.80
C. Hydrogen bond interactions					
D-H···A	d (D-H) Å	d (D-A) Å	d (H-A) Å	(D-H...A)°	
C12-H12···O15(i)	0.93	2.783(3)	2.455(2)	100.74(16)	
C11-H113···O15(iv)	0.96	3.272(5)	2.569(2)	130.24(26)	
C22-H222···O15(v)	0.96	3.589(4)	2.683(2)	157.52(19)	
C26-H261···O15(vi)	0.96	3.443(4)	2.563(2)	152.38(19)	

Symmetry code: (i) x, y, z; (ii) $\frac{1}{2}-x$, $-\frac{3}{2}-y$, $2-z$; (iii) $\frac{1}{2}-x$, $-\frac{1}{2}-y$, $2-z$; (iv) x, $-1-y$, $-\frac{1}{2}+z$; (v) $\frac{1}{2}-x$, $-\frac{1}{2}+y$, $\frac{3}{2}-z$; (vi) $\frac{1}{2}-x$, $-\frac{1}{2}+y$, $\frac{5}{2}-z$; *Note*: *Cg*(1) and *Cg*(2) represents the centroid of the rings (C1-C6) and (C16-C21) respectively.

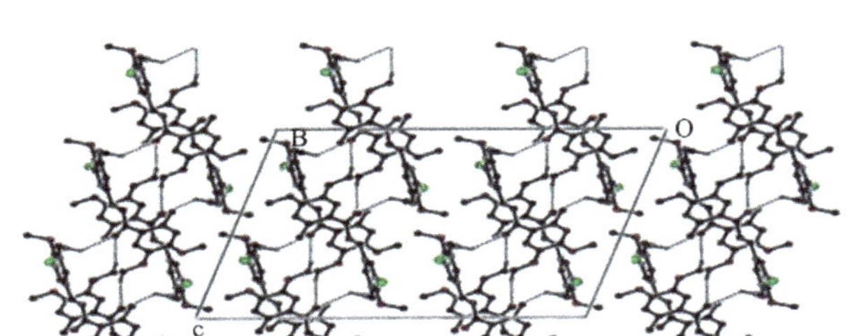

Figure 3. Molecular packing showing C-H···O interactions along ac plane, for a shake of celerity some hydrogen atoms removed.

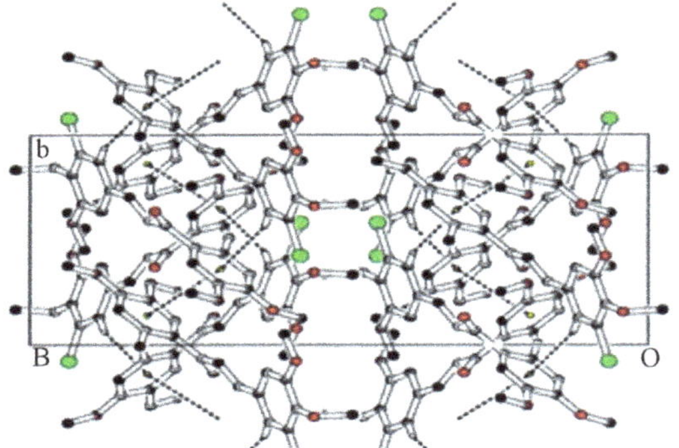

Figure 4. Molecular packing depicting π-π and C-H···π interactions along ab plane, for shake of clarity some hydrogen atoms are removed.

three parameter Lee-Yang-Parr) method with 6-311G* basis set is used to determine optimized bond lengths and angles. **Table 3** compares the experimental data with those obtained by theoretical study which reveals that all optimized bond lengths and bond angles are slightly larger than the experimental values. The highest bond length difference is 0.0979 Å and 0.1238 Å for the C11-O10 bond at RHF and B3LYP respectively, where the biggest bond angle deviation occurred in the C3-O8-C9 angle 4.0313° in RHF method and 4.1322° in B3LYP method. The root mean square error (RMSE) and the correlation coefficient are calculated and the results are presented graphically as shown in **Figure 5**. The RMSE is found to be about 0.022 Å for RHF and 0.038 Å for B3LYP, indicating that the bond lengths obtained by the HF method shows the strongest correlations with the experimental values. The correlation coefficient in bond lengths by RHF and B3LYP are 0.9699 and 0.9277 respectively. For bond angles, the root mean square error is found 1.361° and 1.232° for RHF and B3LYP respectively. The correlation coefficient in bond angles by RHF and B3LYP are 0.8966 and 0.9157 respectively, which reveals that the correlation coefficient for bond angles obtained by RHF method are smaller than those determined by B3LYP method as observed in reported research paper [20].

The molecular conformation described by the torsional angles obtained from X-ray data are C6-C12-C13-C14, C12-C13-C14-C16 and C13-C14-C16-C21 as −176.6(2)° [−178.2701° & −177.5469°], −166.0 (2)° [−172.7265° & −174.4626°] and −142.3 (2)° [−140.9733° & −147.4643°] respectively. The values in the square bracket are those obtained by RHF and B3LYP method respectively. The angle between the mean plane of prop-2-en 1-one group

Table 3. Bond lengths (Å) and angles (°) by X-ray and theoretical calculations for title compound at the RHF/6-311G and B3LYP/6-311G levels of theory.

Bond lengths (Å)	X-ray	RHF	B3LYP
C1-C2	1.383 (4)	1.3840	1.3890
C1-C6	1.408 (4)	1.3854	1.4046
C2-C3	1.385 (4)	1.3751	1.3943
C2-Cl7	1.760 (3)	1.7962	1.8178
C3-O8	1.376 (3)	1.3629	1.3788
C3-C4	1.417 (4)	1.4014	1.4175
C4-O10	1.361 (3)	1.3683	1.3911
C4-C5	1.392 (3)	1.3763	1.3890
C5-C6	1.404 (4)	1.4000	1.4121
C6-C12	1.464 (3)	1.4701	1.4610
C9-O8	1.423 (4)	1.4441	1.4717
C11-O10	1.330 (5)	1.4284	1.4543
C12-C13	1.339 (3)	1.3300	1.4610
C13-C14	1.479 (3)	1.4773	1.4770
C14-O15	1.228 (3)	1.2281	1.2593
C14-C16	1.503 (3)	1.4895	1.4921
C16-C17	1.396 (3)	1.3889	1.4099
C16-C21	1.411 (3)	1.4045	1.4145
C17-C18	1.407 (4)	1.4012	1.4083
C17-C22	1.521 (3)	1.5142	1.5175
C18-C19	1.384 (3)	1.3764	1.3915
C19-O23	1.368 (3)	1.3620	1.3822
C19-C20	1.408 (3)	1.4022	1.4148
C20-C21	1.383 (3)	1.3691	1.3826
C20-O25	1.374 (3)	1.3681	1.3867
C24-O23	1.424 (4)	1.4258	1.4525
C26-O25	1.434 (3)	1.4534	1.4263
Bond Angles (°)	**X-ray**	**RHF**	**B3LYP**
C2-C1-C6	119.5 (3)	119.8049	119.9677
C3-C2-C1	121.3 (3)	122.1467	122.4271
C3-C2-Cl7	120.8 (2)	119.3574	118.9016
C1-C2-Cl7	117.9 (2)	118.4939	118.6711
O8-C3-C2	120.7 (3)	120.9275	199.5017
O8-C3-C4	119.8 (3)	121.0446	122.8734
C2-C3-C4	119.4 (2)	117.9624	117.4753
O10-C4-C5	124.2 (3)	124.0375	124.1057
O10-C4-C3	115.8 (2)	115.3365	115.0927
C5-C4-C3	120.0 (2)	120.6199	120.8002
C4-C5-C6	119.6 (2)	120.6892	120.8309
C5-C6-C1	120.2 (2)	118.7710	118.1946
C5-C6-C12	121.3 (2)	122.8415	123.0725
C1-C6-C12	118.5 (2)	118.3867	118.4328
C3-O8-C9	115.2 (2)	119.2513	119.3522
C11-O10-C4	119.6 (2)	121.4489	119.0727
C13-C12-C6	127.2 (2)	127.4876	127.8938
C12-C13-C14	120.1 (2)	120.965	120.6154
O15-C14-C13	119.8 (2)	119.9883	120.4770
O15-C14-C16	119.6 (2)	119.5906	119.3448
C13-C14-C16	120.5 (2)	120.2911	120.0407
C17-C16-C21	119.6 (2)	119.4958	119.2563
C17-C16-C14	125.1 (2)	125.3151	125.8395
C21-C16-C14	115.3 (2)	115.1763	114.8788
C16-C17-C18	118.4 (2)	117.9234	118.0205
C16-C17-C22	123.9 (2)	124.6705	124.6250
C18-C17-C22	117.6 (2)	117.3577	117.2814
C19-C18-C17	122.0 (2)	122.4108	122.3825
O23-C19-C18	125.2 (2)	124.0269	124.5580
O23-C19-C20	115.4 (2)	116.6251	116.1281
C18-C19-C20	119.4 (2)	119.3440	119.3085
C21-C20-O25	125.2 (2)	124.4546	124.9457
C21-C20-C19	119.2 (2)	118.8226	118.8493
O25-C20-C19	115.7 (2)	116.7227	116.2050
C20-C21-C16	121.5 (2)	121.9925	122.1012
C19-O23-C24	117.9 (2)	121.5507	119.1203
C20-O25-C26	116.7 (2)	120.7125	118.2971

(C12-C14\O15) and the mean plane of the chloro phenyl ring (C1-C6) and methyl phenyl ring (C16-C21) are 29.85 (11)° and 38.46 (9)° respectively. The calculated data are collected in gas phase and no molecular interactions are considered, whereas the experimental data are acquired in the solid state and crystal field interactions, e.g. Van der Waals forces, crystal packing force and hydrogen bond interactions. The observed discrepancies between the theoretical and experimental results may be attributed to different environments of the molecule, being isolated state in gas phase for theoretical study,

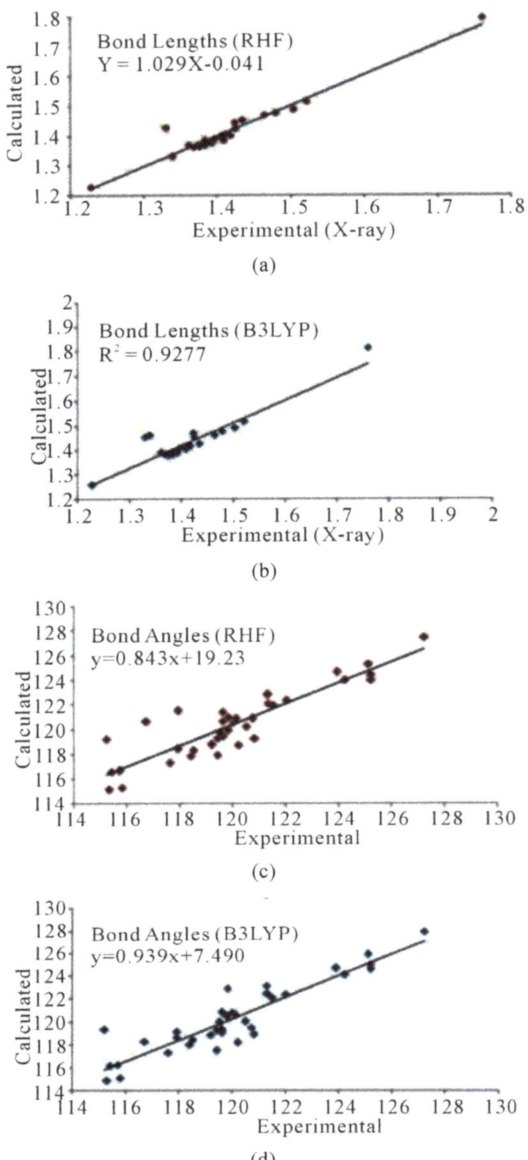

Figure 5. Correlation studies of calculated and experimental bond lengths and bond angles.

whereas the experimental values are of the molecule in solid state there by subjected to the intermolecular forces, which are present into π-π, C-H···π and C-H···O interactions.

3.3. Mulliken Charge Distributions

The Mulliken charge distributions of the title compound have been calculated using RHF and B3LYP methods with 6-311G level (**Table 4**) and graphically shown in **Figure 6**. It may be noted that the all oxygen atoms have negative charge and all hydrogen atoms have positive charge. The oxygen O10 (−0.691300 and −0.531309 from RHF and B3LYP respectively) atom has more

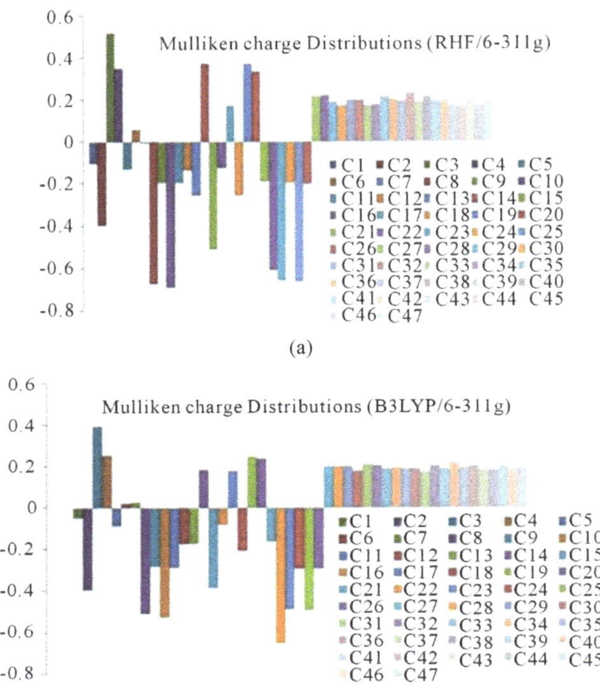

Figure 6. Graphical reorientations of Milliken charge distributions of title compound.

negative charge than other oxygen atoms, whereas the hydrogen atom H21 (0.230926 and 0.204517 from RHF and B3LYP respectively) has more positive charge than the other hydrogen atoms. The results suggest that the oxygen atoms are electron acceptor and charge transfer takes place from H to O. The carbon atom C3 (0.519465 and 0.393703 from RHF and B3LYP respectively) is more positive than the other positive carbon atoms and C22 (−0.611291 and −0.655711 from RHF and B3LYP respectively) is the more negative than the other negative carbons in the title molecule, due to electron-donating substituent at that position. The presence of donor and acceptor atoms suggests the existence of both intra and inter-molecular hydrogen bonding in the crystalline phase.

3.4. Total Energies, Dipole Moments and Molecular Orbitals Analysis

The RHF/6-311G and B3LYP/6-311G methods are used to predict the dipole moment of the title molecule 6.1795 and 6.8948 (**Table 5**) at a total energy of −1602.7509 (RHF) and −1610.6173 (B3LYP) respectively. The high value of the dipole moments suggests that the title compound is reactive and attractive for further interactions with other system. The HOMO and LUMO plots by B3LYP method of the title compound are shown in **Figure 7**. HOMO of the title compound presents a charge density localized mainly on both the phenyl rings and all oxygen atoms where as LUMO is mostly on plane of prop-2-en-1-one group and few carbon atoms of both the phenyl rings. The energy value of HOMO and LUMO is computed at −0.30717 and 0.04286 by RHF method and −0.21689 eV and −0.08978 eV by B3LYP method respectively. The HOMO and its orbital play role of electron donors, and the LUMO and its orbital play role of electron acceptors. The energy gap of HOMO-LUMO explains the charge transfer interaction within the molecule.

4. Study of Antimicrobial Activities

The Minimum Inhibition Concentrations (MICs) of the title compound are determined by the microdilution method as described by the National Committee for Clinical Laboratoty Standard (NCCLS-1992) [21]. The antimicrobial activities tested against the different panel of organisms like *S. aureus* MTCC 96, *S. pyogenus* MTCC 443, *E. coli* MTCC 442 and *P. aeruginosa* MTCC 441 and antifungal strains *C. albicans* MTCC 227 using Gentamycin and *K. nystatin* as reference standards respectively. Each test organism with title compound is run in duplicate. The test plates are incubated at 35°C - 37°C for 24 h. The MIC is taken as the minimum concentration of the dilutions that inhibited the growth of the test microorganism. The concentration of the solvents used in the following assays is maintained at less than 2% so that no inhibition of organisms or interference occurred. The Serial dilution technique is followed by micro method as per NCCLS-1992 manual [21]. The observed MIC values, for bacterial and fungal strains of the title compound are presented in **Table 6**. The data reveals that the response of the drug is significant in *S. aureus* compare to other panel of organisms.

Table 4. Mulliken Charge Distributions of the title molecule.

Atom	Calculated (RHF)	Calculated (B3LYP)	Atom	Calculated (RHF)	Calculated (B3LYP)
C1	−0.103226	−0.049916	O25	−0.665620	−0.498394
C2	−0.399245	−0.399631	C26	−0.203277	−0.296524
C3	0.519465	0.393703	H1	0.214960	0.195785
C4	0.350752	0.253271	H5	0.221216	0.198783
C5	−0.131900	−0.089510	H91	0.188574	0.197300
C6	0.056783	0.020931	H92	0.170305	0.176870
Cl7	−0.003376	0.024023	H93	0.198286	0.208452
O8	−0.677883	−0.510176	H111	0.200044	0.206051
C9	−0.196740	−0.283942	H112	0.173804	0.188522
O10	−0.691300	−0.531309	H113	0.176376	0.190885
C11	−0.197878	−0.291166	H12	0.212435	0.186499
C12	−0.139627	−0.177316	H13	0.201376	0.185665
C13	−0.257158	−0.173867	H18	0.194099	0.173904
C14	0.369580	0.181689	H21	0.230926	0.204517
O15	−0.510804	−0.388171	H221	0.187837	0.187791
C16	−0.126648	−0.082901	H222	0.212743	0.217696
C17	0.172482	0.176513	H223	0.191130	0.191394
C18	−0.258029	−0.210631	H241	0.198432	0.205636
C19	0.374323	0.249912	H242	0.170576	0.184132
C20	0.332634	0.234297	H243	0.168574	0.182609
C21	−0.191313	−0.166480	H261	0.192961	0.199629
C22	−0.611291	−0.655711	H262	0.168619	0.183848
O23	−0.660700	−0.490695	H263	0.196699	0.190735
C24	−0.199975	−0.294656			

Table 5. HOMO-LUMO, total energies and dipole moments of the title compound.

Methods	HOMO (eV)	LUMO (eV)	Energy Band Gap	Total Energy (Hartree)	Dipole Moment (Debye)
RHF	−0.30717	0.04286	0.35003	−1602.7509	6.1795
B3LYP	−0.21689	−0.08978	0.12711	−1610.6173	6.8948

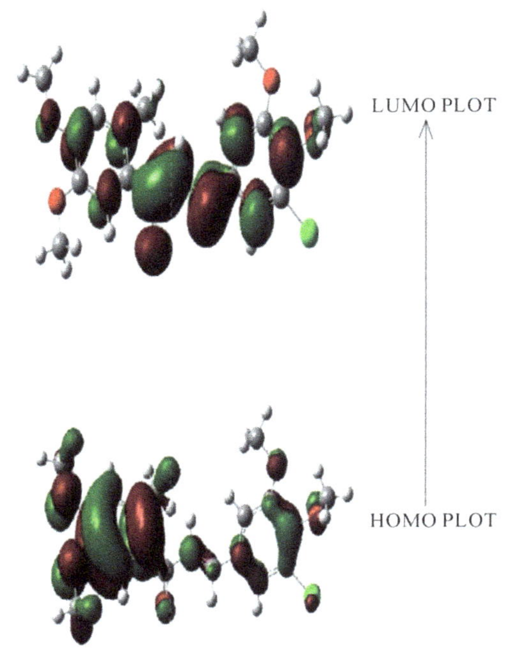

Figure 7. The HOMO and LUMO plots by B3LYP method of the title compound.

Table 6. Minimal inhibition concentrations of bacterial and fungle strains (MIC) in µg/ml.

Bacteria	*S. aureus* MTCC 96	*S. pyogenus* MTCC 443	*E. coli* MTCC 442	*P. aeruginosa* MTCC 441
Std. Drug Gentamycin	0.05	1.0	0.25	0.5
Present study	6.25	25	100	250
Fungal	***C. albicans* MTCC 227**			
Std. Drug K. Nystatin	1000			
Present study	500			

5. Conclusion

A novel chalcone derivative, 3-(3-chloro-4,5-dimethoxy-phenyl)-1-(4,5-dimethoxy-2-methyl phenyl) prop-2-en-1-one, $C_{20}H_{21}ClO_5$, has been synthesized. The synthesized product has been confirmed by chemical analysis, IR, ^1H NMR and ^{13}C NMR, also its three-dimensional structure has been investigated by single crystal X-ray diffraction technique. The predicted optimized molecular conformation is in a good agreement with experimental results. The results of RMSE and correlation coefficient predict the best matching of experimental and theoretical data. The analysis of Milliken charge distributions supports the presence of intra and intermolecular interactions involving the specific atoms. HOMO-LUMO energy band gap further supports the presents of molecular interactions, showing the charge transfer within the molecule. MIC reports reveal that the synthesized drugs are good responses against *S. aureus* bacteria compared to other panel of organisms.

6. Acknowledgements

We are thankful to DST, New Delhi, India for providing the single-crystal X-ray diffractometer (Kappa Apex-II) at Department of Physics, Sardar Patel University, Vallabh Vidyanagar, Gujarat, India, under FIST facility. Sahaj Gandhi is also thankful to UGC, New Delhi, for the financial support (RFSMS) to carry out the research work.

REFERENCES

[1] S. K. Awasthi, N. Mishra, B. Kumar, M. Sharma, A. Bhattacharya, L. C. Mishra and V. K. Bhasin, "Potent Antimalarial Activity of Newly Synthesized Substituted Chalcone Analogs *in Vitro*," *Medicinal Chemistry Research*, Vol. 18, No. 6, 2009, pp. 407-420.

[2] A. Valla, B. Valla, D. Cartier, R. Le Guillou, R. Labia, L. Florent, S. Charneau, J. Schrevel and P. Potier, "New Syntheses and Potential Antimalarial Activities of New 'Retinoid-Like Chalcones,' " *European Journal of Medicinal Chemistry*, Vol. 41, No. 1, 2006, pp. 142-146.

[3] C. Echeverria, J. F. Santibanez, O. Donoso-Tauda, C. A. Escobar and R. R. Tagle, "Structural Antitumoral Activity Relationships of Synthetic Chalcones," *International Journal of Molecular Sciences*, Vol. 10, No. 1, 2009, pp. 221-231.

[4] E. Szliszka, Z. P. Czuba, B. Mazur, L. Sedek, A. Paradysz and W. Krol, "Chalcones Enhance TRAIL-Induced Apoptosis in Prostate Cancer Cells," *International Journal of Molecular Sciences*," Vol. 11, No. 1, 2010, pp. 1-13.

[5] S. Cheenpracha, C. Karalai, C. Ponglimanont, S. Subhadhirasakul and S. Tewtrakal, "Anti-HIV-1 Protease Activity of Compounds from Boesenbergia Pandurata," *Bio-*

organic and Medicinal Chemistry, Vol. 14, No. 6, 2006, pp. 1710-1714.

[6] N. M. Bhatia, K. R. Mahadik and M. S. Bhatia, "QSAR Analysis of 1, 3-Diaryl-2-propen-1-ones and Their Indole Analogs for Designing Potent Antibacterial Agents," *Chemical Papers*, Vol. 63, No. 4, 2009, pp. 456-463.

[7] K. L. Lahtchev, D. I. Batovska, S. P. Parushev, V. M. Ubiyvovk and A. A. Sibirny, "Antifungal Activity of Chalcones: A Mechanistic Study Using Various Yeast Strains," *European Journal of Medicinal Chemistry*, Vol. 43, No. 10, 2008, pp. 2220-2228.

[8] M. Satyanarayana, P. Tiwari, B. K. Tripathi, A. K. Srivastava and R. Pratap, "Synthesis and Antihyperglycemic Activity of Chalcone Based Aryloxypropanolamines," *Bioorganic and Medicinal Chemistry*, Vol. 12, No. 5, 2004, pp. 883-889.

[9] V. V. Serra, F Camoes, S. I. Vieira, M. A. F. Faustino, J. P. C. Tomé, D. C. G. A. Pinto, M. G. P. M. S. Neves, A. C. Tomé, A. M. S. Silva, E. F. da Cruz e Silva and J. A. S. Cavaleiro, "Synthesis and Biological Evaluation of Novel Chalcone-Porphyrin Conjugates," *Acta Chemica Solvenica*, Vol. 56, 2009, pp. 603-611.

[10] U. H. Patel, S. A. Gandhi, V. M. Barot and M. C. Patel, "3-(2-Chloro-3-hydroxy-4-methoxyphenyl)-1-(4,5-dimeth oxy-2-methylphenyl)-prop-2-en-1-one," *Acta Crystallography*, Vol. E68, 2009, pp. o2926-o2927.

[11] U. H. Patel and S. A. Gandhi, "Quantum Chemical Studies on Crystal Structures of Sulfacetamide and Sulfasalazine," *Indian Journal of Pure and Applied Physics*, Vol. 49, 2008, pp. 263-269.

[12] G. M. Sheldrick, "A Short History of SHELX," *Acta Crystallographica*, Vol. A64, No. 1, 2008, pp. 112-122.

[13] A. D. Becke, "A New Mixing of Hartree-Fock and Local Density-Functional Theories," *The Journal of Chemical Physics*, Vol. 98, No. 2, 1993, pp. 1372-1377.

[14] B. G. Jhonson, P. M. Gill and J. A. Pople, "The Performance of a Family of Density Functional Methods," *The Journal of Chemical Physics*, Vol. 98, No. 7, 1993, pp. 5612-5626.

[15] C. Lee, W. Yang and R. G. Parr, "Development of the Colic-Salvetti Correlation-Energy Formula into a Functional of the Electron Density," *Physical Review B*, Vol. B37, No. 2, 1988, pp. 785-789.

[16] M. J. Frisch, G. W. Trucks, H. B. Schlegel, G. E. Scuseria, M. A. Robb, J. R. Cheeseman, G. Scalmani, V. Barone, B. Mennucci, G. A. Petersson, H. Nakatsuji, M. Caricato, X. Li, H. P. Hratchian, A. F. Izmaylov, J. Bloino, G. Zheng, J. L. Sonnenberg, M. Hada, M. Ehara, K. Toyota, R. Fukuda, J. Hasegawa, M. Ishida, T. Nakajima, Y. Honda, O. Kitao, H. Nakai, T. Vreven, J. A. Montgomery, Jr., J. E. Peralta, F. Ogliaro, M. Bearpark, J. J. Heyd, E. Brothers, K. N. Kudin, V. N. Staroverov, R. Kobayashi, J. Normand, K. Raghavachari, A. Rendell, J. C. Burant, S. S. Iyengar, J. Tomasi, M. Cossi, N. Rega, J. M. Millam, M. Klene, J. E. Knox, J. B. Cross, V. Bakken, C. Adamo, J. Jaramillo, R. Gomperts, R. E. Stratmann, O. Yazyev, A. J. Austin, R. Cammi, C. Pomelli, J. W. Ochterski, R. L. Martin, K. Morokuma, V. G. Zakrzewski, G. A. Voth, P. Salvador, J. J. Dannenberg, S. Dapprich, A. D. Daniels, Ö. Farkas, J. B. Foresman, J. V. Ortiz, J. Cioslowski and D. J. Fox, "Gaussian 09, Revision A.1," Gaussian, Inc., Wallingford, 2009.

[17] R. Dennington, T. Keith and J. Millam, "Gauss View Version 5," Semichem Inc., Shawnee Mission, 2009.

[18] J. Bernstein, R. E. Davis and L. Shimoni, "Patterns in Hydrogen Bonding: Functionality and Graph Set Analysis in Crystals," *Angewandte Chemie International Edition in English*, Vol. 34, No. 15, 1995, pp. 1555-1573.

[19] J. F. Malone, C. M. Murray, M. H. Charlton, R. Docherty and A. J. Lavrry, "X-H···π (phenyl) Interactions Theoretical and Crystallographic Observations," *Journal of Chemical Society*, Vol. 93, No. 19, 1997, pp. 3429-3436.

[20] B. Yılmaz, H. Saracoglu, N. Caliskan, I. Yilmaz and A. Cukurovali, "X-Ray Diffraction and Theoretical Approach to the Molecular Structure of (E)-2-(2-(1,3-dioxoisoindolin-2-yl)-1-(3-phenyl-3-methylcyclobutyl)ethyliden e) Hydrazine Carboxamide," *Journal of Chemical Crystallography*, Vol. 42, No. 8, 2012, pp. 897-904.

[21] National Committee for Clinical Laboratory Standards (NCCLS), "Reference Method for Broth Dilution Antifungal Susceptibility Testing of Yeasts," Approved Standard, 2nd Edition, NCCLS Document M27-A2, 2002.

Synthesis and Structure of Dimeric Copper (I) Complex from Bis[(2,2')-dimethyl 2,2'-(1,10-phenanthroline-2,9-diyl) bis(methan-1-yl-1-ylidene)-bis(hydrazinecarbo dithioate)]

Md. Arifuzzaman[1], Tasneem A. Siddiquee[1], Mohammad R. Karim[1*], Aminul H. Mirza[2], Mohamad A. Ali[2]

[1]Department of Chemistry, Tennessee State University, Nashville, USA

[2]Department of Chemistry, Faculty of Science, University of Brunei Darussalam, Gadong, Malaysia

ABSTRACT

Metal complex, bis[(2,2')-dimethyl 2,2'-(1,10-phenanthrline-2,9-diyl)bis(methan-1-yl-1-ylidene)-bis(hydrazinecarrbo dithioate)copper(I)], was synthesized from the reaction of Schiff base, (2,2')-dimethyl 2,2'-(1,10-phenanthroline-2,9-diyl)bis(methan-1-yl-1-ylidene)-bis(hydrazinecarbo dithioate) and $CuCl_2$ at reflux condition in methanol. The copper centers of the complex appear to be reduced. This probably was facilitated by *in situ* oxidative formation of disulfide bond at the uncomplexed ligand moieties. Single crystal X-ray diffraction analysis reveals the distorted tetrahedral geometry around the copper centers. This compound crystallizes in the triclinic space group, *P*-1 with crystallographic parameters: a = 10.006(3) Å, b = 13.272(3) Å, c = 22.123(6) Å, α = 85.656(6)°, β = 81.656(6)°, γ = 73.097(5)°, μ = 1.223 mm^{-1}, V = 2779.5(13) Å3, Z = 2, Dc = 1.437 Mg/m^3, T = 293 (2) K.

Keywords: Schiff Base; Copper Complex; Metal Complex; Phenanthroline; Hydrazinecarbo Dithioate, X-Ray Diffraction

1. Introduction

Schiff bases bearing bidentate N,N' chelating agents such as 1,10-phenanthroline (phen) have been used extensively as a ligand in both analytical and preparative coordination chemistry as well as in the preparation of many mixed-ligand complexes [1]. These complexes have interesting physiochemical properties and have high potential as antibacterial, antifungal and chemotherapeutic agents [2]. Metal complexes derived from 1,10-phenanthroline have also been found to be widely used in areas such as molecular catalysis, solar energy conversion, calorimetric analysis, herbicides, molecular recognition, self-assembly, antineoplastic agents and nucleic acid probes [3]. The key feature of 1,10-phenanthroline is π-electron deficient nitrogen which stabilizes various metal complexes in lowered oxidation states [4]. The copper/phen complexes are able to interact and cleave the DNA/RNA backbone [5,6]. Recently, it has been found that copper(II) complexes of ferrocene(Fc)-conjugated reduced Schiff base of L-tyrosine are cytotoxic in HeLa (human cervical cancer) and MCF-7 (human breast cancer) [7]. 1,10-phenanthroline and its copper (I and II) complexes showed interesting magnetic, electronic and redox process [8].

Schiff base containing amino thiol group (HN_2-NH-CS-R) and their metal complexes exhibit various biological activities. Thiosemicarbazone containing compounds was first used as drugs against tuberculosis, leprosy [9], small pox [10] and tumour [11]. Recently, Triapine (3-amino pyridine-2-carboxaldehyde thiosemicarba zone) has been developed as an anticancer drug and has completed clinical phase II on several cancer types [12, 13].

Schiff bases have been made by reacting 1,10-phenanthroline-2,9-dicarboxaldehydehemi hydrate with sulfur containing amines and studied their antibacterial

*Corresponding author.

properties [14]. However, the synthesis of metal complexes of these Schiff bases has not been studied. We report the synthesis & characterization of an interesting dimeric copper (Cu) complex by using one of these Schiff bases.

2. Materials and Methods

2.1. Materials

The $CuCl_2$ salt and the solvents were commercially available and used without further purification. The Schiff base (2,2')-dimethyl 2,2'-(1,10-phenanthroline-2,9-diyl) bis(methan-1-yl-1-ylidene)-bis(hydrazinecarbodithioate) (1) was prepared according to a published procedure [15]. Infrared spectra were recorded on a Thermo Nicolet iS10 FTIR spectrometer equipped with ATR accessory. NMR spectra were recorded on a Bruker 400 MHz spectrometer. Diffraction study was done on a Rigaku XtaLab mini bench-top instrument. Data collection and data reduction was done with the software of the instrument, "Crystal Clear" [16]. The radiation source was MoKα (λ = 0.71075 Å).

2.2. Synthesis

Copper (II) chloride (0.0134 g, 0.1 mmol) dissolved in methanol (2 ml) was mixed dropwise with boiling solution of the Schiff base **1** (0.044 g, 0.1 mmol) in hot methanol (20 ml). The mixture was refluxed for 1 h and then left to stand overnight (**Scheme 1**). This resulted in precipitate of the copper complex **2**, which was then filtered off and kept in desiccator over silica gel. Then the precipitate was dissolved in DMSO for recrystallization. After 15 days, black crystals of X-ray quality was formed (% yield = 30%). IR: ν (cm^{-1}). 3100, 2900 (C-H aromatic and aliphatic), 1570 (C=N), 1500, 1600 (C=C, aromatic), 1030 (C=S). ^{1}H-NMR (DMSO-d6, ppm): δ H 13.75 (s, 2 NH), 8.65 (d, J = 8.7 Hz, 2H), 8.30 (d, J = 8.7 Hz, 2H), 8.60 (s, 2H), 8.07 (s, 2H), 2.01 (s, 6H).

2.3. Single Crystal X-ray Analysis

Crystals of the complex 2 were grown from DMSO at room temperature. A crystal of 0.17 × 0.13 × 0.09 mm dimension was mounted on a MiTeGen loop with Apiezon grease at ambient temperature. Diffraction study was done on a Rigaku XtaLab mini bench-top instrument. Data collection and data reduction was done with the software of the instrument "Crystal Clear" [17]. The radiation source was MoKα (λ = 0.71075 Å). The compound crystallized in a triclinic crystal system with P-1 space group. Unit cell dimensions: a = 10.006(3) Å, b = 13.272(3) Å, c = 22.123(6) Å, a = 85.656(6)°, b = 81.656(6)°, g = 73.097(5)° and volume = 2779.5(13) Å^3.

Structure Solution and Refinement: The structure was solved by direct methods [16] and expanded using Fourier techniques. The non-hydrogen atoms were refined anisotropically. Hydrogen atoms were refined using the riding model. The final cycle of full-matrix least-squares refinement on F^2 was based on 6595 observed reflections and 607 variable parameters. Which converged with unweighted and weighted agreement factors of: $R1$ = 0.0823 and $wR2$ = 0.2222. This was conducted using the program suite WINGX [18]. The compound crystallized as solvates. Disordered DMSO molecules were present in the lattice that could not be completely modeled. One of the DMSO molecules was located close enough to be within the range of hydrogen bonding interaction. The crystal data and structure refinement details are listed in **Table 1**.

3. Results and Discussion

The coordination sphere around each copper center reveals that copper in the complex was reduced. The ligand in this reaction had only neutral donor atoms as starting material. Hence in order to satisfy the charge neutrality of the complex there needs to be two chloride ions. However, no chloride ion was detected in the structure. This could be rationalized by the fact that thionic sulfur can readily be oxidized to a disulfide form [19,20]. In the process of oxidation of the ligand the copper (II) becomes reduced to copper (I). The C-S bond distance (1.712) of the coordinating sulfur atom in the complex suggests that. this distance is longer that the distance of C=S (1.651)

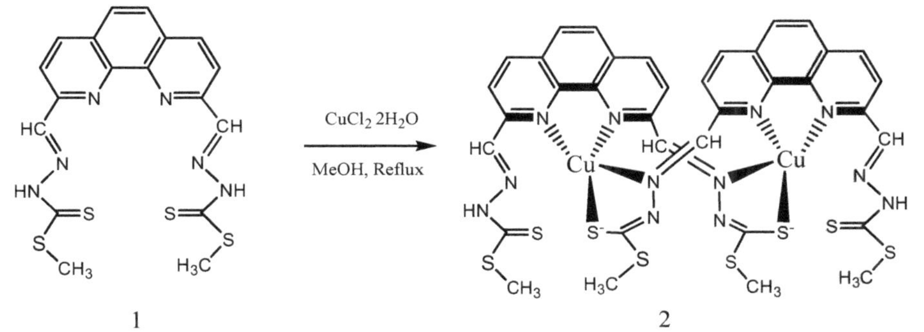

Scheme 1. The synthetic route for the preparation of dimeric copper (II) complex of Schiff base 1.

Table 1. Crystal data and structure refinement for compound 2.

Chemical formula	$C_{36} H_{32} Cu_2 N_{12} S_8$
Formula weight	1016.3396
CCDC deposit	944921
Temperature	293(2) K
Wavelength	0.71075 Å
Crystal system	Triclinic
Space group	P-1
Unit cell dimensions	a = 10.006(3) Å
	b = 13.272(3) Å
	c = 22.123(6) Å
	α = 85.656(6)°.
	β = 81.656(6)°.
	γ = 73.097(5)°.
Volume	2779.5(13) $Å^3$
Z	2
Density (calculated)	1.437 Mg/m^3
Absorption coefficient	1.223 mm^{-1}
F(000)	1232
Crystal size	0.17 × 0.13 × 0.09 mm^3
Theta range for data collection	3.02° to 22.00°.
Index ranges	$-10 \le h \le 10$, $-13 \le k \le 13$, $-23 \le l \le 23$
Reflections collected	16584
Independent reflections	6767 [R(int) = 0.0562]
Completeness to theta = 22.00°	99.3%
Absorption correction	None
Max. and min. transmission	0.8979 and 0.8190
Refinement method	Full-matrix least-squares on F^2
Data/restraints/parameters	6767/0/619
Goodness-of-fit on F^2	1.052
Final R indices [I > 2sigma(I)]	$R1$ = 0.0823, $wR2$ = 0.2222
R indices (all data)	$R1$ = 0.1131, $wR2$ = 0.2586
Largest diff. peak and hole	2.779 d – 0.733 e. $Å^{-3}$

bond in the ligand.

3.1. Bond Lengths and Angles

The ORTEP drawing of the complex is given in **Figure 1**. Cu atom has distorted tetrahedral coordination with sulfur and nitrogen atoms. The bond angles for standard tetrahedral is 109.5° but from the **Figure 1** it has been shown that the angle between S(5)-Cu(1)-N(42) is 86.5 (2)° whereas the angle between S(5)-Cu(1)-N(38) is 131.66 (19)°. The other two angles for central Cu(1) atom are 81.8 (3)° and 100.3 (3)° for angles N(38)-Cu(1)-N(39) and N(39)-Cu(1)-N(42) respectively. However, the Cu(1)-N(38), Cu(1)-N(39), Cu(1)-N(42) and Cu(1)-S(5) distances are 2.038 (6), 2.086 (7), 2.066 (7), and 2.228 (3) Å respectively.

The methyl group attached to S(9) was refined at two positions with site occupancy factors of 0.61 and 0.39 for major and minor conformations, respectively. The methyl group at S(7) atom apparently did not have such freedom probably due to the solvent molecules nested in the lattice around this sulfur atom limiting the free rotation around the single bond. The bond angles and bond distances for other copper center, Cu(2) resembles to that of the first copper center (**Table 2**). Here overall geometry of the complex is cliplike. The clip appearance is provided by semi parallel *phen* moiety. Although the geometry is cliplike, there is not much clearance for another molecule to fit in due to steric hindrance (**Figure 2**).

The average values of bond distances and angles in different atoms are in agreement with the literature values [18]. Selected bond distances and bond angles of molecule **2** are shown in **Table 2**.

Hydrogen Bonds

The bond distance between hydrogen and oxygen, hydrogen and noted noted X-H (Å) and the possible intramolecular interactions by hydrogen bonds noted X-H (Å) of molecule **2** is shown in **Table 3**. **Tables 4** and **5** provide isotropic displacement parameters and Anisotropic displacement parameters of compound 2. These different hydrogen bonds showed in **Figure 3** are responsible of the molecular packing in the unit cell.

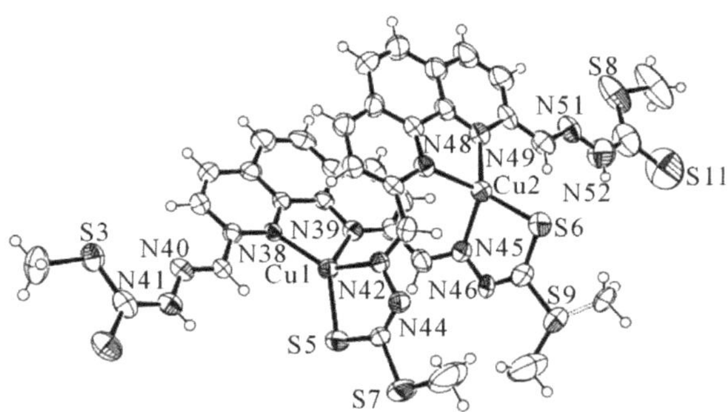

Figure 1. Perspective view of dimer of bis[(2,2')-dimethyl 2,2'-(1,10-phenanthroline-2,9-diyl)bis(methan-1-yl-1-ylidene)-bis(hydrazinecarbodithioate)] Cu(I) complex showing the labelling scheme.

3.2. Crystal Packing

Hydrogen bonds played crucial roles in the crystal pack ing. **Figure 4** shows the presence of two molecules in the unit cell which corresponds to the centrosymmetric triclinic space group *P*-1 with two general positions (x, y, z) and (−x, −y, −z).

4. Conclusion

The new dimeric coordination complex, bis-[(2,2')-dimethyl 2,2'-(1,10-Phenanthroline-2,9-diyl) bis(methan-1-yl-1-ylidene)-bis(hydrazinecarbodithioate) copper(I)], was synthesized with a moderate yield and its crystal structure was determined by X-ray diffraction analysis at room temperature. Structural studies reveal the synergistic multicenter coordination by thiosemicarbazide moiety of the Schiff base ligand. The *phen* moieties of the two ligands are semi-parallel to each other while coordinated to two copper centers. Because of this structural arrangement, the complex looks like a molecular clip. However, the clearance between the two *phen* moiety is

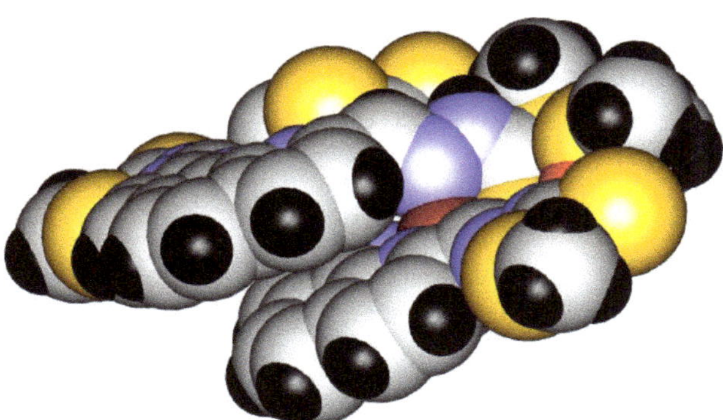

Figure 2. Space filling model of the compound demonstrating limited clearance for clipping in a molecule.

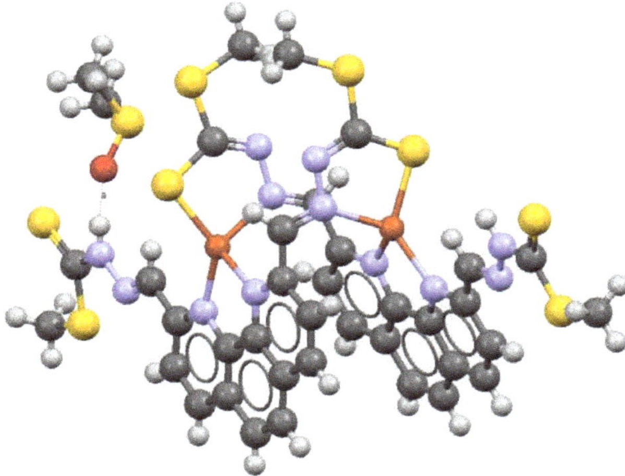

Figure 3. Hydrogen bonding between oxygen of the DMSO solvent molecule and the hydrogen atom of N-H of the ligand moiety.

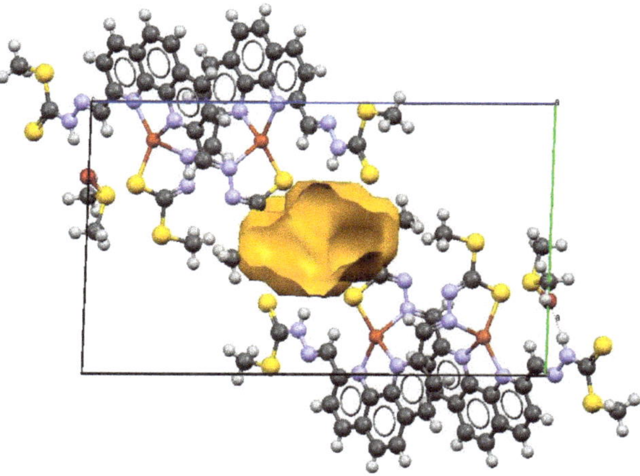

Figure 4. The packing diagram (view along *a*-axis) showing void in the lattice, where disordered solvent molecules are likely to populate.

Table 2. Selected bond angle and bond distances of the compound.

Bond distances	Å	*Bond angles*	°
Cu(1)-N(38)	2.038(6)	N(38)-Cu(1)-N(42)	130.4(3)
Cu(1)-N(42)	2.066(7)	N(38)-Cu(1)-N(39)	81.8(3)
Cu(1)-N(39)	2.086(7)	N(42)-Cu(1)-N(39)	100.3(3)
Cu(1)-S(5)	2.228(3)	N(38)-Cu(1)-S(5)	131.66(19)
Cu(2)-N(49)	2.038(7)	N(42)-Cu(1)-S(5)	86.5(2)
Cu(2)-N(45)	2.063(8)	N(39)-Cu(1)-S(5)	126.6(2)
Cu(2)-N(48)	2.070(7)	N(49)-Cu(2)-N(45)	133.1(3)
Cu(2)-S(6)	2.223(3)	N(49)-Cu(2)-N(48)	80.4(3)
S(3)-C(50)	1.734(11)	N(45)-Cu(2)-N(48)	103.1(3)
S(4)-C(50)	1.651(10)	N(49)-Cu(2)-S(6)	128.5(2)
S(5)-C(43)	1.729(10)	N(45)-Cu(2)-S(6)	85.6(2)
S(6)-C(47)	1.694(12)	N(48)-Cu(2)-S(6)	129.3(2)
S(8)-C(53)	1.693(17)	C(43)-S(5)-Cu(1)	93.7(3)
S(11)-C(53)	1.652(15)	C(47)-S(6)-Cu(2)	94.7(4)
N(40)-C(37)	1.277(11)	C(36)-N(38)-Cu(1)	130.8(5)
N(40)-N(41)	1.368(10)	C(33)-N(38)-Cu(1)	111.5(5)
N(41)-C(50)	1.369(12)	C(25)-N(39)-Cu(1)	129.5(6)
N(42)-C(23)	1.278(11)	C(29)-N(39)-Cu(1)	111.2(5)
N(42)-N(44)	1.403(10)	C(23)-N(42)-Cu(1)	124.4(6)
N(44)-C(43)	1.287(12)	N(44)-N(42)-Cu(1)	116.8(5)
N(45)-N(46)	1.384(11)	C(24)-N(45)-Cu(2)	122.9(7)
N(51)-N(52)	1.340(13)	N(46)-N(45)-Cu(2)	117.5(6)
N(52)-C(53)	1.347(15)	C(20)-N(48)-Cu(2)	129.0(6)
		C(16)-N(48)-Cu(2)	112.6(5)

Table 3. Possible hydrogen bonds.

Donor	H	Acceptor	D....A	D-H	H...A	D-H...A
N(41)	H(41)	N(40)	1.3675(3)	0.86	1.95774(19)	36.632(11)
N(41)	H(41)	O(40)	2.7231(6)	0.86	1.87416	168.890(3)
N(52)	H(52)	N(51)	1.3403(3)	0.86	1.92982(18)	36.527(11)

Table 4. Atomic coordinates ($\times 10^4$) and equivalent isotropic displacement parameters ($Å^2 \times 10^3$) for compound 2. U(eq) is defined as one third of the trace of the orthogonalized Uij tensor.

	x	y	z	U(eq)
Cu(1)	4792(1)	8600(1)	1368(1)	43(1)
Cu(2)	5408(1)	8590(1)	3686(1)	49(1)
S(3)	483(3)	11013(2)	−1033(1)	59(1)
S(4)	92(3)	8883(2)	−1217(1)	66(1)
S(5)	5485(3)	6965(2)	1031(1)	67(1)
S(6)	5205(4)	7129(2)	4205(1)	71(1)
S(7)	4819(5)	5108(2)	1615(2)	91(1)
S(8)	9418(4)	9709(4)	5933(2)	104(1)
S(9)	6324(8)	4950(3)	3787(2)	174(3)
S(11)	10601(5)	7381(6)	6051(3)	173(3)
N(38)	4287(7)	10031(5)	919(3)	32(2)
N(39)	5968(7)	9322(5)	1803(3)	38(2)
N(40)	2165(7)	9968(6)	−204(3)	45(2)
N(41)	1645(8)	9276(6)	−457(3)	49(2)
N(42)	3726(7)	8015(5)	2118(3)	41(2)
N(44)	3834(9)	6935(6)	2132(3)	50(2)
N(45)	6651(8)	7594(6)	3023(3)	51(2)
N(46)	6791(11)	6527(7)	3118(4)	74(3)
N(48)	3974(7)	9668(6)	3200(3)	41(2)
N(49)	5519(7)	9954(6)	4014(3)	43(2)
N(51)	7899(9)	9247(8)	5122(4)	62(2)
N(52)	8724(11)	8363(9)	5356(5)	91(3)
C(10)	7142(10)	9164(9)	4711(4)	57(3)
C(11)	6278(10)	10105(8)	4446(4)	48(2)
C(12)	6232(11)	11096(9)	4613(5)	61(3)
C(13)	5428(14)	11969(10)	4337(5)	72(3)
C(14)	4613(11)	11866(8)	3898(4)	56(3)
C(15)	4680(10)	10832(7)	3766(4)	44(2)
C(16)	3892(9)	10681(7)	3303(4)	40(2)
C(17)	3777(13)	12712(8)	3582(5)	67(3)
C(18)	3018(12)	12588(8)	3149(5)	68(3)
C(19)	3101(10)	11535(7)	2980(4)	49(2)
C(20)	3309(9)	9500(7)	2746(4)	39(2)
C(21)	2521(9)	10311(7)	2400(4)	42(2)
C(22)	2391(10)	11345(8)	2532(4)	50(2)
C(23)	3465(9)	8393(7)	2651(4)	45(2)
C(24)	6811(10)	7878(8)	2455(4)	51(2)
C(25)	6702(9)	8966(7)	2260(4)	43(2)
C(26)	7259(9)	9611(9)	2560(4)	51(2)
C(27)	7075(10)	10631(9)	2374(4)	54(3)
C(28)	6264(9)	11062(7)	1903(4)	45(2)
C(29)	5732(9)	10354(6)	1624(4)	37(2)
C(30)	5970(11)	12111(8)	1680(5)	56(3)
C(31)	5204(11)	12469(8)	1230(5)	59(3)
C(32)	4586(9)	11784(7)	949(4)	46(2)
C(33)	4828(9)	10727(6)	1153(4)	36(2)
C(34)	3726(11)	12124(7)	477(4)	54(3)
C(35)	3233(10)	11412(7)	248(4)	51(2)
C(36)	3503(9)	10377(6)	475(4)	36(2)
C(37)	2949(9)	9628(7)	217(4)	40(2)
C(43)	4590(11)	6471(7)	1656(4)	51(2)
C(47)	6179(15)	6308(9)	3652(5)	81(4)
C(50)	769(10)	9669(8)	−892(4)	53(2)
C(53)	9554(13)	8457(13)	5763(6)	95(4)
C(55)	10574(15)	9523(18)	6502(6)	150(9)
C(56)	−720(12)	11280(10)	−1582(5)	79(4)
C(57)	3850(20)	4810(11)	2299(7)	158(9)
S(10)	1925(5)	6556(3)	380(2)	90(1)
O(10)	2573(11)	7253(7)	−21(5)	116(3)
C(78)	2910(20)	5241(12)	184(11)	171(9)
C(79)	380(20)	6542(15)	7(11)	168(9)
S(14)	−229(12)	5663(8)	−1671(10)	343(10)
O(1)	−1140(30)	6190(20)	−1824(17)	305(14)
C(1)	500(40)	4490(20)	−1812(15)	330(30)
C(2)	1150(40)	6340(40)	−2110(30)	580(60)
S(15)	10120(14)	7030(20)	3770(9)	594(19)
S(10')	2080(60)	6330(40)	−310(30)	250(20)
C(58)	6850(30)	4277(18)	3141(14)	144(17)
C(58')	6030(40)	4810(20)	4511(13)	91(16)

Table 5. Anisotropic displacement parameters (Å2 × 10^3) compound 2. The anisotropic displacement factor exponent takes the form: −2,2[*h2a* * 2U11 + ··· + 2 *hka* * b* U12].

	U11	U22	U33	U23	U13	U12
Cu(1)	57(1)	35(1)	41(1)	1(1)	−13(1)	−18(1)
Cu(2)	60(1)	46(1)	41(1)	−3(1)	−13(1)	−12(1)
S(3)	56(2)	69(2)	55(2)	0(1)	−21(1)	−19(1)
S(4)	58(2)	90(2)	62(2)	−18(2)	−14(1)	−34(2)
S(5)	111(2)	42(2)	45(2)	−6(1)	0(1)	−22(2)
S(6)	109(2)	54(2)	47(2)	3(1)	−7(2)	−19(2)
S(7)	157(3)	38(2)	79(2)	−5(2)	−1(2)	−34(2)
S(8)	75(2)	205(5)	60(2)	−28(2)	−8(2)	−74(3)
S(9)	379(9)	43(2)	73(3)	9(2)	−50(4)	−11(3)
S(11)	111(4)	224(6)	154(5)	−3(4)	−82(3)	29(4)
N(38)	39(4)	34(4)	28(4)	−3(3)	−8(3)	−16(3)
N(39)	32(4)	44(5)	39(4)	−4(3)	−6(3)	−9(3)
N(40)	44(4)	57(5)	36(4)	−11(4)	−2(4)	−18(4)
N(41)	53(5)	53(5)	45(4)	−10(4)	−10(4)	−19(4)
N(42)	44(4)	40(4)	38(4)	−3(3)	−14(3)	−8(3)
N(44)	78(6)	35(4)	46(5)	0(4)	−18(4)	−25(4)
N(45)	58(5)	44(5)	44(5)	1(4)	−13(4)	0(4)
N(46)	115(8)	50(6)	36(5)	8(4)	−23(5)	15(5)
N(48)	45(4)	46(5)	37(4)	−2(3)	−6(3)	−21(3)
N(49)	42(4)	54(5)	37(4)	−2(4)	−4(3)	−19(4)
N(51)	54(5)	92(7)	45(5)	−6(5)	−10(4)	−23(5)
N(52)	85(7)	95(8)	90(7)	0(6)	−49(6)	−7(6)
C(10)	53(6)	77(7)	49(6)	−11(5)	−16(5)	−24(5)
C(11)	52(6)	73(7)	27(5)	−4(5)	−5(4)	−29(5)
C(12)	74(7)	72(8)	51(6)	−10(6)	−11(5)	−40(6)
C(13)	113(10)	71(8)	50(6)	−12(6)	−9(6)	−51(7)
C(14)	80(7)	51(6)	45(6)	−2(5)	−10(5)	−31(5)
C(15)	54(6)	51(6)	36(5)	−3(4)	−8(4)	−27(5)
C(16)	42(5)	50(6)	32(5)	−4(4)	−3(4)	−19(4)
C(17)	104(9)	46(6)	62(7)	1(5)	−22(6)	−34(6)
C(18)	91(8)	38(6)	68(7)	4(5)	−8(6)	−9(5)
C(19)	56(6)	46(6)	44(5)	7(5)	−7(5)	−17(5)
C(20)	40(5)	47(6)	31(5)	−8(4)	0(4)	−15(4)
C(21)	42(5)	49(6)	41(5)	−2(4)	−12(4)	−21(4)
C(22)	48(6)	54(6)	45(5)	2(5)	−8(4)	−9(5)
C(23)	43(5)	50(6)	44(6)	11(5)	−11(4)	−19(4)
C(24)	53(6)	63(7)	37(6)	−4(5)	−12(4)	−10(5)
C(25)	37(5)	53(6)	38(5)	2(4)	−7(4)	−9(4)

Continued

C(26)	43(5)	81(8)	32(5)	−6(5)	−11(4)	−19(5)
C(27)	54(6)	76(8)	48(6)	−12(5)	−13(5)	−36(5)
C(28)	43(5)	56(6)	44(5)	−14(5)	1(4)	−27(5)
C(29)	43(5)	38(5)	34(5)	−5(4)	−2(4)	−19(4)
C(30)	63(7)	58(7)	56(6)	−24(5)	3(5)	−30(5)
C(31)	80(7)	49(6)	58(6)	−3(5)	−5(6)	−34(5)
C(32)	51(6)	43(6)	39(5)	−6(4)	4(4)	−11(4)
C(33)	43(5)	36(5)	29(4)	−2(4)	−2(4)	−11(4)
C(34)	73(7)	39(5)	46(6)	1(5)	−11(5)	−9(5)
C(35)	60(6)	51(6)	40(5)	−3(5)	−13(4)	−10(5)
C(36)	42(5)	36(5)	30(4)	2(4)	−2(4)	−15(4)
C(37)	50(5)	38(5)	30(5)	−5(4)	−10(4)	−8(4)
C(43)	85(7)	31(5)	41(6)	1(4)	−12(5)	−23(5)
C(47)	127(11)	44(7)	61(8)	2(6)	−37(7)	3(6)
C(50)	45(5)	79(7)	40(5)	−6(5)	−5(4)	−25(5)
C(53)	60(8)	144(13)	77(8)	−2(9)	−24(7)	−17(8)
C(55)	81(10)	330(30)	72(9)	−48(13)	−20(8)	−94(14)
C(56)	65(7)	103(10)	83(8)	24(7)	−43(6)	−35(7)
C(57)	260(20)	48(8)	125(14)	12(9)	73(15)	−31(11)
S(10)	108(3)	62(2)	99(3)	−6(2)	−18(3)	−20(2)
O(10)	121(8)	77(6)	150(9)	17(6)	−3(7)	−40(6)
C(78)	181(19)	58(10)	230(20)	−2(12)	−2(17)	24(11)
C(79)	145(18)	130(16)	250(30)	0(17)	−54(17)	−61(14)
S(14)	184(9)	144(7)	680(30)	−66(12)	−152(13)	50(6)
O(1)	230(30)	190(20)	490(40)	−20(30)	−100(30)	−32(19)
C(1)	350(50)	210(30)	240(30)	40(30)	70(30)	160(30)
C(2)	190(30)	560(90)	1080(160)	−580(100)	190(60)	−220(50)
S(15)	186(11)	1140(60)	490(30)	−280(30)	−102(14)	−140(20)
C(58)	200(40)	55(16)	170(30)	−10(16)	−40(20)	−19(17)
C(58')	170(40)	32(17)	60(20)	18(13)	10(20)	−34(18)

inadequate for the complex to serve as a functional molecular clip.

REFERENCES

[1] J. Reedijk, "Comprehensive Coordination Chemistry, Vol. 2," In: G. Wilkinson, R. D. Dillard and J. A. McCleverty, Eds., *Comprehensive Coordination Chemistry*, Pergamon, Oxford, 1987, p. 73.

[2] S. Kumar, D. N. Dhar and P. N. Saxena, "Applications of Metal Complexes of Schiff Bases—A Review," *Journal of Scientific and Industrial Research*, Vol. 68, No. 3, 2009, pp. 181-187.

[3] Y. Shen and B. P. Sullivan, "Versatile Preparative Route to 5-Substituted-1,10-phenanthroline Ligands via 1,10-Phenanthroline-5,6-epoxide," *Inorganic Chemistry*, Vol. 34, No. 25, 1995, pp. 6235-6236.

[4] Md. A. Masood and P. S. Zacharias, "Electrochemical Reversibility of the Cu^{I}-Cu^{0} Couple in Bis[2,9-di(o-substituted phenyl)-1,10-phenanthroline]-copper(I) Complexes. Stability of the Corresponding Copper(0) Species," *Journal of the Chemical Society*, *Dalton Transactions*, No. 1, 1991, pp. 111-114.

[5] A. Zaid, J.-S. Sun, C.-H. Nguyen, E. Bisagni, T. Garestier, D. S. Grierson and R. Zain, "Triple-Helix Directed Cleavage of Double-Stranded DNA by Benzoquinoquinoxa-

line-1,10-phenanthroline Conjugates," *ChemBioChem*, Vol. 5, No. 11, 2004, pp. 1550-1557.

[6] H. Amiri, N. Nekhotiaeva, J.-S. Sun, C.-H. Nguyen, D. S. Grierson, L. Good and R. Zain, "Benzoquinoquinoxaline Derivatives Stabilize and Cleave H-DNA and Repress Transcription Downstream of a Triplex-forming Sequence," *Journal of Molecular Biology*, Vol. 35, No. 4, 2005, pp. 776-783.

[7] T. K. Goswami, S. Gadadhar, A. A. Karande and A. R. Chakravarty, "Photocytotoxic Ferrocene-Appended (l-Tyrosine)copper(II) Complexes of Phenanthroline Bases," *Polyhedron*, Vol. 52, 2013, pp. 1287-1298.

[8] M. A. Masood and D. J. Hodgson, "Synthesis and Characterization of the Multidentate Ligand 2,9-Bis(N-pyrazolylmethyl)-1,10-phenanthroline (bpmp) and Its Copper(I) and Copper(II) Complexes," *Inorganic Chemistry*, Vol. 32, No. 22, 1993, pp. 4839-4844.

[9] E. M. Bavin, R. J. W. Reea, J. M. Robson, M. Seiler, D. E. Seymour and D. Suddaby, "The Tuberculostatic Activity of Some Thiosemicarbazones," *Journal of Pharmacy and Pharmacology*, Vol. 2, No. 1, 1950, pp. 764-772.

[10] G. A. Kune, "To-Day's Drug: Methisazone," *British Medical Journal*, Vol. 2, 1964, p. 621.

[11] A. C. Sartorelli and B. A. Booth, "Inhibition of the Growth of Sarcoma 180 Ascites Cells by Combinations of Inhibitors of Nucleic Acid Biosynthesis and the Cupric Chelate of Kethoxal Bis-(thiosemicarbazone)," *Cancer Research*, Vol. 27, No. 9, 1967, pp. 1614-1619.

[12] C. M. Nutting, C. M. L. Herpan and A. B. Miah, "Phase II Study of 3-AP Triapine in Patients with Recurrent or Metastatic Head and Neck Squamous Cell Carcinoma," *Annals of Oncology*, Vol. 20, No. 7, 2009, pp. 1275-1279.

[13] S. H. N. Moorthy, M. F. S. A. Cerqueira, J. Ramos and A. Farnandes, "Aryl- and Heteroaryl-Thiosemicarbazone Derivatives and Their Metal Complexes: A Pharmacological Template," *Anticancer Drug Discovery*, Vol. 8, No. 2, 2013, pp. 148-152

[14] B. Ma, B. C. Goh and E. H. Tan, "A Multicenter Phase II trial of 3-Aminopyridine-2-carboxaldehyde Thiosemicarbazone (3-AP, Triapine®) and Gemcitabine in Advanced Non-Small-Cell Lung Cancer with Pharmacokinetic Evaluation Using Peripheral Blood Mononuclear Cells," *Investigational New Drugs*, Vol. 26, No. 2, 2008, pp. 169-173.

[15] Md. Arifuzzaman, M. R. Karim, T. A. Siddiquee, A. H. Mirza and M. A. Ali, "Synthesis and Characterization of New Schiff Bases Formed by Condensation of 2,9-Phenathroline-1,10-dialdehyde with Sulfur-Containing Amines," *International Journal of Organic Chemistry*, Vol. 3, No. 1, 2013, pp. 81-86.

[16] G. M. Sheldrick, "Program for the Solution and Refinement of Crystal Structures," University of Göttingen, Göttingen, 1997.

[17] CrystalClear: Rigaku Corporation, "CrystalClear Software User's Guide, Molecular Structure Corporation," Rigaku Corporation, 1999, pp. 1718-1725.

[18] Crystal Crystallography, "WinGX Suite for Small-Molecule Single-Crystal Crystallography," *Journal of Applied Crystallography*, Vol. 32, 1999, pp. 837-838.

[19] S. K. Sahoo, N. Khatun, A. Gogoi, A. Deb and B. K. Patel, "Cu(II) Catalysed Chemoselective Oxidative Transformation of Thiourea to Thioamidoguanidine/2-aminobenzothiazole," *RSC Advances*, Vol. 3, No. 2, 2013, pp. 438-446.

[20] A. Thapper, A. C. Rizzi, C. D. Brondino, A. G. Wedd, R. J. Pais, B. K. Maiti, I. Moura and S. R. Pauelta, "Copper-Substituted Forms of the Wild Type and C42A Variant of Rubredoxin," *Journal of Inorganic Biochemistry*, Vol. 127, 2013, pp. 232-237.

24

Crystal and Molecular Structure of 2-Amino-3-Ethyl Carboxamido-4-Metyl-5-Carboxy Ethyl Thiophene

Dhananjay Dey[1], Venugopal Prakash[2], Vasu[3], Janardhanan Saravanan[4], Deepak Chopra[1*]

[1]Department of Chemistry, Indian Institute of Science Education and Research, Bhopal, India

[2]Shirdi Sai Engineering College, Bangalore, India

[3]Vivekananda Degree College, Bangalore, India

[4]PES College of Pharmacy, Bangalore, India

ABSTRACT

The crystal and molecular structure of 2-Amino-3-ethyl carboxamido-4-methyl-5-carboxy ethyl thiophene ($C_{11}H_{16}N_2O_3S$) has been investigated from single crystal X-ray diffraction data. The primary focus is to investigate the molecular geometry of this compound in the solid state along with the associated inter and intra-molecular hydrogen bonding and related weak interactions present in this molecule. This compound crystallizes in the monoclinic space group $P2_1/c$ with cell parameters, a = 8.1344(3) Å, b = 13.7392(4) Å, c = 11.4704(4) Å, β = 100.769(2)°, V = 1259.36 (7) Å^3, D = 1.352 g·cm^{-3}, Z = 4. The molecular geometry is stabilized by intra-molecular N-H...O=C and C-H...O interactions along with intramolecular C-H...N and C-H...O interactions which contribute towards the stability of the crystal packing.

Keywords: Crystal; Molecular Conformation; Intermolecular Interactions; Spectroscopy; Diffraction

1. Introduction

Thiophene derivatives [1] are of importance in medicinal chemistry and have recently been incorporated into new pharmaceutical and chemical compounds tested as anti-inflammatory agents [2]. This class of compounds exhibit pharmacological activity [3-5]. These are also useful in polymer chemistry because of their mechanical strength, ease of fabrication, flexibility in design, stability, resistance to corrosion and low cost [6]. In view of the importance of this class of heterocycles from a biological and pharmaceutical perspective, we report in this manuscript the synthesis of 2-Amino-3-ethyl carboxamido-4-methyl-5-carboxy ethyl thiophene. The compound has been purified and characterized spectroscopically using FT-IR, ^{1}H and ^{13}C NMR techniques. The purity of the phase has been established by powder X-Ray diffraction. Structural characterization of this compound has been achieved *via* single crystal X-ray diffraction study. Finally, an investigation of the CSD for related compounds containing the thiophene core has also been performed to compare the changes in geometry which accompany the introduction of a 3-ethyl carboxamide and 5-carboxy ester moiety on the thiophene ring.

2. Experimental

2.1. Synthesis of 2-Amino-3-Ethyl Carboxamido-4-Methyl-5-Carboxy Ethyl Thiophene

A mixture of ethyl acetoacetate (5.2 g; 0.04 mol), ethyl cyanocetate (4.52 g; 0.04 mol) and sulphur powder (1.28 g; 0.04 mol) in ethanol (40 ml) were added in a round bottomed flask. To this, morpholine (4.0 ml) was added dropwise with stirring [**Scheme 1**]. The mixture was stirred further for 1 h at 45°C - 50°C, cooled overnight in ice and the solid product obtained was filtered, washed and recrystallised from ethanol. Pink coloured crystals were obtained and these were used for diffraction purposes. Melting point: 106°C.

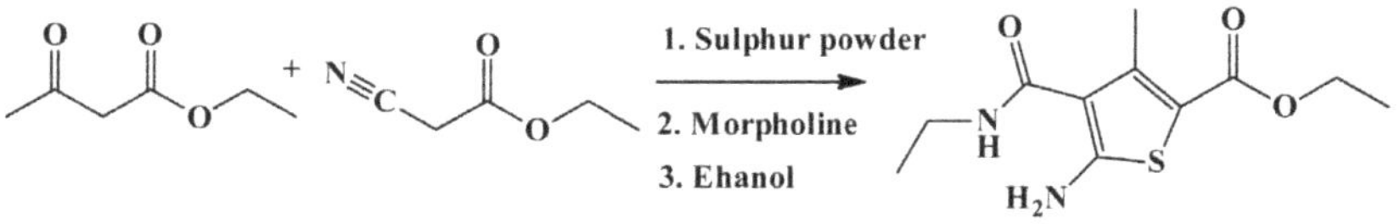

1. Sulphur powder
2. Morpholine
3. Ehanol

Scheme 1

*Corresponding author.

2.2. Spectroscopic Characterization (FTIR, ^{1}H and ^{13}C NMR) of the Synthesized Compound

FTIR (in cm^{-1}: KBr): 3408, 3290, 1681, 1660. ^{1}H NMR (400 MHz, $CDCl_3$): δ 6.43 (s, 2H), 4.24 (q, J = 7.13 Hz, 2H), 4.19 (q, J = 7.12 Hz, 2H), 2.63(s, 3H), 1.30 (t, J = 7.12 Hz, 3H), 1.26 (t, J = 7.12 Hz, 3H). ^{13}C NMR (400 MHz, $CDCl_3$): δ 166.13, 166.09, 162.88, 148.03, 108.53, 108.60, 60.41, 60.09, 16.12, 14.40, 14.34.

2.3. X-Ray Crystallography

Single-crystal X-ray diffraction data were collected on a three circle Bruker APEX-II diffractometer equipped with a CCD area detector using graphite monochromator and Mo-Kα radiation (λ = 0.70173 Å) in φ and ω scan modes. The crystal structure of this compound was refined by least-squares method on the basis of all observed reflections using SHELXL-97 [7] present in WinGx [8] (version 1.80). Empirical absorption correction was applied using SADABS [9]. All hydrogen atoms are fixed in geometrical positions. Non-hydrogen atoms are refined with anisotropic displacement parameters. The molecular connectivity was drawn using ORTEP [10] and the crystal packing diagram was drawn using Mercury (CCDC) program [11]. Geometrical calculations were done using PARST [12] and PLATON [13]. The geometrical optimization of the molecule was performed at the B3LYP/6-31G** level of calculation using TURBOMOLE [14]. The details of the crystal data, data collection and structure refinements are shown in **Table 1**.

3. Results and Discussion

This compound (**Figure 1**) crystallizes in the monoclinic centro-symmetric space group $P2_1/c$ with four asymmetric units in one unit cell. The crystal structure of the compound ($C_{11}H_{16}N_2O_3S$) contains one thiophene moiety. One ethyl amide group is connected with C(3) atom and one ethyl carboxyl group is attached with C(1) atom. The core structure of the molecule is approximately planar. The geometrical restrictions placed on the intermolecular H-bonds are the sum of the van der Waals radii + 0.4 Å and the directionality is greater than 110° [15]. **Table 2** lists all the intra-molecular and intermolecular interactions. The two intra-molecular C(11)-H(11A)...O(2) and N(2)-H(2A)...O(1) hydrogen bonds stabilize the molecular conformation. Strong N(2)-H(2B)...O(2) hydrogen bonds forms molecular chains along the crystallographic b-axis utilising the screw axis as a symmetry element, whereas weak C(10)-H(10A)...N(1) and C(6)-H(6B)... O(3) intermolecular interactions pack the molecules along the c-axis utilizing the glide plane. These interactions have been recognized as key elements for supramolecular association in the solid state [16-21]. The hydrogen bonding capacity of O(1) atom is more than the other oxygen atom present in the title molecule. All intermolecular interactions are shown in the packing diagram (**Figure 2**).

Selected bond distances are shown in **Table 3**. In **Table 4** the experimental torsion angles have been reported.

Table 1. Crystallographic and refinement data of T1.

Empirical formula	$C_{11}H_{16}N_2O_3S$
Formula weight	256.32 g/mole
Crystal colour	Pink
Temperature	298 K
Wavelength	0.71073 Å
Crystal system	Monoclinic
Space group	$P2_1/c$
Unit cell dimensions	a = 8.1344(3) Å, b = 13.7392(4) Å, c = 11.4704(4) Å, β = 100.769(2)°
Volume	1259.36 (7) Å^3
Z, Calculated density	4, 1.352 Mg·m^{-3}
Absorption coefficient	0.256 mm^{-1}
F(000)	544
Crystal size	0.2, 0.1, 0.1 mm
Theta range for data collection	2.34° to 27.43°
Limiting indices	$-10 \le h \le 10, -16 \le k \le 17, -14 \le k \le 14$
Reflections collected/unique	10475/2853[R(int) = 0.0275]
Completeness to theta = 27.43	0.99%
Max. and min. transmission	0.9749, 0.9506
Refinement method	Full-matrix least-squares on F^2
Data/restraints/parameters	2368/0/158
Goodness-of-fit on F^2	1.076
Final R indices [I > 2 sigma(I)]	R_1 = 0.0475, wR_2 = 0.1449
R indices (all data)	R_1 = 0.0655, wR_2 = 0.1554
Largest diff. peak and hole	0.563 e·Å^{-3}, and −0.637 e·Å^{-3}

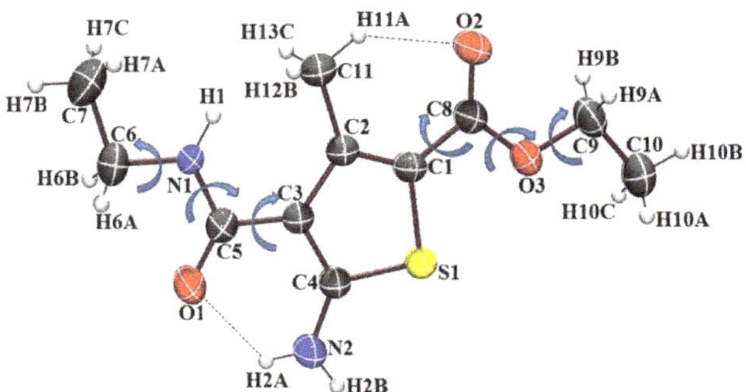

Figure 1. ***ORTEP*** **of the synthesized molecule drawn with 50% ellipsoidal probability. The dotted lines indicates in- tra-molecular N(2)-H(2A)...O(1) and C(11)-H(11A)...O(2) hydrogen bonds. Bending arrows are showing the torsion angles in the asymmetric unit.**

Table 2. Intra- and Intermolecular Interactions in the compound.

D-H...A	D-H (Å)	D...A (Å)	H...A (Å)	D-H...A (°)	Symmetry Code
C(11)-H(11A)...O(2)	1.08	3.027(3)	2.21	131	x, y, z
N(2)-H(2A)...O(1)	1.03	2.685(2)	1.99	122	x, y, z
C(10)-H(10A)...N(1)	1.08	3.593(3)	2.75	134	x + 1, −y + 1/2, z + 1/2
C(6)-H(6B)...O(3)	1.08	3.762(3)	2.79	150	x − 1, −y + 1/2, z − 1/2
C(10)-H(10B)...O(1)	1.08	3.762(3)	2.71	162	x + 1, y, z + 1
N(2)-H(2B)...O(2)	1.03	2.931(2)	1.95	158	−x, y + 1/2, −z + 1/2

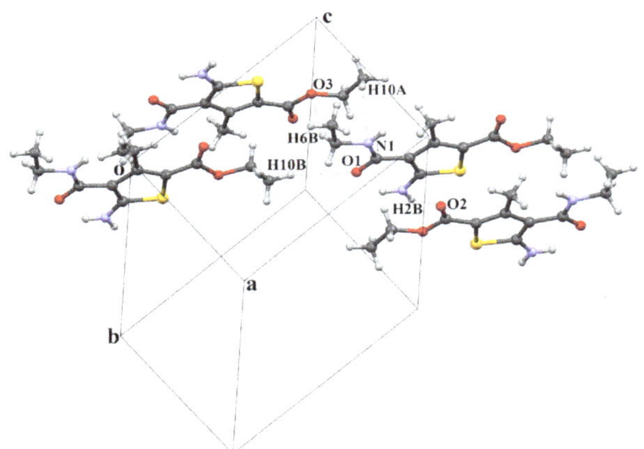

Figure 2. Packing diagram and intermolecular H bonds.

Table 3. Selected bond distances.

Bond	Distance (Å)
C(3)-C(5)	1.46(3)
C(5) = O(1)	1.21(3)
C(5)-N(1)	1.34(3)
N(1)-C(6)	1.45(3)
C(1)-C(8)	1.45(3)
C(8) = O(2)	1.21(3)
C(8)-O(3)	1.34(3)
O(3)-C(9)	1.44(2)

Table 4. Selected torsion angles in degree (°).

Torsion	Angles (°)
C(5)-N(1)-C(6)-C(7)	178.4(2), 159.6(1)[a]
C(6)-N(1)-C(5)-C(3)	179.0(2), 176.1(1)
C(4)-C(3)-C(5)-N(1)	179.8(2), 171.6(1)
C(2)-C(1)-C(8)-O(3),	177.3(2), 179.4(1)
C(9)-O(3)-C(8)-C(1)	180.0(2), 179.8(1)
C(8)-O(3)-C(9)-C(10)	176.3(2), 179.4(1)

([a]: Italicised values obtained from theoretical B3LYP/6-31G** calculation).

It is of interest to note that the torsion C(2)-C(1)-C(8)-O(3) and C(4)-C(3)-C(5)-N(1) are 177.2(2)° and 179.8(2)° indicating planarity with the thiophene ring assisted by delocalisation between the carboxy and carboxamide groups at C(1) and C(3) respectively. The theoretical B3LYP/6-31G** calculations, after geometrical optimization of the molecule, reveal torsion angles and these have been compared with the experimental values. In most of the cases the experimental torsion angles are compareable with the theoretical values. But for C(5)-N(1)-C(6)-C(7), the difference in torsion angle is approximately 18° - 19°, signifying the importance of crystal forces in the packing of molecules. In **Table 5** the search information, retrieved from the Cambridge Structural Database [22] on the presence of specific functional groups on the thiophene moiety has been presented. Search numbers 1 and 2 for the presence of carboxy ester and carboxamide moiety only on the thiophene ring revealed 1 hit only [Structures (A) and (B), **Table 5**]. Search numbers 3 and 4 revealed no hits. It is of interest to compare the torsion angles C(2)-C(1)-C(8)-O(3) and C(4)-C(3)-C(5)-N(1) in the present compounds with those in (A) [23] and (B) [24] respectively. These values are 179.9° and 170.2° respectively.

The phase purity of the compound has been verified by powder X-ray diffraction. It is of interest to note that the experimental and simulated powder patterns (generated from crystallographic coordinates) have a one-to-one correspondence, thereby confirming the single phase behaviour of the compound (**Figure 3**).

4. Conclusion

The title compound is of biological importance and the synthesis of related thiophene compounds is of significance. This is reflected from the CSD wherein related compounds having different functionalities are scarce and hence new compounds can be synthesised, characterized and investigated for their crystal structures. It is of interest to investigate polymorphism in such solids and screen such compounds for their medicinal property. These are expected to have concomitant commercial ramifications in the pharmaceutical industry.

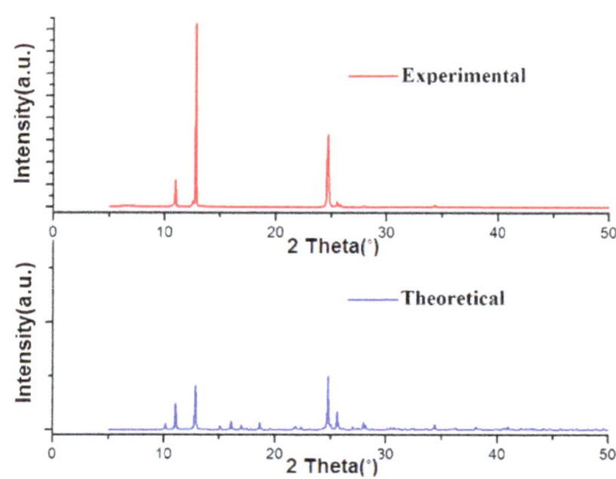

Figure 3. Experimental and theoretical powder pattern for the title compound.

Table 5. CSD [17] search information.

Search number	Moiety	No. of hits	Structure	Details
1	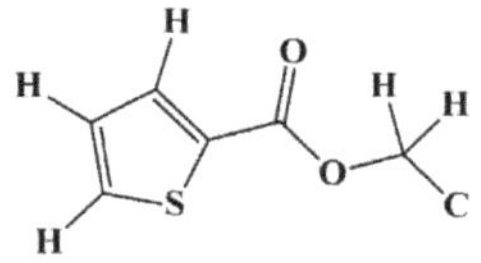	1	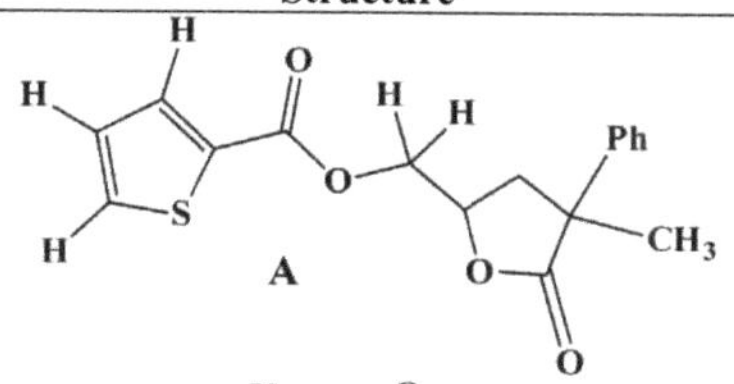 A	Space group = $P2_1/n$ Cell: a = 6.398 Å, b = 23.237 Å, c = 10.314 Å, β = 90.37° [23].
2	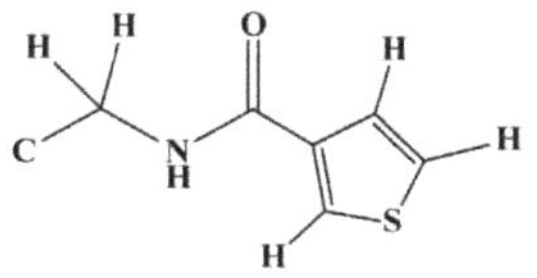	1	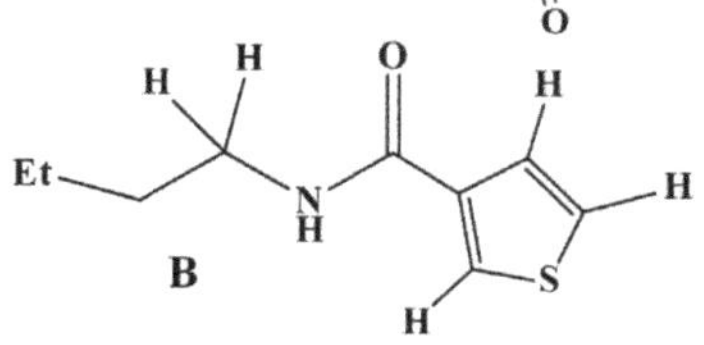 B	Space group = $Pna2_1$ Cell: a = 9.723(1) Å, b = 9.251(1) Å, c = 10.618(1) Å [24].
3	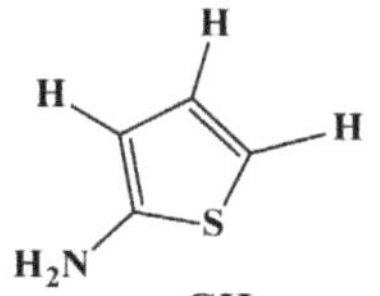	0	-	-
4	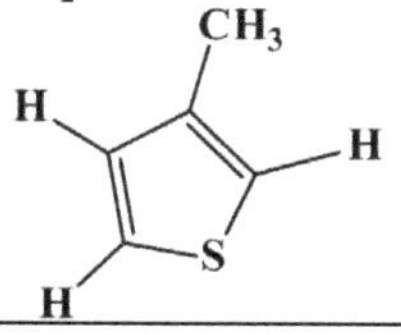	0	-	-

5. Acknowledgements

DC thanks IISER Bhopal for research facilities and DST-Fast track scheme for research funding.

REFERENCES

[1] R. Mishra, K. K. Jha, S. Kumar and I. Tomer, "Synthesis, Properties and Biological Activity of Thiophene: A Review," *Der Pharma Chemica*, Vol. 3, No. 4, 2011, p. 38.

[2] G. Darshan, D. Chopra and J. Saravanan, "N-(X-Methylphenyl)-2-{(Z)-[(2,3,4-trimethoxyphenyl)methylidene]amino}-4,5,6,7-tetrahydro-1-benzothiophene-3-carboxamide, where X = 2 and 3," *Acta Crystallographica*, Vol. C66, No. 12, 2010, p. 571.

[3] Vasu, K. A. Nirmala, D. Chopra, S. Mohan and J. Saravanan, *Acta Crystallographica*, Vol. C60, No. 11, 2004, p. 786.

[4] Vasu, K. A. Nirmala, D. Chopra, S. Mohan and J. Saravanan, *Acta Crystallographica*, Vol. C60, No. 9, 2004, p. 636.

[5] Vasu, K. A. Nirmala, D. Chopra, S. Mohan and J. Saravanan, *Acta Crystallographica*, Vol. E60, No. 2, 2004, p. 236.

[6] H. S. O. Chan and S. C. Ng, "Synthesis, Characterization and Applications of Thiophene-Based Functional Polymers," *Progress in Polymer Science*, Vol. 23, No. 7, 1998, pp. 1167-1231.

[7] G. M. Sheldrick, "SHELXL97, Program for Crystal Structure Refinement," University of Gottingen, 1997.

[8] L. J. Farrugia, "WinGX Suite for Small-Molecule Single-Crystal Crystallography," *Journal of Applied Crystallography*, Vol. 32, No. 4, 1999, pp. 837-838.

[9] G. M. Sheldrick, "A Short History of SHELX," *Acta Crystallographica*, Vol. A64, No.1, 2008, p. 112.

[10] L. J. Farrugia, "ORTEP-3 for Windows—A Version of ORTEP-III with a Graphical User Interface (GUI)," *Journal of Applied Crystallography*, Vol. 30, No. 5, 1997, p. 565.

[11] C. F. Macrae, I. J. Bruno, J. A. Chisholm, P. R. Edgington, P. McCabe, E. Pidcock, L. Rodriguez-Monge, R. Taylor, J. Streek and P. A. Wood, "Mercury CSD 2.0—New Features for the Visualization and Investigation of Crystal Structures," *Journal of Applied Crystallography*, Vol. 41, No. 2, 2008, pp. 466-470.

[12] M. Nardelli, "PARST95—An Update to PARST: A System of Fortran Routines for Calculating Molecular Structure Parameters from the Results of Crystal Structure Analyses," *Journal of Applied Crystallography*, Vol. 28, No. 5, 1995, p. 569.

[13] A. L. Spek, "PLATON-a Multipurpose Crystallographic Tool," Utrecht University, 2002.

[14] I. Dance, "Distance Criteria for Crystal Packing Analysis of Supramolecular Motifs," *New Journal of Chemistry*, Vol. 27, No. 1, 2003, pp. 22-27.

[15] Turbomole V6.3 2011, "A Development of University of Karlsruhe and Forschungszentrum Karlsruhe GmbH, 1989-2007, Turbomole GmbH," 2007. http://www.turbomole.com.

[16] G. R. Desiraju and T. Steiner, "The Weak Hydrogen Bond in Structural Chemistry and Biology," Oxford University Press, Oxford, 1999.

[17] G. R. Desiraju, "C–H...O and Other Weak Hydrogen Bonds. From Crystal Engineering to Virtual Screening," *Chemical Communication*, 2005, pp. 2995-3001.

[18] G. R. Desiraju, "Hydrogen Bridges in Crystal Engineering: Interactions without Borders," *Accounts of Chemical Research*, Vol. 35, No. 7, 2002, pp. 565-573.

[19] S. S. Kuduva, D. C. Craig, A. Nangia and G. R. Desiraju, "Cubanecarboxylic Acids. Crystal Engineering Considerations and the Role of C–H...O Hydrogen Bonds in Determining O–H...O Networks," *Journal of the American Chemial Society*, Vol. 121, No. 9, 1999, p. 1936.

[20] T. Steiner, *Chemical Communication*, No. 8, 1997, pp. 727-734.

[21] G. R. Desiraju, "The C–H···O Hydrogen Bond: Structural Implications and Supramolecular Design," *Accounts of Chemical Research*, Vol. 29, No. 9, 1996, pp. 441-449.

[22] CSD Version 5.33, "The Following Constraints Were Applied: No Refcode Restrictions Applied, 3D Coordinates Determined, R Factor ≤ 0.1, No Errors, Not Polymeric, No Ions, Only Organics," 2011.

[23] I. F. Cottrell, A. R. Cowley, L. J. Croft, L. Hymns, M. G. Moloney, E. J. Nettleton, H. K. Smithies and A. L. Thompson, "Acyloxylactonisations Mediated by Lead Tetracarboxylates," *Tetrahedron*, Vol. 65, No. 12, 2009, pp. 2537-2550.

[24] R. W. Gable, M. J. Laws, C. H. Schiesser, *Acta Crystallographica*, Vol. C53, No. 5, 1997, p. 641.

Analysis of Inhomogeneities in Hydrogen Storage Alloys: A Comparison of Different Methods

Francesco Massimino[1,2]
[1]Università degli Studi, Dipartimento di Chimica IFM, Turin, Italy
[2]NTT—New Tera Technology, Turin, Italy

ABSTRACT

In this work we have realized a simplified model to analyze compositional inhomogeneities in commercial hydrogen storage alloys. We have used it to evaluate the effect of the thermal annealing, together with calorimetric, PCIs and XRD measurement. We have compared results with composition distribution histograms based on Rietveld refinement of XRD patterns. Finally we studied the variation of α and β phases crystallographic parameters with H2 pressure.

Keywords: Hydrogen Storage; Alloys; Inhomogeneities; Sloping Plateau

1. Introduction

Pressure-Composition Isotherms (PCIs) are one of the most informative data when performances of hydrogen storage alloys need to be determined. Basically these curves should be composed, at least for simple systems, of three branches: a stiff rise at the beginning, a flat and wide plateau central region and a second stiff rise at the end. The first and the last part are related to hydrogen solubility, respectively in metal and in hydride. The central flat part is due to transition of metal into hydride phase: as every two-phase equilibrium at a certain temperature it should lay at one well-defined pressure. It is common to obtain a sloping plateau instead of a flat one. Many hypotheses have been proposed to explain it, including kinetic effects and establishment of local equilibriums [1], microstructure and internal stresses [2] and inhomogeneities of the alloy: Park *et al.* [3] proposed to use this feature of the PCIs as a diagnostic method for inhomogeneities.

In the following we tried to implement a simplified and easily applicable version of this model in the analysis of a commercial $LaNi_{4.8}Al_{0.2}$ alloy and we compared results with the ones we obtained from other adhoc experiments based on Rietveld refinement of XRD pattern of both metal powder and partially hydrided powder.

2. Experimental

Experiments have been carried on $LaNi_{4.8}Al_{0.2}$ from Palcan Energy Corporation and 5.0 grade hydrogen has been used.

As-received alloy has been observed in SEM and EDS microanalysis has been used to determine coarsely the average composition of the alloy.

Part of the alloy has been annealed in Ar for one week at 1000°C.

XRD patterns of as-received and annealed alloys have been collected both in vacuum and at different hydrogen pressures, PCIs and calorimetric curves have been measured (see below for further details of each part).

Alloys have been activated following this general procedure for hydrogen applications:

- Preliminary coarse grinding;
- In-vacuum heating to 100°C;
- Soak in high hydrogen pressure (15 - 50 bar depending on the apparatus used);
- Cooling down to room temperature after one hour.

This procedure has been repeated twice after each air exposure of samples.

2.1. Pressure-Composition Curves

PCIs curves have been measured with two different apparatus.

The first one is an automatic Sievert AMC Unit, used to trace PCI curves at temperatures in the range of 25°C - 120°C. The instrument is equipped with an oven and a temperature controller and provides corrections for ΔT between hydrogen reservoir and specimen region. Hydrogen sorbed amounts are calculated using modified Benedict-Webb-Rubin equation for real gases.

The second apparatus is a simplified version of the first one: it consists of a manually operated high pressure steel line, in which ideal gas law is used to evaluate hydrogen sorbed amount. In this case there is no ΔT correction: it is accurate for measurement at room temperature, but it tends to overestimate/underestimate sorbed amounts at temperatures lower/higher than room temperature. The same apparatus is equipped with a C80 calorimeter that allows to measure reaction enthalpies. Pressure is measured with a capacitive transducer.

Reaction enthalpies and entropies can also be estimated from Van't Hoff equation, mathematically operating on PCI curves at different temperatures.

2.2. C80 Calorimeter

Setaram C80 DTA calorimeter is used together with the steel high pressure line previously mentioned. The reference cell is identical to the specimen one; the two cells are connected, so they undergo the same pressure conditions. Signal has been amplified and collected through a LabView custom interface. All measures have been done at constant temperature, thanks to a heater and a temperature controller.

A considerable quantity of sample (6 g) has been used in experiment: in this way it was possible to observe good S/N ratio in the thermal signal and to minimize error in determining sorbed amount, without changing too much hydrogen concentration in the material. In order to improve reproducibility and standardization of the measurement, a fixed cut-off has been introduced in the thermal peak: this avoids as much as possible the effect of thermal oscillation but it may introduce some error, especially in measurements in the plateau region, that is very temperature-dependent; the recovery of the initial temperature may take a really long time due to thermal ballast action operated by hydrides: cut-off may eliminate part of the signal, but at least this error is systematic for all measurement.

Ratio between the integral of thermal peak and the sorbed quantity of hydrogen, that represents enthalpy of reaction, is converging with time and someway reduces the negative effect of the introduction of the cut-off.

Empty cell tests confirmed that heat involved with gas expansion in and from the cell region is negligible in those experiments.

Generally each part of the specimen is considered as it has sorbed the same hydrogen amount at the end of each sorption step; this fact may not be true: due to hysteresis phenomena and really quick reaction rate, for instance, upper part of the sample may absorb more than the lower one and not returning hydrogen at the considered pressure, causing non-homogeneities along the vertical axis.

2.3. X-Ray Diffraction

XRD has been performed with a Phillips X'pert in Bragg-Brentano configuration, equipped with plate sample-holder or XRK900 reactive stage, with Cu X-ray tube; XRK900 has Be windows, Ni filter has been removed in this case to maximize signal and improve acquisition speed.

The sample region is linked to a high pressure and vacuum line that allows imposing a specific hydrogen pressure. Pressure is measured with a capacitive transducer; we have estimated a pressure drop between transducer and cell of 0.5 bar in the 1 - 3 bar region due to resistive effect of the pipes.

We have performed XRD at different fractions of the plateau region. Measures have been executed at the same temperature of the calorimetric one.

For each sorption step several XRD pattern has been collected quickly (0.75°/min) and consequently in a short degree range (2θ = 38° ~ 44°) containing the two major peaks of α and β phases; when XRD pattern was not changing any more, a large degree range (2θ = 25° ~ 75°) XRD patterns have been collected, also with longer acquisition time (0.33°/min).

With plate sample-holder we averaged four acquisitions in an even wider range (2θ = 20° ~ 100°) with longer acquisition time (0.33°/min).

Rietveld analysis has been accomplished using MAUD software [4] on large XRD patterns. Instrumental broadening function has been built using standard LaB6. Due to the removal of Ni filter, also kβ has been introduced in the Rietveld simulation for XRD pattern acquired with reactive chamber stage.

First we have refined Rietveld simulation of a single phase for in vacuum XRD pattern and from this step we have obtain single phase cell parameters. Then we have introduced in the simulation more metal phases with slightly different Al contents. In the second case we have kept constant B factors and microstructure parameters in addiction to previous ones. Relative phase fraction has been achieved from the computation.

In the case of reactive chamber XRD patterns, Rietveld refinement has been performed introducing two phases, α-solution and β-hydride; Al occupancy has been fixed to reach nominal stoichiometry of the alloy. Cell parameters and phase ratio have been obtained from the computation.

3. Data Treatment and Calculation

For the following calculations we needed functions describing $p_{pl}(x_{Al})$, $a(x_{Al})$ and $c(x_{Al})$ and where p_{pl} is plateau pressure, a and c are cell parameters and x_{Al} is the Al content. We have described p_{pl} with Van't Hoff equation.

$$\ln p_{pl} == -\frac{|\Delta H|}{RT}+\frac{|\Delta S|}{R} \tag{1}$$

so we had also to determine $\Delta H(x_{Al})$ and $\Delta S(x_{Al})$.

We data mined in literature [5-8] and in our previous lab measurements for these parameters and we elaborated linear expressions for desired functions. Available data allowed us to determine solidly $\Delta H(x_{Al})$ and $\Delta S(x_{Al})$ for absorption process only.

Linear parameters are shown in **Table 1**.

Low resolution PCIs, hydrogen wt% vs. $\ln(p)$, have been fitted with a sigmoidal curve to better obtain their derivative curve: with this assumption we hypothesize gaussian distribution of inhomogeneities as Fujitani *et al.* [9] did.

The complete expression (2) Park *et al.* [3] proposed has been significantly simplified, once they have verified the distribution character of their function.

$$f(\ln p) = \frac{\mathrm{d}\theta / \mathrm{d}\ln p - k(\theta)}{\theta_\beta - \theta_\alpha}$$

In this expression θ indicates H/M atom ratio, α and β are metal and hydride phase and θ_α and θ_β are phase boundaries.

$$k(\theta) = \frac{k_b(\theta-\theta_\alpha)+k_a(\theta_\beta-\theta)}{\theta_\beta-\theta_\alpha} = \frac{\theta(k_b-k_a)+k_a\theta_\beta-k_b\theta_\alpha}{\theta_\beta-\theta_\alpha} \quad (3)$$

In Equation (3) k_b and k_a are the slopes of the $\theta(\ln p)$ PCI in first and last branch; because of the almost horizontality of these branch in our case the numerical value is very low and even more their difference that may be considered zero. $K(\theta)$ is almost constant and it lowers the value of $f(\ln p)$, amplifying the central values of the distribution and narrowing the function. If $K(\theta)$ is ne--glected the curve obtained may result broadened proportionally to the values of k_b and k_a in respect to the one obtained from the full calculation.

The denominator $(\theta_\beta - \theta_\alpha)$ assumes just a normalizing role and the derivative $\mathrm{d}\theta/\mathrm{d}\ln p$ remains the only fundamental factor; the normalization of the distribution can be imposed *a posteriori* at the end of the data treatment.

Note that the usage of H/M atom ratio is no longer required and one can choose his preferred unit, e.g. hydrogen wt%.

The last step was converting $f(\ln p)$ into $f(x_{Al})$ using the Van't Hoff equation and linearized $\Delta H(x)$ and $\Delta S(x)$ expressions.

Table 1. Cell and thermodinamic parameters vs. Al contents, linearized expressions.

	a	c	ΔH	ΔS
Intercept	5.0175(8)	3.976(3)	–31.10(6)	–109.0(1.6)
Slope	0.0362(6)	0.0889(6)	16.6(4)	–9.7(3)

For what concerns the calorimetric curve, we have defined an average weight percent to avoid problems cited at the end of C80 paragraph: we have associated each measured ΔH to the mean value between current hydrogen wt% and the one related to previous step, that is lower for absorption and higher for desorption.

In the multi-phase Rietveld refinement we have avoided to simulate the presence of too many phases, this would have introduced too many degree of freedom. We have used a maximum of five phases, fixed in Al content and cell parameters; they have been "prepared" using the linearized function for $a(x)$ and $c(x)$. The phases differ by a constant Δx_{Al} value and each one is representative of a class of the same width. We have repeated the refinement using a different sampling of the phase space, namely "preparing" other phases differing $\Delta x_{Al}/2$ from each one of the previous. After that we averaged the partially superposed classes, obtaining a histogram composed of classes with half broadness.

Finally in order to build a distribution based on the reactive chamber experiments, we used hydride-metal phase ratio obtained from the Rietveld computations and we associated each increment in the hydride quantity to the reached pressure. Then we converted pressure into x_{Al} as done in the previous. The classes obtained in this way are not of the same wideness and cannot be used to plot a distribution histogram: for that reason we have normalized the value of each class dividing it by the class wideness and we normalized the whole distribution a posteriori.

4. Results

SEM imaging could not highlight inhomogeneities both in the as-received and annealed alloy. Also EDS analysis variability stays widely inside experimental uncertainly and gives atomic composition of La 18.6%, Ni 78.8% and Al 2.6%, corresponding to an Al/(Al + Ni)*5 ratio of 0.16, a bit lower than nominal one of 0.2.

XRD peaks, as shown in **Figure 1**, broaden for as-received alloy: this is particularly evident for (00*l*) peaks. Sharper peaks may be an index of less strained microstructure, as expected after annealing.

PCIs plateau flatten after annealing; PCIs of as-received and annealed alloys cross exactly at midpoint of plateau and hysteresis remains of the same magnitude. In **Figure 2** PCIs collected at 313 K with manual instrument for both alloys are shown. The flattening of the plateau after annealing is a clear clue suggesting the presence of inhomogeneities in the received alloy.

As shown in **Figure 3**, reaction enthalpy, determined with C80, is constant for the annealed alloy and it is about 33.8 kJ/mol, while the result for the as-received alloy is higher for first-absorbed (or last-desorbed) hy

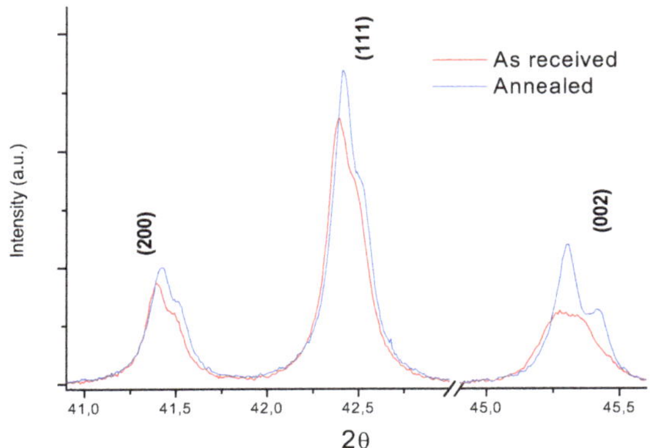

Figure 1. XRD peaks of as-received and annealed alloys.

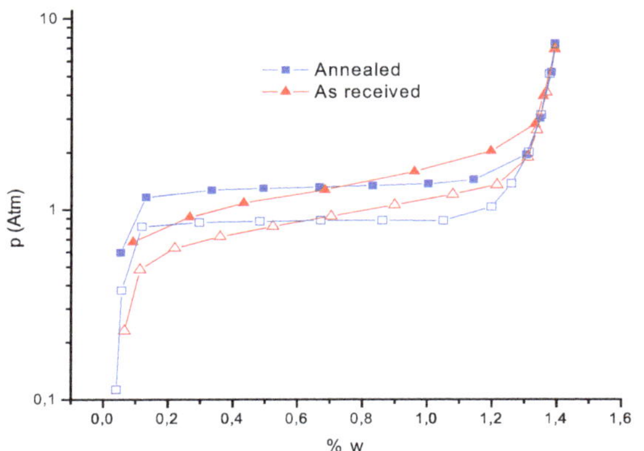

Figure 2. PCIs at 313 K of as-received and annealed alloys.

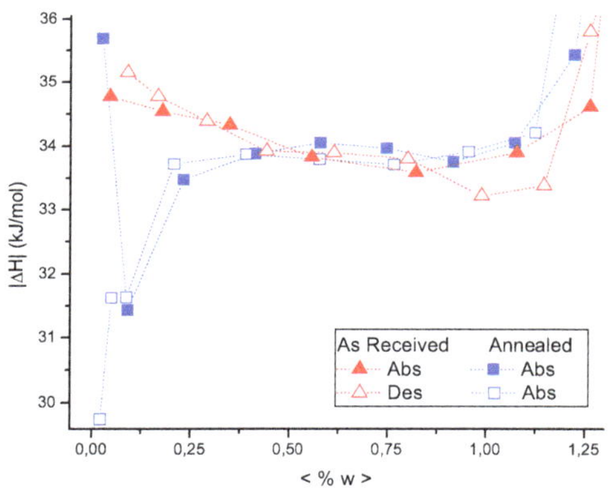

Figure 3. Calorimetric determined |ΔH| at 313 K as-received and annealed alloys; this data have been collected contemporary to the PCIs of Figure 4.

drogen molecules and lower toward the final part of plateau. High heat of reaction is related to higher Al-content. Calorimetric curve is confirming that the modification of PCIs after annealing is not a merely kinetic effect.

Execution of the experiment with reactive chamber was greatly more problematic in the case of the annealed alloy: in fact the presence of a sloping plateau allowed us to explore more systematically the sample in the whole central range, while the flat plateau forces the alloy to pass abruptly from an almost pure α solid solution to a completely hydrided material β. For this reason there is a gathering of data to the sides of the plateau.

Figure 4 sums up the results of Rietveld refinement coming from this experiment.

Cell volume of the α phase suddenly increases after the first hydrogen load; α cell volume of as-received alloy shrinks with the rise of hydrogen pressure after the same initial expansion. In the case of β phase, for as-received there is a gradual increment of volume, while in the annealed alloy it remains almost constant until a stiff rise in the last part.

The same trend can be directly and qualitatively observed from the XRD pattern; the main peaks of both phases are shown in **Figure 5** that underlines also the high variability of (002) hydride peak in relation to the increasing H_2 pressure.

Figure 6 presents the distributions and the histograms obtained from the various techniques previously described.

Every distribution shows a narrowing effect of the annealing process. Manual and automatic PCI instrument results are almost perfectly superposing, in particular for as-received sample. Centers of XRK and sloping plateau distributions coincide while the multi-phase Rietveld refinement gives results at substantially higher x_{Al} values.

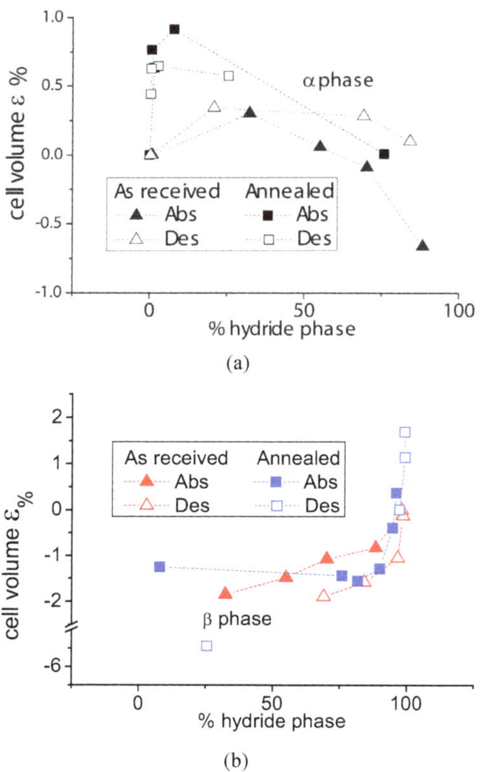

Figure 1. Cell volume variation: α phase take 0 bar phase as reference (a); β phase take 10 bar phase as reference (b).

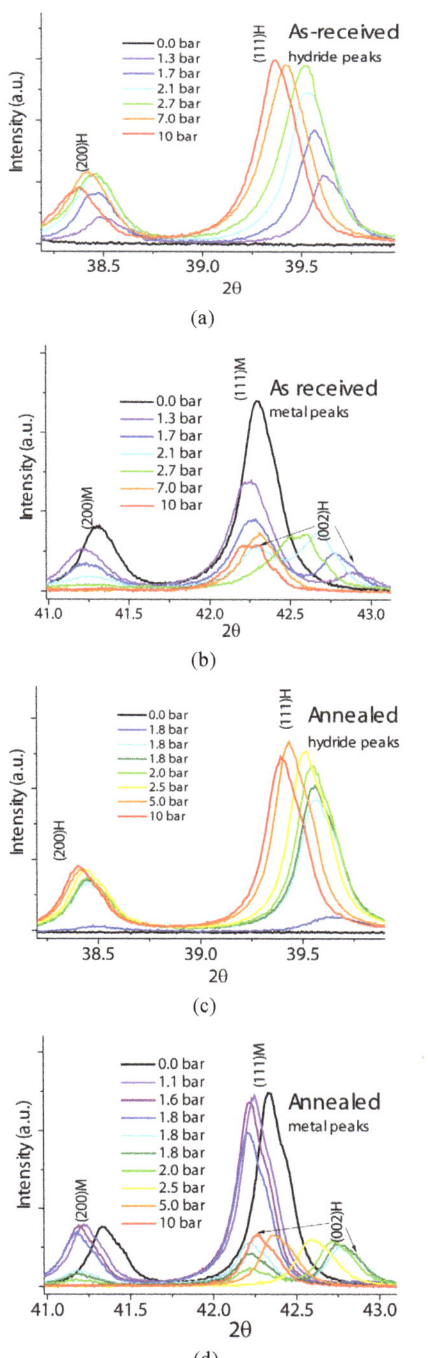

Figure 5. Main XRD peaks of α and β phase during the reactive chamber experiment.

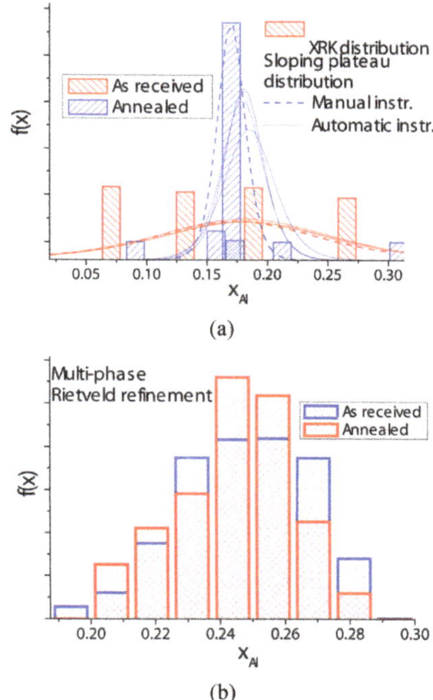

Figure 6. Distributions from various techniques: (a) histograms from reactive chamber experiment, lines from sloping plateau simplified model, dashed and continuous ones come from manual and automatic instrument data respectively; (b) histograms come from multi-phase Rietveld refinement.

5. Discussion

Almost all the analyses performed on the bought alloy suggest that the material is not homogeneous; variations in Al content, and consequently on Ni one, change molecular weight of the unit formula of the alloy of 33 g/mol per unitary Δx_{Al} : even hypothesizing a variation of 0.3 on xA_l it is not enough to appear with SEM imaging or to be clearly detected with EDS micro-analysis. For this reason we have no information about the spatial distribution of the inhomogeneities. Repeating of the PCIs measurement on specimen taken from other parts of the ingot brings to the same results: inhomogeneities are substantially equally spread all over the material.

The source of inhomogeneities is still not clear, but one hypothesis may be segregation acting on each grain during solidification from the melt.

c cell parameter exhibits a greater dependency from both x_{Al} and hydrogen concentration (*i.e.* pressure), as shown by the higher value of slope for linearized cx_{Al} and shift of the (00*l*) peaks. Notice that percent variation of $c(x_{Al})$ is even greater due to smaller dimension of *c* compared to *a*. This relative stability of the *a* parameter allows the growth of phases with slightly different x_{Al} with relative low strain, at least in some direction. As a consequence XRD pattern is not so different after annealing, expecially for (*hk*0) peaks.

Calorimetric measurement confirmed the presence in the as-received alloy of Al-richer and Al-poorer fractions, fraction that tends to the average composition after the annealing: this fact is supported by the midpoint interception of PCIs collected both before and after annealing.

Trend of cell volume of α and β is the result of three different phenomena:

1) H_2 rising pressure increments the gas solubility in both α and β phase, the higher is the hydrogen dissolved in a phase the bigger is its crystalline cell;

2) In inhomogeneous alloys Al-richer fractions are reacting at lower H_2 pressure, and the more pressure rise, the more unreacted fraction is made of Al-poorer phases, which are the last to transform into β phase; Al-poorer has smaller cell parameters, so the average a and c decrease in both α and β phases, increasing H_2 pressure for this effect;

3) As every material, this alloy may shrink due to increasing hydrostatic pressure and re-expand when pressure lowers; considering the low pressures involved this factor may be the less important, at least in the plateau region.

Observing cell volume ε% trends one may conclude that:

1) Effect is very strong for β phase and really low, except for first hydrogen load, for α solid solution; this effect is basically the only one observed in the right absorption branch of annealed alloy; this difference of behavior may correspond to a dissimilarity in Sievert's constants;

2) Effect is present in as-received alloy for α phase but is overcome by *i.* effect for β phase;

3) Effect may explain the increase of hydride cell volume as a relaxation while pressure is lowered after a complete hydrogen load.

There is agreement between sloping plateau model distributions and the ones based on reactive chamber experiments. In those cases the conversion between $\ln p$ and x_{Al} is based on the same thermodynamic literature data ΔH and ΔS. The automatic instrument curve of the annealed alloy is a bit broadened respect to the one coming from manual measurement: in the second apparatus the time between one step and the following is extremely longer than in the first one, due to the waiting needed to have the calorimeter correctly settled; for this reason the automatic measure may be influenced by some kinetic effects and same effects propagate to the distribution curve.

The multi-phase Rietveld distribution overestimates average x_{Al}. This problem seems to be intrinsic of the method. If one tries to simulate a single phase XRD over the same experimental data, obtained cell parameters correspond to a correct x_{Al} (0.17 in this case); once more phases have been introduced, the ones with bigger cell size gain in scale factor during the refinement and higher $\langle x_{Al} \rangle$ is obtained at the end of the process. Al content has not as its only effect the increase of the cell size, but it also changes relative heights of the XRD peaks in a nonlinear way. According to this statement the best agreement between experimental data and simulated curve may be found in different conditions in the single-phase refinement respect to the multi-phase one. Literature data come from a "single-phase library" and it may be not so correct to use it for a multi-phase simulation.

Considering this, the only information this multi-phase simulation is correctly furnishing is that there is an effective narrowing of the inhomogeneities distribution after annealing.

6. Conclusions

Commercial hydrogen storage alloys may have compositional inhomogeneities related to their production processes.

The simplified model we proposed seems to be a quick, good and simple way to estimate the variations in composition of one element x in an alloy belonging to a system in which the function $p_{pl}(x)$ is known. A low resolution PCIs is sufficient for the whole calculation. This simplified model could be extended, with the knowledge of more functions (e.g. $\Delta H(x)$ and $\Delta S(x)$) and the availability of PCIs at different temperatures, to a more complex system with three or more varying atomic species.

Three effects are acting in the variation of the crystallographic parameters with H_2 pressure and their trends may be very complex.

Multi-phase Rietveld refinement is probably not a fully trustable analytical technique with the available data for this kind of material; however it may be an interesting field into systematically inquire to build an alternative data library for future studies.

REFERENCES

[1] W. A. Oates and T. B. Flanagan, "On the Origin of Increasing Hydrogen Pressures in the Two Solid Phase Regions of Intermetallic Compound-Hydrogen Systems," *Scripta Metallurgica*, Vol. 17, No. 8, 1983, pp. 983-986.

[2] D. Wang, T. B. Flanagan and T. Kuji, "Hysteresis Scans for Pd-H and Pd-alloy-H Systems," *Physical Chemistry Chemical Physics*, Vol. 4, No. 17, 2002, pp. 4244-4254.

[3] C. Park, "Analysis of Sloping Plateaux in Alloys and Intermetallic Hydrides I. Diagnostic Features," *Journal of Alloys and Compounds*, Vol. 384, No. 1-2, 2004, pp. 203-207.

[4] L. Lutterotti, "Maud-Materials Analysis Using Diffraction," 2011.

[5] C. Da-li, *et al.*, "Effects of Al Partial Substitution for Ni on Properties of $LaNi_{5-x}Al_x$," *Transactions of Nonferrous Metals Society of China*, Vol. 17, No. S1, 2007, pp. s967-s971.

[6] J. Liu, *et al.*, "Electrochemical Characterization of $LaNi_{5-x}Al_x$ (x = 0.1 - 0.5) in the Absence of Additives," *Journal of Power Sources*, Vol. 161, No. 2, 2006, pp.

1435-1442.

[7] T. Kodama, "The Thermodynamic Parameters for the $LaNi_{5-x}Al_x$-H_2 and $MmNi_{5-x}Al_x$-H_2 Systems," *Journal of Alloys and Compounds*, Vol. 289, No. 1-2, 1999, pp. 207-212.

[8] H. Senoh, *et al.*, "Systematic Investigation on Hydrogen Storage Properties of RNi_5 (R: Rare Earth) Intermetallic Compounds with Multi-Plateau," *Materials Science and Engineering B*, Vol. 108, No. 1-2, 2004, pp. 96-99.

[9] S. Fujitani, *et al.*, "Development of Hydrogen-Absorbing Rare-Earth Ni-Alloys for a 20-Degrees-C Refrigeration System," *Journal of Alloys and Compounds*, Vol. 192, No. 1-2, 1993, pp. 170-172.

Computational Study of Induction Heating Process in Crystal Growth Systems—The Role of Input Current Shape

Mohammad Hossein Tavakoli[1*], Tayebe Nadery Mostagir[2]
[1]Physics Department, Bu-Ali Sina University, Hamedan, Iran
[2]Department of Physics, Kurdistan University, Sanandaj, Iran

ABSTRACT

A set of 2D steady state finite element numerical simulations of electromagnetic fields and heating distribution for an oxide Czochralski crystal growth system was carried out for different input current shapes (sine, square, triangle and sawtooth waveforms) of the induction coil. Comparison between the results presented here demonstrates the importance of input current shape on the electromagnetic field distribution, coil efficiency, and intensity and structure of generated power in the growth setup.

Keywords: Computer Simulation; Induction Heating; Czochralski Method; Growth from Melt; Metals

1. Introduction

Radio frequency induction heating is frequently used in crystal growth technology. The process principle consists of applying an alternating current in a conductor or coil called inductor (RF-coil) that generates an alternating electromagnetic field in the space. The alternating electromagnetic field induces eddy currents in metal crucible where the crystal material is placed and should be to melt. These currents lead to Joulean heating (RI^2) of the crucible in the form of temporal and spatial volumetric heating. Distribution and control of the induced power along the crucible cross-section and length are quite important which result in temperature difference and flow field in the growth setup [1-3].

In order to produce the required heating pattern within a metal crucible and afterheater it is necessary to accurately model and predict the electromagnetic field and the eddy currents distribution produced by the RF-coil under different operating conditions such as geometry and orientation of the metallic parts, cross section of the coil turns, the crucible shape and position, and frequency choice [4-7]. Selection of input current shape is another critical issue, which is particularly important for certain selective heating applications. In this article, we try to investigate the effects of different input voltage shapes, *i.e.*, sine, square, triangle and sawtooth waveforms (**Figure 1**) on the strength and distribution of the electromagnetic fields and heat generation in a Czochralski setup using the mathematical modeling and computer simulation. It should be noted, however, that despite of the differences in the patterns, each pattern is periodic. This point is important for our analysis of the driving current shape, *i.e.*, they can be represented as closely as desired by the combination of a sufficiently large number of sinusoidal patterns that form a harmonic series (Fourier series). Every non-sinusoidal current pattern consists of a fundamental and a complement of harmonics, which can be considered as a superposition of sine pattern of a fundamental frequency ω and integer multiples of that frequency [8].

2. Mathematical Model

2.1. Governing Equations

Since the real induction heating process is very complex, we make some simplifying assumptions in our approach. In our mathematical model used for numerical calculations, we make the following five assumptions: 1) the heating system is rotationally symmetric about the z-axis, so that all quantities are independent of the azimuthal coordinate φ; 2) all materials are isotropic, non-magnetic and have no net electric charge; 3) the displacement current is neglected; 4) the distribution of driving electrical current (also voltage) in the RF-coil is uniform; and 5)

*Corresponding author.

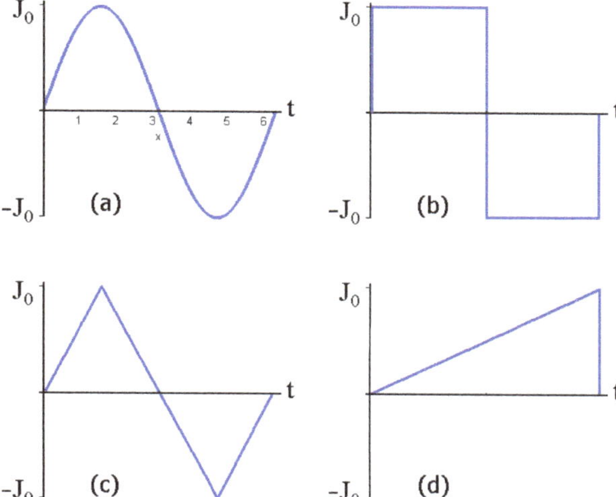

Figure 1. Four input current shapes (a) Sine; (b) Square; (c) Triangle; and (d) Sawtooth waveforms of the induction coil.

the driving and induced currents have only one angular component (*i.e.*, φ-direction). Under these assumptions, the governing equations are [4];

$$\frac{\partial}{\partial r}\left(\frac{1}{r}\frac{\partial \psi_B}{\partial r}\right)+\frac{\partial}{\partial z}\left(\frac{1}{r}\frac{\partial \psi_B}{\partial r}\right)=\hat{\mathcal{L}}\psi_B=-\mu_0 J_\varphi \quad (1)$$

where

$$\hat{\mathcal{L}}=\frac{\partial}{\partial r}\left(\frac{1}{r}\frac{\partial \psi_B}{\partial r}\right)+\frac{\partial}{\partial z}\left(\frac{1}{r}\frac{\partial \psi_B}{\partial r}\right) \quad (2)$$

and

$$J_\varphi=\begin{cases} J_d^{\text{coil}}+J_e^{\text{coil}}=J_d^{\text{coil}}-\dfrac{\sigma_{co}}{r}\dfrac{1}{r}\dfrac{\partial \psi_B}{\partial t} \\ \qquad \text{driving and eddy currents in the coil} \\ J_e^{\text{crucible}}=-\dfrac{\sigma_{cr}}{r}\dfrac{1}{r}\dfrac{\partial \psi_B}{\partial t} \text{ eddy currents in the crucible} \end{cases} \quad (3)$$

in which ψ_B is the magnetic stream function defined by $\psi_B(r,z,t)\equiv rA_\varphi(r,z,t)$, where A_φ is the azimuthal component of the vector potential, (r,z) the cylindrical coordinates, J the charge current density, σ the electrical conductivity, μ_0 the magnetic permeability of free space and t the time.

The energy dissipation rate in all metallic parts (coil, crucible and afterheater) is computed as

$$P(r,z,t)=\frac{J_\varphi^2}{\sigma} \quad (4)$$

Finally we average over one period to obtain the volumetric heat generation rate (*i.e.*, the time averaged quantity),

$$q(r,z)=\frac{\omega}{2\pi}\int_0^{\frac{2\pi}{\omega}} \mathrm{P}(\mathrm{r},\mathrm{z},\mathrm{t})\,\mathrm{dt} \quad (5)$$

where ω is the frequency of the electrical current in the induction coil.

a) Sine Waveform

Assuming the driving current in the RF-coil as a sine form $J_d=J_0\sin\omega t$, we can consider a solution of the form

$$\psi_B^{\text{sine}}=S(r,z)\sin wt+C(r,z)\cos wt \quad (6)$$

where $S(r,z)$ is the in-phase component and $C(r,z)$ is the out-of-phase component of the solution.

Now the coupled set of elliptic PDE's for $S(r,z)$ and $C(r,z)$ is:

$$\hat{\mathcal{L}}S^{\text{sine}}=\begin{cases} -\mu_0\left(J_0+\dfrac{\sigma_{co}\omega}{r}C\right) \text{coil} \\ -\mu_0\dfrac{\sigma_{co}\omega}{r}C \text{ crucible} \\ 0 \text{ elsewhere} \end{cases} \quad (7)$$

$$\hat{\mathcal{L}}C^{\text{sine}}=\begin{cases} \mu_0\dfrac{\sigma_{co}\omega}{r}S \text{ coil} \\ \mu_0\dfrac{\sigma_{co}\omega}{r}S \text{ crucible} \\ 0 \text{ elsewhere} \end{cases} \quad (8)$$

where $\hat{\mathcal{L}}$ is the linear operator defined in (2).

After solving (7) and (8) for $S(r,z)$ and $C(r,z)$, the eddy currents distribution and the energy dissipation rate can be computed via

$$J_e^{\text{sine}}=-\frac{\sigma}{r}\frac{\partial \psi_B}{\partial t}=-\frac{\sigma\omega}{r}\left[S(r,z)\cos wt-C(r,z)\sin wt\right] \\ =J_C\cos wt+J_S\sin wt \quad (9)$$

and

$$P(r,z,t) \\ =\frac{J_\varphi^2}{\sigma} \\ =\begin{cases} \dfrac{\sigma_{co}\omega^2}{r^2}\left[S^2+\left(\dfrac{J_{0r}}{\sigma_{co}\omega}+C\right)^2+S\left(\dfrac{J_{0r}}{\sigma_{co}\omega}+C\right)+\sin\omega t\right] \text{coil} \\ \dfrac{\sigma_{cr}\omega^2}{r^2}\left(S^2+C^2-CS\sin 2\omega t\right) \text{crucible} \end{cases} \quad (10)$$

Consequently, the volumetric heat generation rate is

$$q^{\text{sine}}(r,z)=\begin{cases} \dfrac{\sigma_{co}\omega^2}{r^2}\left[S^2+\left(\dfrac{J_{0r}}{\sigma_{co}\omega}+C\right)^2\right] \text{coil} \\ \dfrac{\sigma_{cr}\omega^2}{r^2}\left(S^2+C^2\right) \text{crucible} \end{cases} \quad (11)$$

b) Square Waveform

The square waveform of the driving current in the RF-coil can be approximated by a sum of harmonics using Fourier series as

$$J_d = J_0 F^{\text{square}}(t) = \frac{4J_0}{\pi}\sum_{n=1}^{\infty}\frac{\sin(2n-1)\omega t}{2n-1} \tag{12}$$

and then

$$\begin{aligned}\psi_B^{\text{square}} &= \sum_{n=1}^{\infty}\psi_{B(n)}^{\text{square}} \\ &= \sum_{n=1}^{\infty}\left[S_n\sin(2n-1)\omega t + C_n\cos(2n-1)\omega t\right]\end{aligned} \tag{13}$$

$$\begin{aligned}J_e^{\text{square}} = \sum_{n=1}^{\infty}J_{e(n)}^{\text{square}} = \sum_{n=1}^{\infty}\frac{(2n-1)\sigma\omega}{r}\big[&C_n\sin(2n-1)\omega t \\ &-S_n\cos(2n-1)\omega t\big]\end{aligned} \tag{14}$$

$$\hat{\mathcal{L}}S_n^{\text{square}} = \begin{cases} -\mu_{co}\left[\dfrac{4J_0}{(2n-1)\pi} + \dfrac{(2n-1)\sigma_{co}\omega}{r}C_n\right] & \text{coil} \\ -\dfrac{(2n-1)\sigma_{cr}\omega\mu_{cr}}{r}C_n & \text{crucible} \\ 0 & \text{elsewhere}\end{cases} \tag{15}$$

$$\hat{\mathcal{L}}C_n^{\text{square}} = \begin{cases} \dfrac{(2n-1)\sigma_{co}\omega\mu_{co}}{r}S_n & \text{coil} \\ \dfrac{(2n-1)\sigma_{cr}\omega\mu_{cr}}{r}S_n & \text{crucible} \\ 0 & \text{elsewhere}\end{cases} \tag{16}$$

$$q^{\text{square}}(r,z) = \sum_{n=1}^{\infty}q_n^{\text{square}}(r,z)$$

$$= \begin{cases} \displaystyle\sum_{n=1}^{\infty}\frac{(2n-1)^2\sigma_{co}\omega^2}{2r^2}\left[S_n^2 + \left(\frac{4J_0 r}{(2n-1)^2\pi\sigma_{co}\omega} + C_n\right)^2\right] & \text{coil} \\ \displaystyle\sum_{n=1}^{\infty}\frac{(2n-1)^2\sigma_{cr}\omega^2}{2r^2}\left[S_n^2 + C_n^2\right] & \text{crucible}\end{cases} \tag{17}$$

c) Triangle Waveform

The triangle waveform of the input current in the coil is approximated as

$$J_d = J_0 F^{\text{triangle}}(t) = \frac{8J_0}{\pi^2}\sum_{n=1}^{\infty}\frac{(-1)^{n-1}}{(2n-1)^2}\sin(2n-1)\omega t \tag{18}$$

$$\begin{aligned}\psi_B^{\text{triangle}} &= \sum_{n=1}^{\infty}\psi_{B(n)}^{\text{triangle}} \\ &= \sum_{n=1}^{\infty}\left[S_n\sin(2n-1)\omega t + C_n\cos(2n-1)\omega t\right]\end{aligned} \tag{19}$$

$$\begin{aligned}J_e^{\text{triangle}} &= \sum_{n=1}^{\infty}J_{e(n)}^{\text{triangle}} \\ &= \sum_{n=1}^{\infty}\frac{(2n-1)\sigma\omega}{r}\big[C_n\sin(2n-1)\omega t \\ &\qquad -S_n\cos(2n-1)\omega t\big]\end{aligned} \tag{20}$$

$$\hat{\mathcal{L}}S_n^{\text{triangle}} = \begin{cases} -\mu_{co}\left[\dfrac{(-1)^{n-1}8J_0}{(2n-1)^2\pi^2} + \dfrac{(2n-1)\sigma_{co}\omega}{r}C_n\right] & \text{coil} \\ -\dfrac{(2n-1)\sigma_{cr}\omega\mu_{cr}}{r}C_n & \text{crucible} \\ 0 & \text{elsewhere}\end{cases} \tag{21}$$

$$\hat{\mathcal{L}}C_n^{\text{triangle}} = \begin{cases} \dfrac{(2n-1)\sigma_{co}\omega\mu_{co}}{r}S_n & \text{coil} \\ \dfrac{(2n-1)\sigma_{cr}\omega\mu_{cr}}{r}S_n & \text{crucible} \\ 0 & \text{elsewhere}\end{cases} \tag{22}$$

$$q^{\text{triangle}}(r,z) =$$

$$\sum_{n=1}^{\infty}q_{(n)}^{\text{triangle}}\begin{cases} \displaystyle\sum_{n=1}^{\infty}\frac{(2n-1)^2\sigma_{co}\omega^2}{2r^2} \times\left[S_n^2 + \left(\frac{(-1)^{n-1}8J_0 r}{(2n-1)^3\pi^2\sigma_{co}\omega}\right) + C_n^2\right] & \text{coil} \\ \displaystyle\sum_{\infty}^{n=1}\frac{(2n-1)^2\sigma_{cr}\omega^2}{2r^2}\left(S_n^2 + C_n^2\right) & \text{crucible}\end{cases} \tag{23}$$

d) Sawtooth Waveform

The related equations of the sawtooth waveform of the input current can be written similar to the square and triangle waveforms. They are

$$J_d = J_0 F^{\text{sawtooth}}(t) = -\frac{J_0}{\pi^2}\sum_{n=1}^{\infty}\frac{\sin n\omega t}{n} \tag{24}$$

$$\psi_B^{\text{sawtooth}} = \sum_{n=1}^{\infty}\psi_{B(n)}^{\text{sawtooth}} = \sum_{n=1}^{\infty}\left[S_n\sin n\omega t + C_n\cos n\omega t\right] \tag{25}$$

$$J_e^{\text{sawtooth}} = \sum_{n=1}^{\infty}J_{e(n)}^{\text{sawtooth}} = \sum_{n=1}^{\infty}\frac{n\sigma\omega}{r}\left[S_n\sin n\omega t + C_n\cos n\omega t\right] \tag{26}$$

$$\hat{\mathcal{L}}S_n^{\text{sawtooth}} = \begin{cases} -\mu_{co}\left[\dfrac{-J_0}{n\pi} + \dfrac{n\sigma_{co}\omega}{r}C_n\right] & \text{coil} \\ -\dfrac{n\sigma_{cr}\omega\mu_{cr}}{r}C_n & \text{crucible} \\ 0 & \text{elsewhere}\end{cases} \tag{27}$$

$$\hat{\mathcal{L}}C_n^{\text{sawtooth}} = \begin{cases} \dfrac{n\sigma_{co}\omega\mu_{co}}{r} S_n \text{ coil} \\ \dfrac{n\sigma_{cr}\omega\mu_{cr}}{r} S_n \text{ crucible} \\ 0 \text{ elsewhere} \end{cases} \quad (28)$$

$$q^{\text{sawtooth}}(r,z) = \sum_{n=1}^{\infty} q_{(n)}^{\text{sawtooth}} \begin{cases} \sum_{n=1}^{\infty} \dfrac{n^2\sigma_{co}\omega^2}{2r^2}\left[S_n^2 + \left(\dfrac{-J_0 r}{\pi n^2 \sigma_{co}\omega}\right) + C_n^2\right] \text{coil} \\ \sum_{n=1}^{\infty} \dfrac{n^2\sigma_{cr}\omega^2}{2r^2}\left(S_n^2 + C_n^2\right) \text{crucible} \end{cases} \quad (29)$$

2.2. The Calculation Conditions

The driving current density in the induction coil is calculated by $J_0 = \sigma_{co}V_{\text{coil}}/(2\pi R_{\text{coil}}N)$, where V_{coil} is the total voltage of the coil, R_{coil} is the mean value of the coil radius and N is the number of coil turns. The boundary conditions are $\psi_B = 0$; both in the far field $(r, z \to \infty)$ and at the axis of symmetry ($r = 0$).

Values of electrical conductivity employed for our calculations are presented in [9], operating parameters are listed in **Table 1** and the geometry of the growth heating system is shown in **Figure 2**. The fundamental partial equations require using a numerical discretization method to solve them. Calculation of the equations with boundary conditions has been made by 2D finite element method.

The two-dimensional computational domain with the finite element triangle mesh is shown in **Figure 3**. In the space close to and in the metal parts (*i.e.* crucible, afterheater and RF-coil) the mesh is denser because of the high gradients of the electromagnetic fields. After solving the set of equations, we can obtain the electromagnetic field structure in the system as well as the volumetric power distribution in the crucible, afterheater and RF-coil.

Table 1. Operating parameters used for calculations.

Description (units)	Symbol	Value
Crucible inner radius (mm)	r_c	50
Crucible wall thickness (mm)	l_c	2
Crucible inner height (mm)	h_c	100
Baffle inner radius (mm)	r_b	35
Bottom heater height (mm)	h_{bh}	50
Height of the thick bottom (mm)	h_{tb}	10
Radius of the round bottom corner (mm)	r_{cb}	10
Coil inner radius (mm)	r_{co}	78
Coil width (mm)	l_{co}	13
Coil wall thickness (mm)	l_{co}	1.5
Height of coil turns (mm)	h_{co}	20
Distance between coil turns (mm)	d_{co}	3
Total voltage of the RF-coil (v)	V_{coil}	200
Current frequency of RF-coil (kHz)	f	10

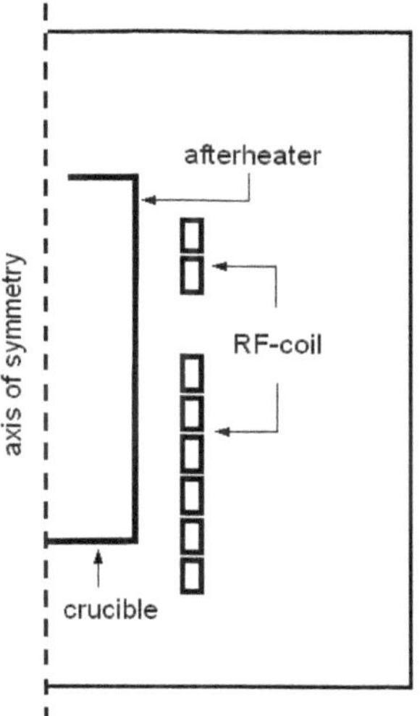

Figure 2. Sketch of an oxide Czochralski growth heating.

Figure 3. The finite element mesh structure of the calculation domain.

3. Results and Discussion

We explain the results of electromagnetic field and heating pattern in an oxide CZ setup including a cylindrical metal crucible, active afterheater and RF-coil corresponding to a real growth situation with different shapes of driving current in the RF-coil and with unique amplitude and frequency.

3.1. Electromagnetic Fields

Figures 4-7 show the distribution of in-phase component and out-of-phase component of the magnetic stream function (ψ_B) for the cases of sine, square, triangle, and sawtooth waveforms, respectively, in the growth setup.

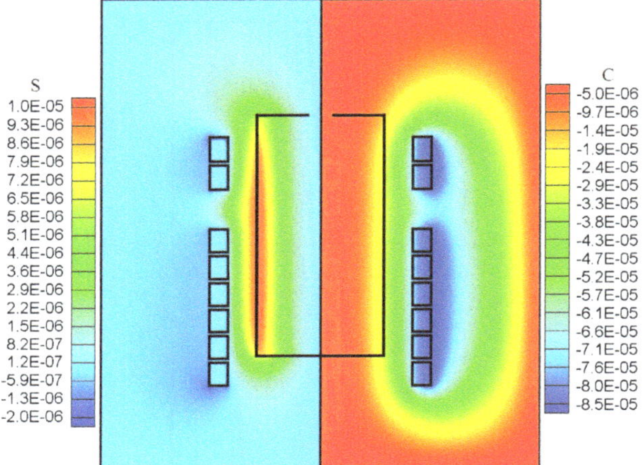

Figure 4. Components of the magnetic stream function (ψ_B) calculated for the case of sine waveform. The left hand side shows the in-phase component (*S*) and the right hand side shows the out-of-phase component (*C*) in the setup.

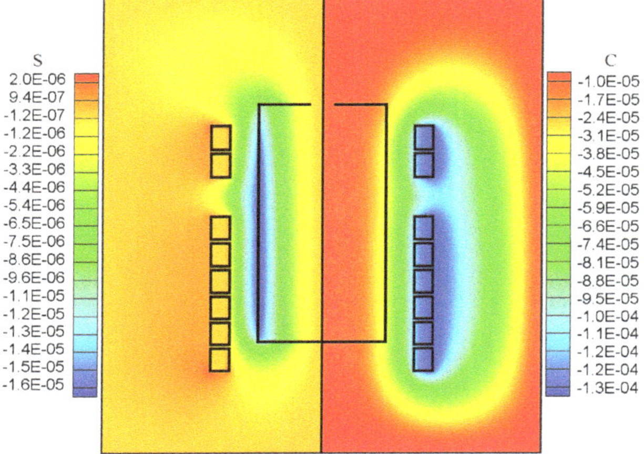

Figure 5. Components of the magnetic stream function (ψ_B) calculated for the case of square waveform. The left hand side shows the in-phase component (*S*) and the right hand side shows the out-of-phase component (*C*) in the heating setup.

The maximum of in-phase component (S_{max}) is located at the lowest and top edges of the RF-coil while the minimum (S_{min}) is located on the middle of the outer surface of crucible and afterheater wall, for the square and triangle waveforms. But for the sine and sawtooth waveforms, it is vice versa, that is, the positions of the (S_{max}) and (C_{max}) are replaced. For the out-of-phase component (*C*), the minimum is located on the outer surfaces of the induction coil turns for the cases of sine, square and triangle waveforms while the maximum is placed there for the case of sawtooth waveform. The distribution of *C*-component has a linear gradient in the space between the coil and the crucible and afterheater wall for all cases. The crucible and afterheater wall squeezes this compo-

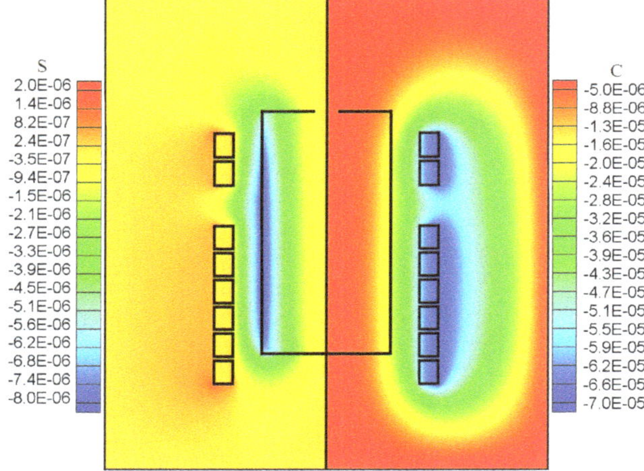

Figure 6. Components of the magnetic stream function (ψ_B) calculated for the case of triangle waveform. The left hand side shows the in-phase component (*S*) and the right hand side shows the out-of-phase component (*C*) in the heating setup.

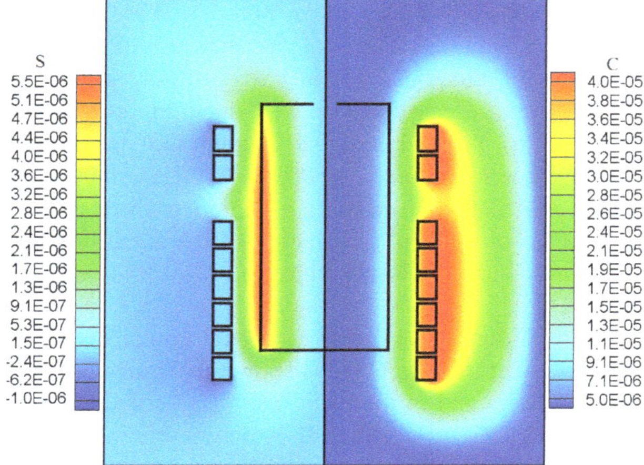

Figure 7. Components of the magnetic stream function (ψ_B) calculated for the case of sawtooth waveform. The left hand side shows the in-phase component (*S*) and the right hand side shows the out-of-phase component (*C*) in the heating setup.

nent to the area between the crucible and afterheater, and the RF-coil. Some interesting advantages are:

- The gradient of the *S*-component is too high in the area close to the maximum and minimum points, which is not true for other parts of the system;
- Deformation and distortion of the *S*-component in the area close to the extreme edges of the RF-coil are particularly visible for all cases (edge effect). For the crucible and afterheater, spatial distribution of the *S*-component is along and parallel to their sidewall;
- The strongest electromagnetic fields belong to the square waveform while the weakest fields are produced by the sawtooth waveform.

3.2. Heat Generation

The volumetric heat generation rate (q) in the crucible and afterheater has been shown for all cases in **Figure 8**. The power intensity is at its maximum value at the middle portion of the outer surface of the crucible sidewall, which arises from the skin effect.

The most important features are:

- The heating structure of the crucible and afterheater is the same for all cases expect for their intensity. The most powerful energy is produced by square, sine, triangle and sawtooth waveform, respectively, **Figure 9**. This feature is predictable from the related electromagnetic fields distribution;
- The spatial distribution of heat generation in the induction coil is mostly uniform with local "hot spots" (highly heated areas) at the lowest and upper edges, which is shown in **Figure 10**. The skin effect and proximity effect are responsible for appearance of these undesirable overheating because the induced eddy currents are concentrated on the top and lowest corners of the RF-coil [10,11];
- It is worth to note that despite of different total power generation, the coil efficiency (*i.e.*, the part of the energy delivered to the coil that is transferred to the workpiece) does not change and is approximately the same for all cases (**Table 2**).

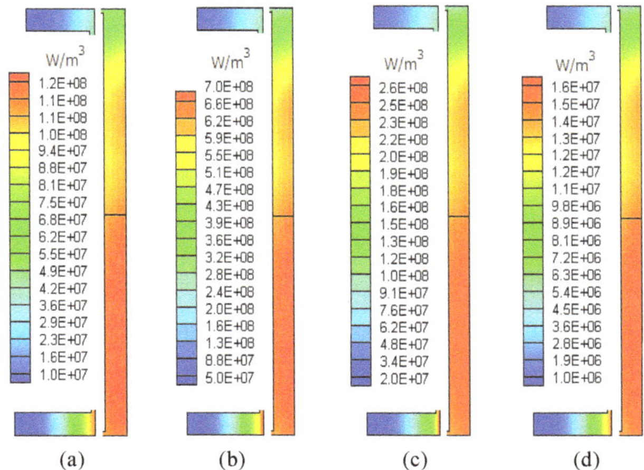

Figure 8. Volumetric power distribution (q) in the crucible and afterheater computed for (a) sine; (b) square; (c) triangle; and (d) sawtooth waveform of the driving current (for a better demonstration the crucible and afterheater sidewall and bottom are separately magnified).

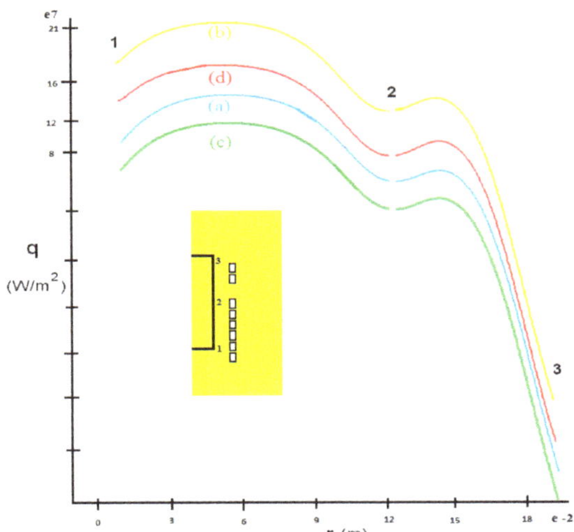

Figure 9. Profiles of the heat generated along the outer surface of the crucible and afterheater side wall calculated for (a) sine; (b) square; (c) triangle; and (d) sawtooth waveform of the input current.

4. Conclusions

To study the dependence of electromagnetic distribution and heating pattern on the input current shape (sine, square, triangle and sawtooth waveforms) of the induction coil, a set of 2D numerical calculations was performed.

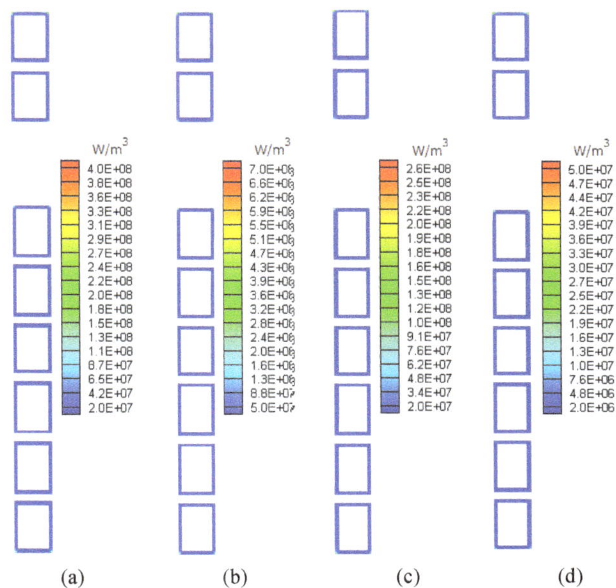

Figure 10. Volumetric power distribution (q) in the induction coil calculated for (a) Sine; (b) Square; (c) Triangle; and (d) Sawtooth waveform of the input current.

Table 2. Detail information about the heat generated in the CZ coil (Heating efficiency is the part of the energy delivered to the coil that is transferred to the crucible and afterheater).

Waveform	Crucible and afterheater (kW)	Induction coil (kW)	Heating efficiency (%)
Sine	15	1.4	91.5
Square	26	2.4	91.6
Triangle	10	0.9	91.5
Sawtooth	1.9	0.2	92.1

From the computational results described above, we can conclude:

- The spatial structure of electromagnetic fields and generated heat is a complex function of several parameters such as setup geometry and driving current shape;
- The electromagnetic fields distribution within the crucible and afterheater as well as the RF-coil is not uniform. This electromagnetic fields nonuniformity causes a nonuniform heating pattern in the crucible and afterheater, which in turn leads to a nonuniform temperature profile in the growth system.

A square input current results in a high intense heating of the crucible and afterheater while a sawtooth waveform leads to a low heating intensity in that part of the system. Different amount of produced energy in the setup is due to differences in the intensity and distribution of the electromagnetic fields. Understanding the physics of these properties is important during designing of an induction system for certain crystal growth applications.

REFERENCES

[1] J. J. Derby, L. J. Atherton and P. M. Gresho, "An Integrated Process Model for the Growth of Oxide Crystals by the Czochralski Method," *Journal of Crystal Growth*, Vol. 97, No. 3-4, 1989, pp. 792-826.

[2] D. T. J. Hurle, "Crystal Pulling from the Melt," Springer-Verlag, Berlin, Heidelberg, 1993.

[3] O. Klein and P. Philip, "Transient Numerical Investigation of Induction Heating during Sublimation Growth of Silicon Carbide Single Crystals," *Journal of Crystal Growth*, Vol. 247, No. 1-2, 2003, pp. 219-235.

[4] M. H. Tavakoli, F. Samavat and M. Babaiepour, "Influence of Active Afterheater on the Induction Heating Process in Oxide Czochralski Systems," *Crystal Research and Technology*, Vol. 43, No. 2, 2008, pp. 145-151.

[5] M. H. Tavakoli, A. Ojaghi, E. Mohammadi-Manesh and M. Mansour, "Influence of Coil Geometry on the Induction Heating Process in Crystal Growth Systems," *Journal of Crystal Growth*, Vol. 311, No. 6, 2009, pp. 1594-1599.

[6] M. H. Tavakoli, E. Mohammadi-Manes and A. Ojaghi, "Influence of Crucible Geometry and Position on the Induction Heating Process in Crystal Growth Systems," *Journal of Crystal Growth*, Vol. 311, No. 17, 2009, pp. 4281-4288.

[7] M. H. Tavakoli, H. Karbaschi, F. Samavat and E. Mohammadi-Manesh, "Numerical Study of Induction Heating in Melt Growth Systems—Frequency Selection," *Journal of Crystal Growth*, Vol. 312, No. 21, 2010, pp. 3198-3203.

[8] G. H. Hardy and W. W. Rogosinski, "Fourier Series," Dover, 1999.

[9] M. H. Tavakoli, "Modeling of Induction Heating in Oxide Czochralski Systems Advantages and Problems," *Crystal Growth Design*, Vol. 8, No. 2, 2007, pp. 483-488.

[10] S. Zinn and S. L. Semiatin, "Elements of Induction Heating," ASM International, Cleveland, 1988.

[11] V. Rudnev, D. Loveles, R. Cook and M. Black, "Handbook of Induction Heating," CRC Press, New York, 2003.

Crystal Structure and Solution Structural Dynamic Feature of 1,8-Dibenzoyl-2,7-Dimethoxynaphthalene

Akiko Okamoto, Shoji Watanabe, Kosuke Nakaema, Noriyuki Yonezawa
Department of Organic and Polymer Materials Chemistry, Tokyo University of Agriculture and Technology, Tokyo, Japan

ABSTRACT

The crystal structure and the dynamic feature of molecular structure in solution for 1,8-dibenzoyl-2,7-dimethoxynaphthalene are revealed by X-ray crystallographic analysis and VT-NMR measurements. In crystal, the molecule of the title compound is located on a twofold rotation axis. The two benzoyl groups are situated in an opposite direction. The dihedral angle between the mean planes of the phenyl ring and the naphthalene ring system is 80.25(6)°. The benzene ring and carbonyl moiety in each benzoyl group are almost coplanar. The molecular packing is stabilized by weak C–H...O hydrogen bonds and a π-π stacking interaction between the benzene rings [centroid-centroid and interplanar distances of 3.6383(10) and 3.294 Å, respectively]. In solution, the temperature-dependent rotation behavior of the C–C bond between the benzene ring and the ketonic carbonyl group has been observed by ^{1}H VT-NMR measurements. Furthermore, comparison of the C–C bond rotation behavior between the benzene ring and the carbonyl group with 1-benzoyl-2,7-dimethoxynaphthalene has clarified that the C–C bond between the ketonic carbonyl group and the naphthalene ring rotates slower than the 1,8-dibenzoylated homologue.

Keywords: Non-Coplanarly Accumulated Aromatic Rings Molecules; Crystal Structural Features; Bond Rotation Behavior in Solution; X-Ray Crystallography; VT-NMR Spectroscopy

1. Introduction

Non-coplanarly accumulated aromatic rings molecules, such as binaphthyl and biphenyl compounds, have attracted significant attention because of their characteristic properties, unique shapes, and various applications [1-7]. *Peri*-substituted naphthalenes have also received much attention as unique structured aromatic core compounds for variety of the functional materials [8-12]. Therefore, the structural analyses have been actively performed [13-15]. Recently, the authors' group has revealed that diaroylation at 1,8-positions of 2,7-dimethoxynaphthalene smoothly proceeds [16,17]. According to X-ray crystal structural study, the obtained 1,8-diaroylnaphthalene has unique non-coplanar alignment of aromatic rings [18]. The curious reversible aroylation behavior of the naphthalene derivative [16] and chemospecific and regioselective ethereal alkyl-oxygen bond cleavage reaction of aroylated naphthalenes [19] can be explained on the basis of the structural features of the aroylated naphthalenes. Under these circumstances, the authors have undertaken the structural studies of the aroylnaphthalne compounds [20-26] with investigation of the formation and the related reaction behaviors.

In this article, the crystallographical structural characteristics and the dynamic feature in solution structure of a 1,8-diaroylated naphthalene derivative having two methoxy groups at the 2,7-positions are described as the most simple homologue of monoaroylated derivative.

2. Experimental

All reagents were of commercial quality and were used as received. Solvents were dried and purified using standard techniques. Pentoxide-methanesulfonic acid (P_2O_5-MsOH) was prepared according to literature [27].

2.1. Measurements

^{1}H NMR spectra were recorded on a JEOL JNM-AL300 spectrometer (300 MHz) and a JEOL ECX400 spectrometer (400 MHz). Chemical shifts are expressed in ppm relative to internal standard of Me_4Si (δ 0.00). ^{13}C NMR spectra were recorded on a JEOL JNM-AL300 spectrometer (75 MHz). Chemical shifts are expressed in ppm relative to internal standard of $CDCl_3$ (δ 77.0). IR spectra were recorded on a JASCO FT/IR-4100 spectrometer. High-resolution FAB mass spectra were recorded on a JEOL MStation (MS700) ion trap mass spectrometer in positive ion mode.

2.2. Synthetic Procedure of the Title Compound

The title compound was synthesized via direct condensation mediated by P_2O_5-MsOH of 2,7-dimethoxynaphthalene with benzoic acid (**Scheme 1**). To a mixture of 2,7-dimethoxynaphthalene (0.200 mmol, 37.6 mg) and benzoic acid (0.440 mmol, 174 mg), P_2O_5-MsOH (0.88 mL) was added by portions at rt. After the reaction mixture was stirred at 60℃ for 3 h, it was poured into iced water (20 mL) and the mixture was extracted with $CHCl_3$ (15 mL × 3). The combined extracts were washed with 2 M NaOH aq and followed by sat NaCl aq. The organic layers thus obtained were dried over anhydrous sodium sulfate. The solvent was removed under reduced pressure to give powdery product. Isolation of the title compound was carried out by column chromatography [hexane: AcOEt = 2:1] (1,8-diaroylnaphthalene 63%; 3-monoaroylnaphthalene 19%; 1-monoaroylnaphthalene 3%). Colorless single crystals suitable for X-ray diffraction were obtained by recrystallization from ethanol.

1,8-benzoyl-2,7-dimethoxynaphthalene: Colorless needle (EtOH); m.p. = 530 K; IR (KBr): 1665, 1626 cm^{-1}; ^{1}H NMR (400 MHz, $CDCl_3$): 3.68 (6H, s), 7.21 (2H, d, J = 9.2 Hz), 7.34 (4H, dd, J = 7.6, 7.6 Hz), 7.49 (2H, t, J = 7.4 Hz), 7.70 (4H, d, J = 7.4 Hz), 7.95 (2H, d, J = 9.2 Hz) ppm; ^{13}C NMR (75 MHz, $CDCl_3$):56.40, 111.24, 121.47, 125.55, 127.95, 129.09, 129.84, 132.03, 132.64, 138.61, 156.28, 196.875 ppm. The above melting point and spectral data are compatible with the literature [28].

1-benzoyl-2,7-dimethoxynaphthalene:Colorless plate (Hexane + methylenechloride); m.p. = 358.5 - 362 K; IR (KBr): 1663, 1627 cm^{-1}; ^{1}H NMR δ (400 MHz, $CDCl_3$): 3.71 (3H, s), 3.79 (3H, s), 6.80 (1H, d, J = 2.8 Hz), 7.01 (1H, dd, J = 9.2, 2.8 Hz), 7.17(1H, d, J = 9.2 Hz), 7.43 (2H, t, J = 8.0 Hz), 7.57 (1H, t, J = 8.2 Hz), 7.72 (1H, d, J = 8.0 Hz), 7.84 - 7.89 (3H, m) ppm; ^{13}C NMR (75 MHz, $CDCl_3$): 55.09, 56.24, 102.02, 110.17, 117.01, 121.67, 124.29, 128.48, 129.44, 129.62, 130.96, 132.98, 133.31, 137.98, 154.93, 158.77, 198.07 ppm; HRMS (FAB; *m*-nitrobenzyl alcohol [*m*-NBA]) m/z: $[M + H]^+$; Calcd for $C_{19}H_{17}O_3$, 293.3365; found, 293.1185.

2.3. X-Ray Crystallography

For the crystal structure determination, the single-crystal of the compound $C_{26}H_{20}O_4$ was used for data collection on a four-circle Rigaku R-AXIS RAPID diffractometer (equipped with a two-dimensional area IP detector). The graphite-mono-chromated Cu Kα radiation (λ = 1.54187 Å) was used for data collection. The lattice parameters were determined by the least-squares methods on the basis of all reflections with $F^2 > 2\sigma(F^2)$. The data collection and cell refinement were performed using *PROCESS-AUTO* [29] software. The data reduction was performed using *CrystalStructure* [30]. The structures were solved by direct methods using *SIR*2004 [31] and refined by a full-matrix least-squares procedure using the program *SHELXL*97 [32]. All H atoms were found in a difference map and were subsequently refined as riding atoms, with the aromatic C–H = 0.95 Å and methyl C–H = 0.98 Å, and with $U_{iso}(H) = 1.2U_{eq}(C)$. Molecular structure of the compound showing the atomic numbering scheme is shown in **Figure 1**. The crystallography details for the structures determination of the compound are displayed presented in **Table 1**.

MeO OMe P_2O_5-MsOH 60°C, 3 h MeO OMe 63% MeO OMe 3% MeO OMe 19%

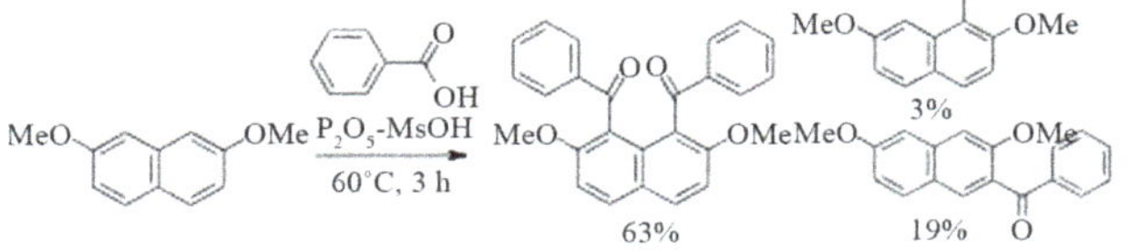

Scheme 1. Synthetic reaction of the title compound: P_2O_5-MsOH mediated direct condensation of 2,7-dimethoxynaphthalene with benzoic acid.

2.4. Variable Temperature NMR

Variable temperature ^{1}H NMR spectra were recorded on an FT-NMR operating at 400 MHz. Chemical shift values were reported in parts per million (ppm) relative to $(CH_3)_4Si$ (TMS). The solvent used in all cases was $CDCl_3$ and CS_2 (1:4 v/v) solution. Low temperature spectra were obtained with the use of a JEOL cooling system.

3. Results and Discussion

The title compound was synthesized via direct condensation

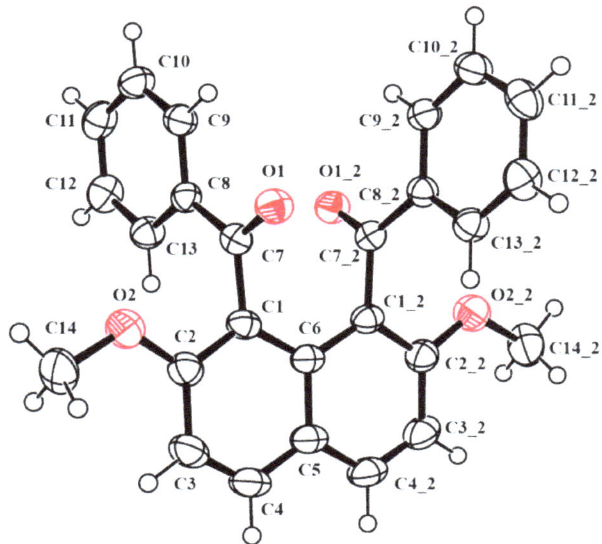

Figure 1. Molecular structure of the title compound, with the atom-labeling scheme and displacement ellipsoids drawn at the 50% probability level. The symbol "_2" refers to symmetry code: −x, y, −z + 1/2.

mediated by P_2O_5—MsOH of 2,7-dimethoxynaphthalene with benzoic acid (see Experimental section). **Table 1** shows the crystallographic data of the title compound. **Table 2** shows selected bond lengths and angles. **Table 3** gives selected torsion angles. **Figure 1** gives *ORTEP* representation of the molecular structure of the title compound, as determined by the structured X-ray analysis [33]. The molecule of the title compound lies across a crystallographic 2-fold axis so that the asymmetric unit contains one half of the molecules. Thus, the two benzoyl groups are situated in an opposite direction (anti-orientation). The benzoyl groups are twisted away from the naphthalene moiety, and the dihedral angle is 80.25(6)°. The torsion angle between the carbonyl group and the naphthalene ring is –76.73(16)° [C6–C1–C7–O1] and that between the carbonyl group and the phenyl group is 179.76(13)° [O1–C7–C8–C13].

Table 1. Crystallographic data and structure refinement parameters.

Empirical formula	$C_{26}H_{20}O_4$
Formula weight	396.43 g·mol^{-1}
Crystal shape, colour	Needle, colorless
Temperature	93(2) K
Wavelength	1.54187 Å
Crystal sytem	Monoclinic
Space group	C 2/c
Unit cell dimensions	a = 13.9677 (4) Å b = 10.2145 (3) Å c = 14.6966 (4) Å b = 109.711(2)°
Volume	1973.95 (10) Å^3
Z, calculated density	4, 1.334 Mg·m^{-3}
Absorption coefficient	0.72 mm^{-1}
F(000)	832
Crystal size	0.50 × 0.10 × 0.10 mm
Theta range for data collection	3.2° to 68.1°
Limiting indices	$-16 \le h \le 16$ $-12 \le k \le 12$ $-17 \le l \le 17$
Reflections collected/unique	17362/1807 [Rint = 0.027]
Completeness to theta = 68.21°	99.6%
Max. and min transmission	0.930 and 0.838
Refinement method	Full-matrix least-squares on F^2
Data/restraints/parameters	1807/0/139
Goodness-of-fit on F^2	1.08
Final R indices [I > 2 sigma (I)]	R1 = 0.039, wR2 = 0.107
R indices (all data)	R1 = 0.047, wR2 = 0.115
Largest diff. peak and hole	0.19 e·Å^{-3} and −0.21 e·Å^{-3}

Table 2. Selected bond lengths (Å) and angles (°).

O1—C7	1.2197 (16)	C2—C1—C7	115.70 (13)
C1—C2	1.382 (2)	C6—C1—C7	123.36 (13)
C1—C6	1.4264 (17)	O1—C7—C8	121.63 (13)
C1—C7	1.5158 (19)	O1—C7—C1	118.49 (12)
C7—C8	1.4814 (19)	C8—C7—C1	119.88 (12)
C8—C13	1.3908 (19)	C13—C8—C7	122.02 (13)
C8—C9	1.396 (2)	C9—C8—C7	118.88 (13)

Table 3. Selected torsion angles (°).

O1—C7—C8—C13	179.76 (13)
O1—C7—C8—C9	0.4 (2)
C2—C1—C7—O1	97.99 (16)
C6—C1—C7—O1	−76.73 (16)

In the crystal structure, the molecular packing of the title compound is mainly stabilized by van der Waals interaction. In addition, the packing of the molecule is stabilized by relatively weak C–H...O hydrogen bonding, namely, C12–H12...O1^{i} [symmetry code: i) x, –y + 1, z + 1/2], C14–H14B...O1^{ii} [symmetry code: ii) –x + 1/2, y – 1/2, –z + 1/2], and a π-π stacking interaction [centroid-centroid and interplanar distances of 3.6383 (10) and 3.294 Å, respectively]. In the packing, the molecules are arranged by C–H...O hydrogen bonding along the *c* axis of the unit cell, and by a π-π stacking interaction perpendicular the bc plane of the unit cell (**Figures 2** and **3**). The nonbonding distances are listed in **Table 4.**

The dynamic behavior of the title compound in solution was estimated by variable temperature NMR (VT-NMR) measurement performed in chloroform-*d* and CS_2 (1:4, v/v) solution from 293 to 173 K. **Figure 4(a)** shows the ^{1}H VT-NMR spectra of the title compound in the aromatic region. At 293 K, two signals of δ 7.21 and 7.95 ppm are assigned the protons at 3(6)- and 4(5)-positions of the naphthalene ring, respectively. The signals of δ 7.34, 7.49, and 7.70 ppm are assigned to the protons at *m*-positions, *p*-position, and *o*-positions of benzoyl groups. The signals of δ 7.21 and 7.95 ppm (at 293 K) are scarcely changed from 293 to 173 K. On the other hand, the figures of the three signals of δ 7.34, 7.49, and 7.70 ppm (at 293 K) are drastically changed. The signal of δ 7.49 ppm (at 293 K) is broadened from 293 to 233 K then sharpened from 213 to 173 K. The signals of δ 7.34 and 7.70 ppm (at 293 K) are broadened from 293 to 213 K and each of them splits into two broad signals at 193 K. The both of the pairs of broad signals are sharpened again at 173 K.

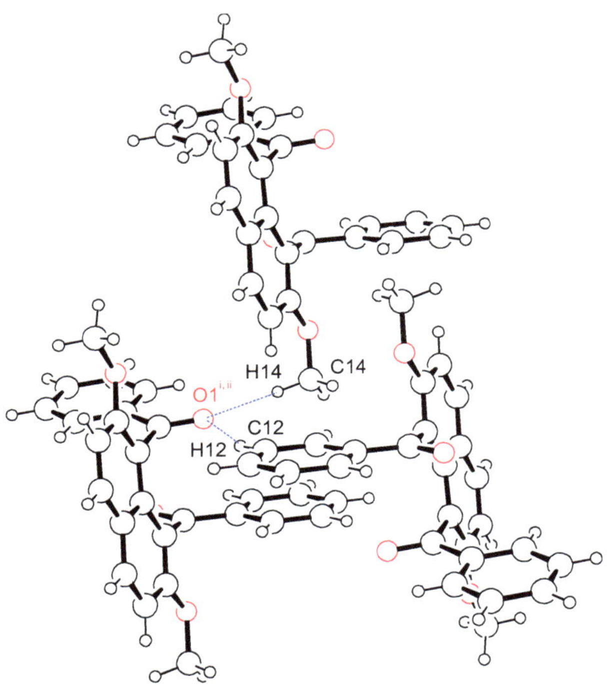

Figure 2. C–H…O interactions of methoxy group and benzene ring with carbonyl group [Symmetry codes: (i) x, −y + 1, z + 1/2; (ii) −x + 1/2, y − 1/2, −z + 1/2].

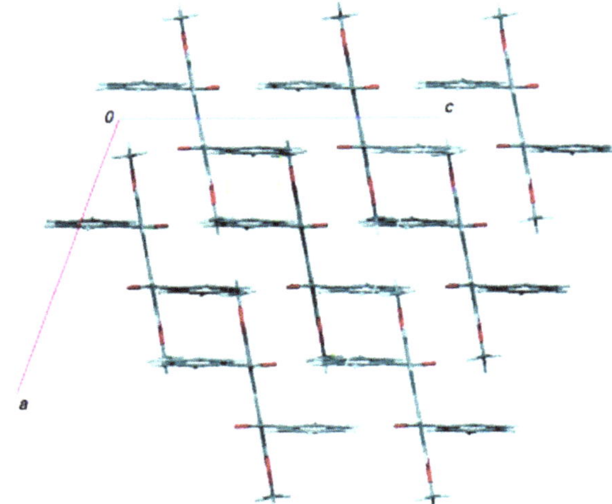

Figure 3. A partial packing diagram of the title compound, viewed down the b axis. The dashed lines indicate hydrogen bonds (blue dashed lines) and π-π stacking interactions (green lines).

Table 4. Nonbonding distances and related geometrical parameters in 1,8-dibenzoyl-2,7-dimethoxynaphthalene (°).

D—H···A	D—H (Å)	H···A (Å)	D···A (Å)	D—H···A (deg.)
C12—H12···O1[(i)]	0.95	2.60	3.4987 (19)	159
C14—H14B···O1[(ii)]	0.98	2.39	3.344 (2)	164

Symmetry codes: (i) x, −y + 1, z + 1/2; (ii) −x + 1/2, y − 1/2, −z + 1/2.

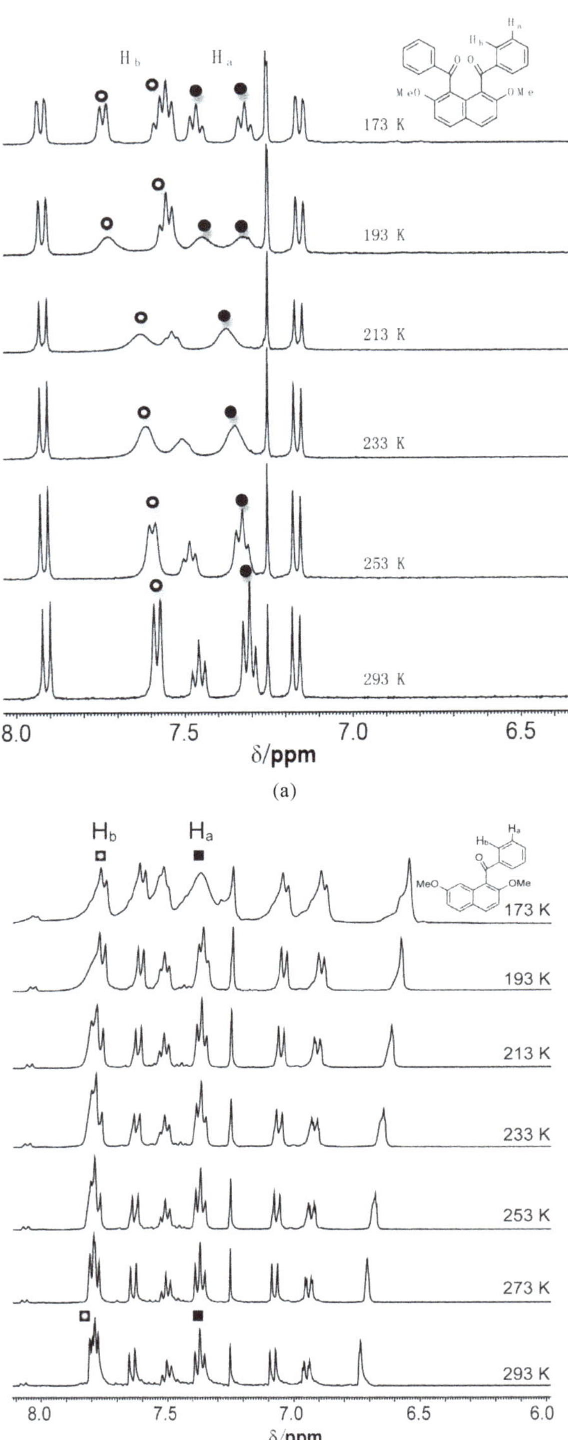

Figure 4. Variable-temperature ^{1}H NMR study in $CDCl_3$/CS_2 (1:4 v/v): (a) the title compound (6.5 - 8.0 ppm); (b) 1-benzoyl-2,7-dimethoxynaphthalene (6.0 - 8.0 ppm).

The sequent signal changes for the protons at *o*- and *m*-positions of the benzoyl group as shown in the VT-NMR spectra of the title compound suggest that the C–C bond rotation between benzene ring and ketonic carbonyl group is gradually slowed down with decreasing temperature then the magnetically non-equivalent environment is made on the benzene ring. The rotation barrier of the C(benzene)—C(carbonyl) bond was calculated as ΔG = 9.63 kcal/mol on the basis of the detailed VT-NMR measurements (**Figure 5**) [34].

In a similar manner, VT-NMR measurement of the homologous compound of 1-benzoyl-2,7-dimethoxynaphthalene was carried out (**Figure 4(b)**). For spectrum at 293 K, the signals of δ 6.80, 7.01, 7.17, and 7.72 ppm are assigned to the protons at 8-, 6-, 5-, and 3-positions of the naphthalene ring, respectively. The two signals of δ 7.43 and 7.57 ppm are assigned to the protons at *m*-positions and *p*-position of the benzoyl groups. The signals from δ 7.84 to 7.89 ppm are overlapped by the two kinds of signals, the proton at 4-position of the naphthalene ring and the protons at *o*-positions of the benzoyl group. The signals assigned to the protons of the benzoyl group are broadened in the temperature range from 293 to 173 K. However, no split behavior is observed at 173 K.

The title compound and the homologue have two kinds of the C–C bonds allowed to rotate, *i.e.*, C(benzene)–C(carbonyl) and C(naphthalene)–C(carbonyl) bonds. The steric environment around the neighboring carbonyl groups in the title compound is highly congested as shown in **Figure 1**. Naturally, the rotation behavior of C(naphthalene)–C(carbonyl) bond should be slower than 1-benzoylnaphthalene homologue. In the consequence, the temperature-dependent C(benzene)–C(carbonyl) bond rotation behavior might be observed. In other words, C(naphthalene)–C(carbonyl) bond in the title compound rotates slowly in the temperature range that the C(benzene)–C(carbonyl) bond is allowed to rotate freely.

4. Conclusion

Conclusively, the crystal structural shapes and the dynamic feature of solution structure of 1,8-dibenzoyl-2,7-dimethoxynaphthalene are clarified. In crystal, the aroyl groups of the compound are perpendicularly attached to the naphthalene ring core and situated in an opposite direction. The two types of C–H…O interactions of ketonic carbonyl group with benzene ring and methoxy group and π-π interactions between benzene rings mainly stabilize the molecular packing. According to VT-NMR study, the split signals assigned to the *o*- and *m*-protons of the benzene ring are observed from 193 to 173 K. Comparison of the dynamic behavior with 1-benzoylnaphthalene homologue shows that the C–C bond rotation between ketonic carbonyl group and naphthalene ring in the title compound is enough slow to detect the rotation of the C–C bond rotation between benzene ring and ketonic carbonyl group. The structural information of the title compound in solid state and solution affords some hitherto-unknown aspects in molecular formula and intramolecular rotation properties relationship of these non-coplanarly accumulated aromatic rings molecules.

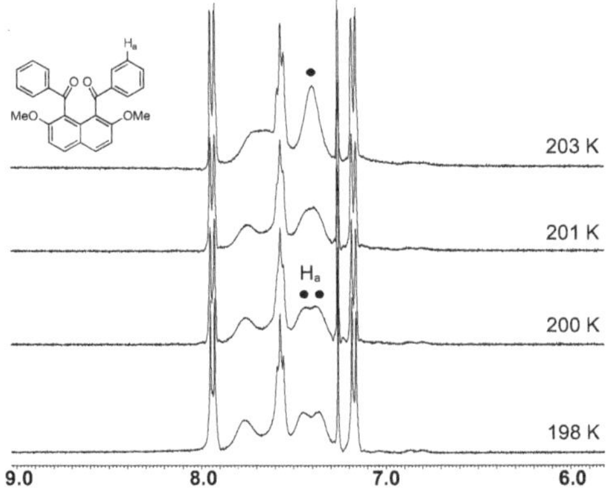

Figure 5. The temperature-dependence of ^{1}H NMR signals of the title compound in $CDCl_3/CS_2$ (1:4 v/v) from 203 to 198 K (6.0 - 9.0 ppm).

5. Acknowledgements

This work was partially supported by the Shorai Foundation for Science and Technology (Tokyo, Japan) and the Ogasawara Foundation for the Promotion of Science and Engineering (Tokyo, Japan).

REFERENCES

[1] F. Gasparrini, M. Pierini, C. Villani, A. Filippi and M. Speranza, "Induced-Fit in the Gas: Phase Conformational Effects on the Enantioselectivity of Chiral Tetra-Amide Macrocycles," *Journal of American Chemical Society*, Vol. 130, No. 2, 2008, pp. 522-534.

[2] N. H. Shah, G. L. Butterfoss, K. Nguyen, B. Yoo, R. Bonneau, D. L. Rabenstein and K. Kirshebaum, "Oligo(N-aryl Glycines): A New Twist on Structured Peptoids," *Journal of American Chemical Society*, Vol. 130, No. 49, 2008, pp. 16622-16632.

[3] I. Alfonso, M. I. Burguete, F. Galindo, S. V. Luis and L. Vigara, "Molecular Rotors as Simple Models to Study Amide NH-Aromatic Interactions and Their Pole in the Folding of Peptide-Like Structures," *Journal of Organic Chemistry*, Vol. 72, No. 21, 2007, pp. 7947-7956.

[4] F. R. Zhang, H. B. Song and G. F. Zi, "Synthesis and Catalytic Activity of Group 5 Metal Amides with Chiral Biaryldiamine-Based Ligands," *Dalton Transactions*, Vol. 40, No. 7, 2011, pp. 1547-1566.

[5] J.-P. Genet, "Asymmetric Catalytic Hydrogenation. Design of New Ru Catalysts and Chiral Ligands: From Laboratory to Industrial Applications," *Accounts of Chemical Research*, Vol. 36, No. 12, 2003, pp. 908-918.

[6] O. Lucchi, "High Symmetry Chiral Auxiliaries Containing Heteroatoms," *Pure and Applied Chemistry*, Vol. 68, No. 4,

1996, pp. 945-950.

[7] K. Maruoka, "Asymmetric Phase Transfer Catalysis," Wiley-VCH, Weinheim, 2008.

[8] M. T. Scerba, C. M. Leavitt, M. E. Diener, A. F. DeBlase and T. Lectka, "NH^+-F Hydrogen Bonding in a Fluorinated "Proton Sponge" Derivative: Integration of Solution, Solid-State, Gas-Phase, and Computational Studies," *Journal of Organic Chemistry*, Vol. 76, No. 19, 2011, pp. 7975-7984.

[9] S. Shinamura, E. Miyazaki and K. Takiyama, "Synthesis, Properties, Crystal Structures, and Semiconductor Characteristics of Naphtho[1,2-b:5,6-b']Dithiophene and Diselenophene Derivatives," *Journal of Organic Chemistry*, Vol. 75, No. 4, 2010, pp. 1228-1234.

[10] Z. Y. Wang and A. L. Guen, "Synthesis and Properties of Poly(arylene ether)s Containing 1,8-Dibenzoylnaphthalene Units," *Macromolecules*, Vol. 28, No. 10, 1995, pp. 3728-3732.

[11] Y. L. Jiang, X. N. Gao, G. N. Zhou, A. Patel and A. Javer, "Selective Recognition of Uracil and Its Derivatives Using a DNA Repair Enzyme Structual Mimic," *Journal of Organic Chemistry*, Vol. 75, No. 2, 2010, pp. 324-333.

[12] X. F. Mei, R. M. Martin and C. Wolf, "Synthesis of Sterically Crowded Atropisomeric 1,8-Diacridylnaphthalene for Dual-Mode Enantioselective Fluorosensing," *Journal of Organic Chemistry*, Vol. 71, No. 7, 2006, pp. 2854-2861.

[13] S. Cohen, M. Thirumalaikumar, S. Pogodin and I. Agranat, "*Peri* Interactions in Naphthalene Diketones: A Preference for (Z,Z) Conformations," *Structure Chemistry*, Vol. 15, No. 4, 2004, pp. 339-346.

[14] L.-H. Jing, D.-B. Qin, L. He, S.-J. Gu, H.-X. Zhang and G. Lei, "Dimethyl Naphthalene-1,8-Dicarboxylate," *Acta Crystallographica Section E*, Vol. 61, No. 11, 2005, pp. o3595-o3596.

[15] P. H. Gore and K. Henrick, "1,8-Dibenzoyl-2,7-Dimethylnaphthalene," *Acta Crystallographica Section B*, Vol. B36, No. 10, 1980, pp. 2462-2465.

[16] A. Okamoto and N. Yonezawa, "Reversible ArS_E Aroyla tion of Naphthalene Derivatives," *Chemistry Letters*, Vol. 38, No. 9, 2009, pp. 914-915.

[17] A. Okamoto, R. Mitsui, H. Oike and N. Yonezawa, "Lewis Acid-Mediated ArS_E Aroylation of Naphthalene Derivative: Distinct Second Aroylation Behavior of Naphthyl Ketone," *Chemistry Letters*, Vol. 40, No. 11, 2011, pp. 1283-1284.

[18] K. Nakaema, A. Okamoto, K. Noguchi and N. Yonezawa, "1,8-Bis(4-Chlorobenzoyl)-2,7-Dimethoxynaphthalene," *Acta Crystallographica Section E*, Vol. E63, No. 10, 2007, p. o4120.

[19] A. Okamoto, R. Mitsui, S. Watanabe, T. Tsubouchi and N. Yonezawa, "Chemospecific and Regioselective Etheral Methyl-Oxygen Bond Cleavage Behavior of Aroylated Dimethoxynaphthalenes by Combined Action of $AlCl_3$ and Aroyl Group," *International Journal of Organic Chemistry*, Vol. 2, No. 3, 2012, pp. 194-201.

[20] T. Tsumuki, A. Isogai, A. Nagasawa, A. Okamoto and N. Yonezawa, "2,7-Dimethoxy-1-(2-Naphthoyl) Naphthalene," *Acta Crystallographica Section E*, Vol. E68, No. 8, 2012, p. o2595.

[21] R. Sakamoto, K. Sasagawa, D. Hijikata, A. Okamoto and N. Yonezawa, "1,8-Dibenzoylnaphthalene-2,7-Diyl Dibenzate," *Acta Crystallographica Section E*, Vol. E68, No. 8, 2012, p. o2454.

[22] K. Sasagawa, D. Hijikata, R. Sakamoto, A. Okamoto and N. Yonezawa, "[2,7-Dimethoxy-8-(4-Propylbenzoyl)Naphthalen-1-yl](4-Propylphenyl)Methanone," *Acta Crystallographica Section E*, Vol. E68, No. 8, 2012, p. o2596.

[23] K. Sasagawa, D. Hijikata, T. Kusakabe, A. Okamoto and N. Yonezawa, "4-{[8-(4-Acetyloxybenzoyl)-2,7-Dimethoxynaphthalen-1-yl]Carbonyl}Phenyl Acetate," *Acta Crystallographica Section E*, Vol. E68, No. 8, 2012, p. o2503.

[24] T. Muto, K. Sasagawa, A. Okamoto, H. Oike and N. Yonezawa, "[8-(4-Chlorobenzoyl)-2,7-Dimethoxy-Naphthalen-1-yl](2,4,6-Trimethylphenyl)Methanone," *Acta Crystallographica Section E*, Vol. E68, No. 3, 2012, p. o906.

[25] T. Muto, K. Sasagawa, A. Okamoto, H. Oike and N. Yonezawa, "(3,5-Dimethylphenyl)[8-(3,5-Dimethylbenzoyl)-2,7-Dimethoxynaphthalen-1-yl] Methanone," *Acta Crystallographica Section E*, Vol. E68, No. 4, 2012, p. o1200.

[26] T. Tsumuki, S. Murohashi, A. Nagasawa, A. Okamoto and N. Yonezawa, "(3,6-Dimethoxynaphthalen-2-yl) (Naphthalen-2-yl)Mathanone," *Acta Crystallographica Section E*, Vol. E68, No. 9, 2012, p. o2653.

[27] P. Eaton, G. R. Carlson and J. T Lee, "Phosphorus Pentoxide-Methane Sulfonic Acid. Convenient Alternative to Polyphosphoric Acid," *Journal of Organic Chemistry*, Vol. 38, No. 23, 1973, pp. 4071-4073.

[28] A. M. Gorelik, A. V. Reznichenko, N. A. Andronova, E. A. Luk'yanets, "Acylation of 2,7-Dimethoxynaphthalene," *Zhurnal Organicheskoi Khimii*, Vol. 19, No. 1, 1983, pp. 199-206.

[29] Rigaku, "PROCESS-AUTO," Rigaku Corporation, Tokyo, 1998.

[30] Rigaku/MSC, "CrystalStructure," Rigaku/MSC, The Woodlands, Houston, 2004.

[31] M. C. Burla, R. Caliandro, M. Camalli, B. Carrozzini, G. L. Cascarano, L. De Caro, C. Giacovazzo, G. Polidori and R. Spagna, "*SIR*2004: An Improved Tool for Crystal Structure Determination and Refinement," *Journal of Applied Crystallography*, Vol. 38, No. 2, 2005, pp. 381-388.

[32] G. M. Sheldrick, "A Short History of SHELX," *Acta Crystallographica Section A*, Vol. A64, No. 1, 2008, pp. 112-122.

[33] K. Nakaema, S. Watanabe, A. Okamoto, K. Noguchi and N. Yonezawa, "1,8-Dibenzoyl-2,7-Dimethoxy-Naphthalene," *Acta Crystallographica Section E*, Vol. E64, No. 5, 2008, p. o807.

[34] H. S. Gutowsky and C. H. Holm, "Rate Processes and Nuclear Magnetic Resonance Spectra II. Hindered Internal Rotation of Amides," *Journal of Chemical Physics*, Vol. 25, 1956, pp. 1228-1234.

Permissions

The contributors of this book come from diverse backgrounds, making this book a truly international effort. This book will bring forth new frontiers with its revolutionizing research information and detailed analysis of the nascent developments around the world.

We would like to thank all the contributing authors for lending their expertise to make the book truly unique. They have played a crucial role in the development of this book. Without their invaluable contributions this book wouldn't have been possible. They have made vital efforts to compile up to date information on the varied aspects of this subject to make this book a valuable addition to the collection of many professionals and students.

This book was conceptualized with the vision of imparting up-to-date information and advanced data in this field. To ensure the same, a matchless editorial board was set up. Every individual on the board went through rigorous rounds of assessment to prove their worth. After which they invested a large part of their time researching and compiling the most relevant data for our readers. Conferences and sessions were held from time to time between the editorial board and the contributing authors to present the data in the most comprehensible form. The editorial team has worked tirelessly to provide valuable and valid information to help people across the globe.

Every chapter published in this book has been scrutinized by our experts. Their significance has been extensively debated. The topics covered herein carry significant findings which will fuel the growth of the discipline. They may even be implemented as practical applications or may be referred to as a beginning point for another development. Chapters in this book were first published by Scientific Research Publishing Inc.; hereby published with permission under the Creative Commons Attribution License or equivalent.

The editorial board has been involved in producing this book since its inception. They have spent rigorous hours researching and exploring the diverse topics which have resulted in the successful publishing of this book. They have passed on their knowledge of decades through this book. To expedite this challenging task, the publisher supported the team at every step. A small team of assistant editors was also appointed to further simplify the editing procedure and attain best results for the readers.

Our editorial team has been hand-picked from every corner of the world. Their multi-ethnicity adds dynamic inputs to the discussions which result in innovative outcomes. These outcomes are then further discussed with the researchers and contributors who give their valuable feedback and opinion regarding the same. The feedback is then collaborated with the researches and they are edited in a comprehensive manner to aid the understanding of the subject.

Apart from the editorial board, the designing team has also invested a significant amount of their time in understanding the subject and creating the most relevant covers. They scrutinized every image to scout for the most suitable representation of the subject and create an appropriate cover for the book.

The publishing team has been involved in this book since its early stages. They were actively engaged in every process, be it collecting the data, connecting with the contributors or procuring relevant information. The team has been an ardent support to the editorial, designing and production team. Their endless efforts to recruit the best for this project, has resulted in the accomplishment of this book. They are a veteran in the field of academics and their pool of knowledge is as vast as their experience in printing. Their expertise and guidance has proved useful at every step. Their uncompromising quality standards have made this book an exceptional effort. Their encouragement from time to time has been an inspiration for everyone.

The publisher and the editorial board hope that this book will prove to be a valuable piece of knowledge for researchers, students, practitioners and scholars across the globe.

List of Contributors

Ikram Dhouib, Tahar Mhiri and Zakaria Elaoud
Laboratoire de l'Etat Solide, Département de Chimie, Faculté des Sciences de Sfax, Sfax, Tunisia

Salih Al-Juaid
Chemistry Department, Faculty of Science, King Abdulaziz University, Jeddah, KSA

Yuxin Song
Department of Microtechnology and Nanoscience, Photonics Laboratory, Chalmers University of Technology, Gothenburg, Sweden
State Key Laboratory of Functional Materials for Informatics, Shanghai Institute of Microsystem and Information Technology, Chinese Academy of Sciences, Shanghai, China

Shumin Wang
Department of Microtechnology and Nanoscience, Photonics Laboratory, Chalmers University of Technology, Gothenburg, Sweden
State Key Laboratory of Functional Materials for Informatics, Shanghai Institute of Microsystem and Information Technology, Chinese Academy of Sciences, Shanghai, China

Carl Asplund, Rickard Marcks von Würtemberg and Hedda Malm
IRnova AB, Kista, Sweden

Amir Karim
Acreo AB, Kista, Sweden

Xiang Lu, Jun Shao
National Laboratory for Infrared Physics, Shanghai Institute of Technical Physics, CAS, Shanghai, China

Ching-Hwa Ho
Graduate Institute of Applied Science and Technology, National Taiwan University of Science and Technology, Taipei, Taiwan

Kiran K. Pujar and Manohar V. Kulkarni
Department of Chemistry, Karnatak University, Dharwad, India

G. N. Anilkumar
M S Ramaiah Institute of Technology, Department of Physics, Bangalore, India

Satoshi Yamauchi
Department of Biomolecular Functional Engineering, Ibaraki University, Hitachi, Japan

Yoh Imai
Department of Electric and Electronic Engineering, Ibaraki University, Hitachi, Japan

Lubna M. Sharaf El-Deen, Mohamed H. Badr, Abdel-Mageed H. Khafagy, Dalia U. Nassar
Physics Department, Faculty of Science, Minufiya University, Shibin El Kom, Egypt

Kheira Feham and Douniazed El Abed
Laboratoire de Chimie Fine (LCF), Université d'Oran, Es-Senia, Algérie

Abdelkarim Benkadari and Abdelkader Chouaih
Laboratoire SEA2M, Université de Mostaganem, Mostaganem, Algérie

Abdellah Miloudi
Laboratoire de Chimie Fine (LCF), Université d'Oran, Es-Senia, Algérie
Département de Physique-Chimie, ENSET d'Oran, Oran, Algérie

Gérard Boyer
Laboratoire HIT, Université Paul Cezanne, Marseille, France

Oleg D. Linnikov, Irina V. Rodina, Igor G. Grigorov and Evgeniy V. Polyakov
Institute of Solid State Chemistry, Ural Branch of the Russian Academy of Sciences, Ekaterinburg, Russia

Sibi Narayanan and Devadasan Velmurugan
Centre of Advanced Study in Crystallography and Biophysics, University of Madras, Maraimalai (Guindy) Campus, Chennai, India

Delphine Parmentier, Murielle Bertrand and Pascal Baron
Commissariat à l'Energie Atomique et aux Energies Alternatives, Radiochemistry and Process Department, Bagnols-sur-Cèze, France

Edouard Plasari
Reactions and Chemical Engineering Laboratory-CNRS, Nancy, France

Anastasia P. Gazhulina and Mikhail O. Marychev
Lobachevsky State University of Nizhni Novgorod, Nizhni Novgorod, Russia

M. Junaid Bushiri
Department of Physics, Cochin University of Science and Technology, Kochi, India

T. C. Kochuthresia and V. K. Vaidyan
Department of Physics, University of Kerala, Thiruvananthapuram, India

S. Athimoolam
Department of Physics,University College of Engineering, Anna University Tirunelveli Region, Nagercoil, India

V. Ramakrishnan
School of Physics, Madurai Kamaraj University, Madurai, India

Gasidit Panomsuwan
Department of Materials, Physics and Energy Engineering, Graduate School of Engineering, Nagoya, Japan

Nagahiro Saito
Department of Materials, Physics and Energy Engineering, Graduate School of Engineering, Nagoya, Japan
EcoTopia Science Institute, Nagoya University, Nagoya, Japan
Green Mobility Collaborative Research Center, Nagoya, Japan

Devadasan Velmurugan
Centre of Advanced Study in Crystallography and Biophysics, University of Madras, Maraimalai (Guindy) Campus, Chennai, India

Wanqiang Zhang
School of Chemistry and Chemical Engineering, Xuchang University, Xuchang, China

Akiko Okamoto, Atsushi Nagasawa, Siqingaowa and Noriyuki Yonezawa
Department of Organic and Polymer Materials Chemistry, Tokyo University of Agriculture and Technology, Tokyo, Japan

S. Madan Kumar, B. C. Manjunath and N. K. Lokanath
Department of Studies in Physics, University of Mysore, Mysore, India

G. S. Lingaraju and M. P. Sadashiva
Department of Studies in Chemistry, University of Mysore, Mysore, India

M. M. M. Abdoh
Department of Physics, Faculty of Science, An Najah National University, Nabtus West Bank, Palestinian Territories

Shuichi Miyamoto
Faculty of Pharmaceutical Sciences, Sojo University, Kumamoto, Japan

Chieko Naoe and Kazuo T. Nakamura
School of Pharmacy, Showa University, Tokyo, Japan

Masaru Tsunoda
Faculty of Pharmacy, Iwaki Meisei University, Iwaki, Japan

Veerendra Kumar A. Kalalbandi and J. Seetharamappa
Department of Chemistry, Karnatak University, Dharwad, India

Sahaj A. Gandhi
Department of Physics, Sardar Patel University, Vallabh Vidyanagar, India

Vijay M. Barot and Mitesh C. Patel
P. G. Center in Chemistry, Smt. S. M. Panchal Science College, Talod, India

Md. Arifuzzaman, Tasneem A. Siddiquee, Mohammad R. Karim
Department of Chemistry, Tennessee State University, Nashville, USA

Aminul H. Mirza and Mohamad A. Ali
Department of Chemistry, Faculty of Science, University of Brunei Darussalam, Gadong, Malaysia

Dhananjay Dey and Deepak Chopra
Department of Chemistry, Indian Institute of Science Education and Research, Bhopal, India

Venugopal Prakash
Shirdi Sai Engineering College, Bangalore, India

Vasu
Vivekananda Degree College, Bangalore, India

Janardhanan Saravanan
PES College of Pharmacy, Bangalore, India

Francesco Massimino
Università degli Studi, Dipartimento di Chimica IFM, Turin, Italy
NTT—New Tera Technology, Turin, Italy

Mohammad Hossein Tavakoli
Physics Department, Bu-Ali Sina University, Hamedan, Iran

Tayebe Nadery Mostagir
Department of Physics, Kurdistan University, Sanandaj, Iran

Akiko Okamoto, Shoji Watanabe, Kosuke Nakaema, Noriyuki Yonezawa
Department of Organic and Polymer Materials Chemistry, Tokyo University of Agriculture and Technology, Tokyo, Japan

www.ingramcontent.com/pod-product-compliance
Ingram Content Group UK Ltd.
Pitfield, Milton Keynes, MK11 3LW, UK
UKHW052226270726
14059UKWH00003B/142